GREENHOUSE GASES AND ANIMAL AGRICULTURE

Proceedings of the 1st International Conference
on Greenhouse Gases and Animal Agriculture
Obihiro, Japan, 7-11 November, 2001

GREENHOUSE GASES AND ANIMAL AGRICULTURE

Proceedings of the 1st International Conference
on Greenhouse Gases and Animal Agriculture
Obihiro, Japan, 7-11 November, 2001

Edited by

Junichi Takahashi
Obihiro University of
Agriculture and Veterinary Medicine
Obihiro, Hokkaido
Japan

and

Bruce A. Young
The University of Queensland
Gatton, Australia

2002

ELSEVIER

AMSTERDAM – BOSTON – LONDON – NEW YORK – OXFORD – PARIS
SAN DIEGO – SAN FRANCISCO – SINGAPORE – SYDNEY – TOKYO

ELSEVIER B.V.
Radarweg 29, 1043 NX, Amsterdam, The Netherlands

First edition 2002

Library of Congress Cataloging-in-Publication-Data
A catalog record from the Library of Congress has been applied for.

British Library Cataloguing in Publication Data
A catalogue record from the British Library has been applied for.

ISBN: 0-444-51012-5
ISBN 13: 978-0-444-51012-9

The publisher's policy is to use **paper manufactured from sustainable forests**

Transferred to Digital Printing 2007

PREFACE

This book contains a compilation of papers presented at the 1st International Conference on Greenhouse Gases and Animal Agriculture, held in the Tokachi Plaza, Obihiro, Hokkaido Japan from 7-11 November 2001.

This conference brought together scientists, engineers and people involved in practical agriculture to discuss and examine the involvement of Greenhouse Gases in Animal Agriculture. The issues of the involvement of animals in greenhouse gas release is not only important in protecting the atmospheric environment from excessive concentrations of infrared absorbing gases but also to gaining improved efficiency of utilizing agricultural resources and protecting agricultural soils. While carbon is bound in organic forms it can be recycled back into agricultural systems or can be utilized for generation of energy sources to replace the excessive use of fossil fuels.

Nitrogen and other nutrients contained in animal effluent and agricultural by-products are valuable resources that with recycling to the soil can reduce the need for chemical fertilizers and contribute to soil quality. Appropriate recycling of such materials makes use of valuable organic resources, reduces the use of fossil fuels and the potential contributions of nitrous oxide release to the atmosphere and the development of acid rain.

The different approaches used by the broad cross section of disciplines represented at this meeting provided an excellent opportunity for interchange of ideas and the raising of issues that may not come out of a meeting specializing in a single discipline. Scientists from diverse fields such as animal nutrition, feed technology, biotechnology and soil science had the opportunity to mix with technologists from commercial companies working with fuel cells and biogas systems as well as with social economists.

Over 200 scientific delegates from over 20 countries, active in the field of greenhouse gases and animal agriculture, participated in the conference and 100 presentations were made during the meeting, based on recent research findings, which are published herein after a fair selection by the Editorial Committee.

Through programs of international cooperation and sharing of ideas and scientific results, we can better understand the possible contributions of animals in agriculture to the generation of greenhouse gases and the opportunities available to reduce gas emissions. Minimizing the emission of greenhouse gases reduces the risk of accelerated global warming and at the same time we can improve the health of our environment.

Junichi Takahashi

CONTENTS

Part 1

Overview of greenhouse gases and animal agriculture

Greenhouse Gases and Animal Agriculture
J. Takahashi & B. A. Young (Editors)

Sustainable systems of animal agriculture

M. Suzuki

United Nations University, 5-53-70 Jingumae, Shibuya-ku, Tokyo 150-8925 Japan

Intensification of human activities and population increase during the last century brought about conflicts between natural ecosystems and human activities in various areas. For establishing sustainable human activities, it is important to realize the finiteness of the globe system. Zero Emission concept is based on constructing human activities consisting of both minimum exploitation of resources and minimum load to environment. This leads to maximum resources productivities, which is needed for formulating a sustainable system. For stockbreeding, construction of complete cyclic use of materials such as nitrogen is a key factor in selection of systems.

Keywords: sustainability, Zero Emission, environment revolution, nitrogen cycle

1. INTRODUCTION

At the start of a new century, the world is facing a number of problems. Some of the serious problems such as environmental degradations caused by various anthropogenic reasons in various area and expected natural resources degradation and exhaustion, lead to the question of possibility of the human survival. As a matter of fact, the current generation is considered to be the first of the history of mankind that realized the finiteness of the globe. We need to develop a new paradigm for human activities including industrial and domestic activities in order to coexist in a sustainable manner with the natural environment based upon the concept of finiteness. All of the human ways should be reconsidered along this line and stockbreeding industry is of course no exception.

2. ENVIRONMENTAL REVOLUTION

Shuntaro Ito, professor emeritus of University of Tokyo, identified five most significant evolutions in human history and called them revolutions. First revolution happened about five million years ago in East Africa where upheaval of land occurred. Because of forest degradation in the eastern side of the highland range, apes came down from the trees and started to walk upright. This was the start of anthropogenic history. Cerebrum has developed since then and tools, languages, fire cooking, etc. were invented.

The second revolution occurred in various placed in the world at almost the same time, about 10 to 8 thousand years ago when farming and taming of wild species were developed. This was the start of agriculture, stock farming and settlement. This happened in Palestine, Mesopotamia, East and South Asia, West Africa and New Continent independently. About 4 thousand years ago, people acquired flood control technology and developed agricultural irrigation.

This led to the next revolution, urban revolution. Namely, large scale industrialized agriculture brought about excess of agricultural products, which promoted evolution of various occupations other than agriculture. Governor, clerks, soldiers, merchants, etc. evolved to protect and manage the excess wealth within a community – a wall surrounded city. This urban revolution occurred in four different places, that is, Egypt, Sumer, Indus and Huang River.

When peoples lived in enclosed wall cities for numbers of generations, they became intellectually matured. Various philosophers appeared in 4 to 6 century BC. Thales in Greek, Shakamuni in India, Confucius and hundred philosophers in China appeared almost at the same time in different civilizations. Spiritual revolution leads to the evolution of several religions.

Science revolution started only in Europe at the later half of 17 century. Descartes introduced the mechanistic view of nature, Bacon emphasized the empiricism and many pioneering scientists like Newton, Galilei, Copernicus and Boyle initiated innovative simplified views of natural systems. Development of this modern science triggered the idea of human's superiority to the other entire natural existence and possibility of creating and remodeling the world for the sake of humankind's comfort. Subsequent progresses of science led to the gigantic industrial development when combined with the utilization of fossil fuels, which brought about improvement of human welfare and amenity in lifestyles as well as large-scale conversion of natural ecosystems in various parts of the globe.

Now we are facing the counterattack of natural systems through the global climate change, deterioration of hydrosphere, eutrophication or desertification of soil systems, deterioration of air, man-made chemicals pollutions, etc. All of these problems arose from the fact that the environmental system that surrounds human activities has the finite capacity and deterioration of the environment eventually affects human activities. Also ever increasing massive exploitation of natural resources is expected to invite exhaustion of non-renewable resources and diminishing of renewable resources on the earth. Our generation is said to be the first one that faced the limit of the earth systems. We have to consider seriously about finding a new paradigm to cope with the finiteness of both resources and environmental capacities. This is the reason why the present time is defined as the time of environment revolution. The most important issue is therefore how to find a sustainable system of human activities.

3. CONCEPT OF SUSTAINABILITY

There have been numbers of discussions on definition of "sustainability" so far. The sources of problems are clearly derived from the excess intensification of the anthropogenic activities compared with the finite limit of the natural ecosystems on the earth. Thus a practical definition of "sustainability" might be "Optimization of economic level of human activity within the limit of renewable resources supply and the acceptance capacity of natural ecosystems." In order to achieve this state, the working principle should be:

- To maximize resources productivity (product generated per unit resources) and to minimize material/energy load to the environment

Besides, in line with this definition, all the approaches should be of holistic nature. A systematic approach is especially important in order to clarify interlinkage of unit acts in the anthropogenic sphere.

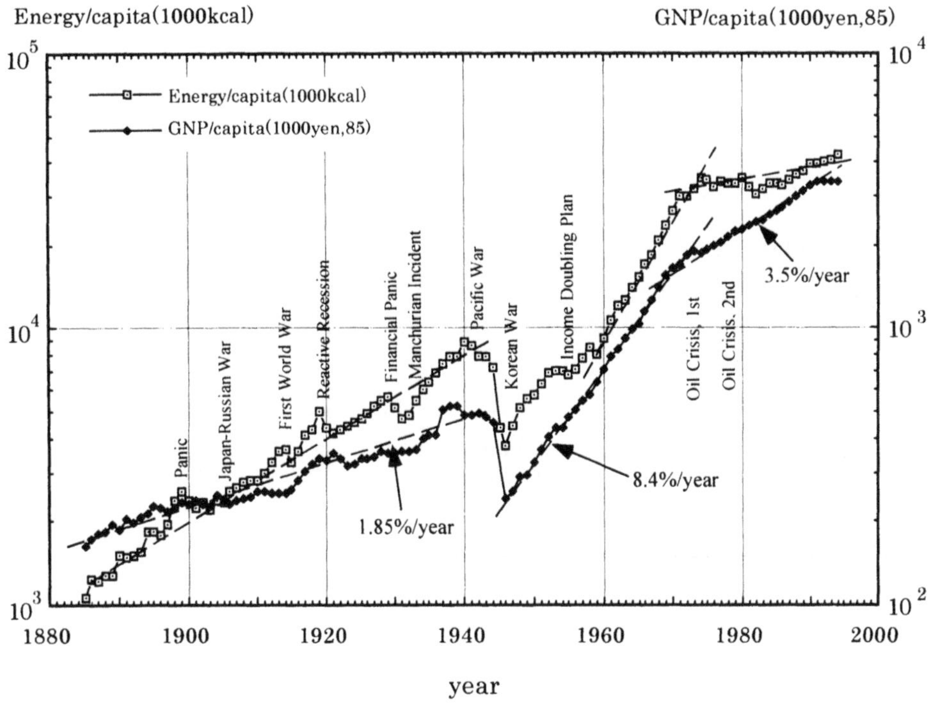

Figure 1. Change of CNP and energy consumption per capita in Japan since modernization starts.

4. PROBLEMS OF INDUSTRIALIZATION—JAPANESE EXPERIENCES

Japan experienced the remarkable growth of economy since the modernization started at late 19C. Figure 1 shows the long-term trends of gross national product (GNP) and energy consumption after the modern revolution of Japan. GNP figures are normalized on the basis of 1985 currency value. Almost a constant growth rate of 1.85%/year had been observed until the start-up of the Second World War. The situation changed after Japan was defeated at the WWII. The recoveries of GNP and energy consumption were so rapid (8.4%/year), which looked as "the miraculous high growth of economy". At 1970s Japan experienced the energy crisis twice, which leveled down the growth rate of energy consumption and at 1990s the bubble economy corruption greatly contributed to the slowing down of the GNP growth.

As illustrated in Figure 2, all the biological growths found in nature start from a logarithmic growth stage. Logarithmic growth implies that an intrinsic growth rate of the system is realized without any restricting conditions. The growth gradually becomes limited by supply of feeds/nutrients, spatial limitations, adverse interactions among components of various natures, etc. At simple systems such as a continuous monoculture in biological reactors, limiting factors of growth can be easily fed back to the growth rate of species. In these cases, stable steady state are rather easily found and achieved by taking a path shown as Soft Landing (I) in Figure 2. In more complicated systems such as biological predator-prey systems that involve plural species, growth of predators can easily eat up preys and extinction of prey species occurs, which eventually affects the existence of predator species.

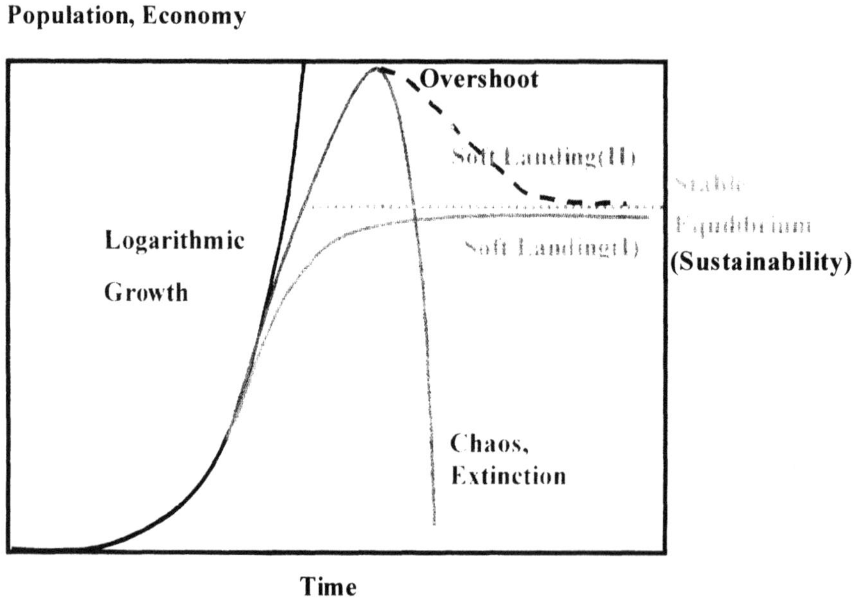

Figure 2. Typical growth patterns in complex systems

In complex systems, such as economic growth of a country and transition of natural ecosystems, limiting conditions of growth or external boundary conditions are sometime difficult to be fed back because of its complex structural natures and its larger time constants of the systems. Also inertial movement of systems often causes a delay in realization of the limit. This causes an overshoot in growth, which may result in chaotic behavior or eventual extinction of the system. The current growth situation of industrialized countries might correspond to these cases. To approach the stable steady state, specific working principles that are completely different from the working principles in the logarithmic growth period should be identified and employed.

A first step for this direction is to make an estimation of a level of sustainable steady state. Establishment of a sound material balance system is the most important basis for the sustainable society. After a goal of growth is visualized, strategies for soft landing on the goal must be discussed in detail.

Environmental problems have been derived from the intensification of anthropogenic activities in many areas. In the case of agricultural environment in Japan, wastes discharged from stock-breeding account for as much as 94.3 million ton per year which is equivalent to the amount of sludge generated from sewage treatment, 85.5 million ton per year. Those wastes if not properly treated or managed, become a heavy burden to the environment.

5. ZERO EMISSION APPROACHES

In considering stock-breeding wastes, traditional ways are to think about how to treat wastes at the exit of a stock yard and to add treatment processes to eliminate organic constituents, nitrogen, phosphorus, odor, color, etc. This typical treatment concept is called "end-of-pipe" method. "End-of-pipe" method has not been successful in treating strong waste stream from stockyards in most cases.

Reutilization is considered as a next step by employing some ways of value addition to the wastes. In fact, anaerobic digestion processes for producing methane gas and various composting processes have been proposed and already applied in some places in Japan. These examples become good examples in limited cases since anaerobic digestion is followed by the problems of residual liquors that again contain biorefractory organics and strong nitrogen and compost products have not established sufficient demand among farmers who have been accustomed to easy agriculture based on chemical fertilizer applications. Also if recycling of nitrogen contained in stock manures to the farmland becomes prevalent, eutrophication of soils may become a serious problem in Japan.

Zero Emission concept is based upon the concept of sustainability, that is as stated before, to maximize resources productivity (product generated per unit resources) and to minimize material/energy load to the environment and networking of various processes based upon a holistic balance is performed. When Zero Emission concept is applied to stockbreeding, nitrogen cycle might become one of the key components to be considered. Not only an end-of-type treatment process is pursued, nitrogen contained in the wastes should be returned to the feed to the stocks. Current stockbreeding depends on the feeds whose 75% are imported from oversea (nitrogen base). Hence it is necessary to reuse nitrogen in the wastes for production of feed cereals and to minimize the import of stock feeds.

As one of the projects of the United Nations University, a small scale Zero Emission experiment illustrated in Figure 3 had been conducted, where material balances and conversion rates in each unit process were determined.

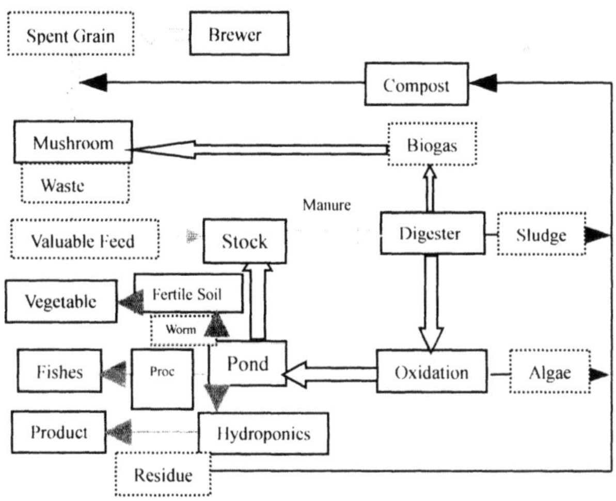

Figure 3 Example of Zero Emission systems based on integrated biological processes

This system introduces the spent grain from a beer brewery to culture mushrooms. Enzymes produced by the mushroom help decomposition of lignin contained in the spent grain, which after mushroom harvesting becomes a good feed to pigs. Manure of the pigs is used to generate biogas methane in a digester, where a half of carbon content is converted to methane gas. Residual liquid phase contains ammonium ion, which is led to waterways where ammonium ion is oxidized to nitrate ion and is taken up by algae growing in the pond.

Hydroponics is another value added product and varieties of fishes are raised in the pond. In this way, the wastes from a beer brewery are converted to various valuable products where solar energy is the only source of driving the system.

Other systems may be designed for stockbreeding where input and output balances in the system must be carefully examined so that accumulation or diminution of specific components is prevented. Also for a holistic approach to design stockbreeding systems, eating habits in a country including supply and demand of meats and dairy products should be designed so that a sustainable country style can be realized.

Greenhouse Gases and Animal Agriculture
J. Takahashi & B. A. Young (Editors)

Greenhouse gases and the animal industries

B. A. Young

Recycled Organics Consortium, The University of Queensland, Gatton, Qld4343, Australia.

The awareness of greenhouse gases and global warming are relative recent issues and there is still much uncertainty, debate and misunderstanding as to their overall importance. The issues are extremely complex and currently there is lack of knowledge as to all factors involved and their relative importance. Domestic livestock have been implemented as significant components from several points of view ranging from land clearance, intensive land-use practices, enteric fermentation, to waste disposal practices. Clearly, we need more research and understanding of factor involved and documentation of their relative importance. Until this information is available, care and attention must be take to ensure farming practices do not contribute unnecessarily to any imbalance in the carbon equations and cause a build up of greenhouse gases. Increasing demands from growing human populations and consumerism are putting pressures on agriculture for increased production and thus challenge livestock producers and scientists to develop sustainable practices which minimize possible harmful effects on the environment.

Keywords: greenhouse gases, reduction strategies, animal industries, community awareness

1. INTRODUCTION

Awareness over the past decade of greenhouse gases and the potential for global warming are of increasing concern to governments, the public and scientists around the world. The issues are truly international and require international cooperation (IPCC 1990). This paper provides an overview the greenhouse gas issues relating to animal agriculture and the Australian situation is analyzed as an example of a country with major livestock industries.

2. WHAT ARE THE GREENHOUSE GASES?

The gases that bring about global warning, via the greenhouse gas effect, are water vapour (H_2O), carbon dioxide (CO_2), methane (CH_4), nitrous oxide (N_2O) and chlorofluorocarbons (CFC's). Concern has arisen from the increasing amounts of these gases from anthropogenic origin (arising from human activities) that could contribute to global warming rather than the natural atmospheric cycles. The natural occurrence of water vapour in the earth's atmosphere is by far the major greenhouse gas contributor accounting for some 14 to 15 C to global warming. Without this warming from water vapour, life, as we know it, would not be possible.

Moss (1993) has documented the atmosphere concentrations, average resident times and trends in concentrations of the greenhouse gases of anthropogenic origin. He estimated carbon dioxide contributed about half of the anthropogenic additions to atmospheric concentrations and that this gas has increased from about 280 ppmv during pre-industrial times to the present levels of about 350 ppmv. The increase has been attributed largely to the increased combustion of fossil fuels, industrialisation and motor vehicles in the advanced nations. Carbon dioxide has a long troposphere resident time (>200 years). Its infrared absorption capacity is, however, restricted to a narrow band around 16 *u*m which is presently near saturation because of the amount of carbon dioxide present. Thus additional increases in levels of atmospheric carbon dioxide may contribute only marginally to further global warming.

In contrast, methane occurs at relatively low concentrations, 0.8 ppmv during pre-industry times but has increased to the current level of about 1.7 ppmv over the past 200 years. Despite its low concentrations; increases in atmospheric methane concentration potentially contributes to increased global warming because of its broad spectrum of infrared absorption and the fact that there is little overlap with the relatively saturated absorption zones of water vapour and carbon dioxide. Any increase in atmospheric concentration of methane has been estimated to be more than 50 times greater than for a similar increase in atmospheric water vapour or carbon dioxide. Therefore, even a small increase in methane concentration in the atmosphere exerts a potentially significant contribution to global warming.

Rates of methane release to the atmosphere are closely associated with natural wetlands, termites and oceans, and also with increasing human activities linked to the petrochemical industries, paddy rice cultivation, enteric fermentation in livestock, and animal and human wastes disposal processes (IPCC 1994). According to the IPCC, the annual release of methane to the atmosphere is about 535Mt wherein about one quarter arises from natural wetlands while, for example, only about 25Mt can be attributed to the breakdown of animal wastes. Depending upon how the calculations are made, 60% or more of the amount of methane released has been directly or indirectly attributed to human activities and the increasing demands arising from the human population. For these reasons methane is currently receiving considerable attention in the greenhouse gas debate and methane production has been very much linked with animal agriculture.

3. AUSTRALIA'S GREENHOUSE STRATEGY

As an example of one nation's actions in the greenhouse debate, Australia has made substantial effort to reduce greenhouse gas emissions (C of A, 1998).

- Australia supports and continues to make strong contributions to international negotiations on reducing greenhouse gas emissions.
- The Australian government has established the Australian Greenhouse Office, the world's first dedicated national greenhouse office.
- Substantial funding for research initiatives and monitoring has been provided to support a range of strategies to reduce greenhouse gas emission.
- The Australian Greenhouse Office is investigating the feasibility of a domestic emissions trading scheme.

Greenhouse strategies are coordinated and financially supported through the Australian Greenhouse Office that has several targeted programs to reduce greenhouse gas emissions (C of A, 1998). These include:-

The *Greenhouse Science Program* is aimed at improving global understanding of climate change. Objectives include improving models for climate prediction and description of regional climate change; gaining better understanding of processes relevant to climate change; improving our understanding of the role of greenhouse gases and the carbon cycle; improving appreciation of links between climate change and ocean circulation; and the development of specific predictive scenarios to counter the possible consequences of climate change in Australia.

The goal of the *Greenhouse-Friendly Power Supply Program* is to achieve an additional two percent uptake of renewable energy use and, as a consequence, Australia's renewable energy generation of electricity will rise from 10 to 12%. Some of the new technologies under development include energy generation and management from hot dry-rock geothermal; wind turbines; solar cell sources, the replacement of the lead-acid battery with more energy dense systems and the generation of electrical energy from municipal and agricultural wastes and by-products. A focused goal of this program is to achieve, through the use of the new technologies, greater renewable energy use in rural and remote areas for both domestic houses and whole communities.

The *Improved Energy Efficiency Programs* are aimed at increasing minimal energy performance standards for household goods, such as refrigerators, electric water heaters, etc., the development of minimal energy performance standards for electric motors, lights and air conditioners, and the adoption of mandatory energy efficiency standards in building codes. Mandatory improved efficiencies in fuel consumption for motor vehicles has be implemented along with encouragement for conversion to alternate fuels such as diesel and natural gas. Natural gas powered cars are 50% cheaper to run than petrol fuelled cars, and natural gas vehicles produce about 18% fewer greenhouse gas emissions.

A program called, *The Greenhouse Gas Challenge*, has been implemented, on a voluntary basis for industry and government partnerships, to assist industry in reducing greenhouse gas emissions. More than three hundred major enterprises have joined this voluntary challenge. As a consequence, some 25Mt of CO_2 equivalent (CO_2 -e) have been eliminated from the emission stream. Even the Sydney Olympic Organising Committee was a challenge member and the recent games demonstrated commitment and carried a clear message of Australia's desire for energy efficiency and greenhouse gas reduction

Of significance, for agriculture and rural communities, is the *Bush for Greenhouse Program* that has attracted much attention, improved research funding for greenhouse related aspects of animal agriculture and is also attracting corporate investments into the plantings of forest trees.

4. ACCOUNTING FOR GREENHOUSE GAS EMISSIONS

The *National Carbon Accounting System* will establish Australia's 1990 emission baseline for land use changes and forestry, and provide the data needed for on-going reporting of the greenhouse sink performance (AGO 2000). The aim is to have this system internationally creditable and to reduce scientific uncertainties surrounding the estimation of emissions and sinks. At the same time, the aim is to have the system transparent and for there to be verification in reporting.

Under the *Framework Convention on Climate Change*, Australia has produced an annual inventory on greenhouse gas emissions from 1990 to 1997 (NGGIC, 1999). In 1997, for example, the Australian Greenhouse Gas Inventory totaling some 431 Mt $CO_{2\text{-}e}$ was based on the following sectors contributions:

Stationary Energy	55%
Transport Energy	16%
Fugitive Energy	7%
Industrial Processing	2%
Agriculture	22%
Forestry	-6%
Waste	4%

It is estimated that approximately one quarter of the total anthropogenic $CO_{2\text{-}e}$ emissions in Australia comes from agriculture while waste disposal contributes less than 5%. In contrast, the equivalent calculated for forestry is currently a sink of about 6%. Within the agriculture sector, where the common types of greenhouse gases are methane and nitrous oxide, the sources of emissions include:-

- *Enteric fermentation* - emissions associated with microbial fermentation during digestion by ruminants and some non-ruminant animals (64.4Mt $CO_{2\text{-}e}$)
- *Manure processing* - emissions associated with decomposition of organic waste of largely animal origin (2.2Mt $CO_{2\text{-}e}$)
- *Rice cultivation* - methane emitted from anaerobic decay of plant and other organic material in rice fields (0.7Mt $CO_{2\text{-}e}$)
- *Agricultural soils* - emissions associated with the disturbance of agricultural land by cropping and application of fertilisers (15.4Mt $CO_{2\text{-}e}$)
- *Clearing and burning of agricultural land* - large segments of Australian savannah country have been cleared in recent years and the consequences are substantial emissions of greenhouse gas equivalents (11.3Mt $CO_{2\text{-}e}$).

Recently there has been a continued increase in emissions from agriculture in Australia but not equally across all areas. Ruminant animals are the largest source of Australia's methane emission. In 1997, domestic livestock generated about 70% of the agricultural sectors emission, representing 15% of the total national emissions (Lumb et al, 1995). There was an estimated peak in livestock-related methane emissions in 1991 with the rate of emissions declining since. Although the numbers of cattle have increased, there has been a large reduction in sheep numbers. These reductions in sheep numbers have been largely responsible for the decline in rural emissions. It is of interest to note that only the domestic livestock are included in the Australian calculations. The contributions of natural ecosystems and native animals are, by convention, excluded from the emission inventory because they are not a human-induced. However, of possible significance, but presently not included in the anthropogenic calculations, are feral animals, imported insects and, of course, the enteric fermentation in the digestive tracts of the 19 million Australians who live mainly in the larger cities.

Agricultural crops, soil and fire-related emissions are the largest source of emission of nitrous oxide in Australia. In 1997, the total nitrous oxide emissions reached 26Mt $CO_{2\text{-}e}$ with nearly 80% from agriculture, mainly from nitrogenous fertiliser plus some associated with the burning of grasslands and clearing of trees. Since 1990 there has been an increase in $CO_{2\text{-}e}$ emissions from increased land clearing and fertiliser use (NGGIC, 1999). Furthermore, rice production has increased in the '90s that has resulted in an increase in emission from this area of agriculture. The area under rice cultivation has increased from one hundred thousand hectares in 1990 to one hundred and fifty hectares in 1997 (NGGIC, 1999).

5. WHAT IS BEING DONE TO REDUCE GREENHOUSE EMISSIONS FROM THE ANIMAL AGRICULTURE?

There is support for research programs to help the animal industries and agriculture to reduce the atmospheric release of greenhouse gases and to achieve increased energy use efficiencies. Some examples of this research were presented at a recent international workshop at our laboratory (Young et al, 1999; Takahashi and Young 2000). In Australia, research programs are specifically targeted at rangeland systems and animal waste processing for intensive livestock holdings. Research is addressing the more efficient use of fertilisers, minimising tillage practices and identification of alternatives to the use of fossil fuels, and the reduced burning of agricultural residues. These activities have potential benefits in encouraging sustainable agricultural practices that go well beyond the direct greenhouse benefits.

Substantial research effort is being given to the manipulation of fermentation in the rumen and the development of methanogenic vaccines to reduce methane release from the rumen and to increase animal production. There is also considerable research effort in bio-gas production and other technologies to reduce greenhouse emission to the environment from intensive animal industries and their effluent.

6. COMMUNITY AWARENESS

One of the more difficult aspects of understanding the situation with respect to greenhouse gases is the sheer magnitude of the masses of gas involved. Few people can, for example, appreciate the magnitude of a Mt of carbon dioxide let alone 450Mt of methane releases into the atmosphere. To ease our understanding of the complexities and confusion arising from the use of very large numbers our laboratory has developed a scheme that reduces the calculation of the amounts of gases to a unit area (square meter or hectare) of land (Young 2000). This allows key issues relating to greenhouse gases emissions to be illustrated on a local area to get the message to persons not familiar with the very complex issues involved at a global level.

Natural physical and biological cycles can be readily represented diagrammatically to illustrate local responsible for the fluxes of materials between the atmosphere, biosphere and the geosphere. As can the effects of increases in demands from human activities and unsustainable practices associated with the intensification of land use, the excessive clearing of natural forest and vegetation are contributing to imbalances. The mining of geo-fossil materials (petrochemical and coal resources) over the past 200 years has resulted in a substantial flow of carbon dioxide equivalents from the geosphere to the atmosphere. As this flow is largely unidirectional, it is a major contributor to the accumulation of atmospheric greenhouse gases and potentially global warming.

7. CONCLUSION

Are we, as a society, too ready to point the finger at others and their use of fossil fuels, the coal industry or domestic livestock as the course of the greenhouse problems. The cities and the large numbers people therein with high consumer lifestyles and the ever increasing demands for food supply, electricity and energy for heating and transport that must be recognised as important driving forces to anthropogenic greenhouse gas generation. Recent publicity and environmental movements have made city communities much more aware of the potential ill effects of greenhouse gases and global warming. This awareness has had a significant effect to change the attitudes of people and in the operations of some industries. The consequences have been changes in the way people think about the environment and the development of better attitudes towards protecting the environment for future generations.

REFERENCES

AGO, 2000: http://www.greenhouse.gov.au/pubs/internationalsinks.

Commonwealth of Australia, 1998. The National Greenhouse Strategy-Strategic Framework for Advancing Australia's Greenhouse Response. Australian Greenhouse Office, Canberra.

IPCC, 1990. Climate Change: the IPCC Scientific Assessment, Cambridge University Press, Cambridge, UK.

IPCC, 1994. Climate Change in 1994 (J.H.Houghton etal eds), Cambridge University Press, Cambridge, UK.

Moss, A.R., 1993. Methane. Chalcombe Publications, Kingston, Canada.

NGGIC, 1999. National Greenhouse gas inventory – Analysis of Trends 1990 to 1997 and National Greenhouse Response Strategy – Selected indicators 1990 to 1997, National Greenhouse Gas Inventory Committee, Canberra.

Lumb, J.M., Pears, A., Beer, T. and Galbally I., 1995. Performance Indicies for the National Greenhouse Response Strategy. Report to the Department of the Environment, Sports and territories, May 1996.

Takahashi, J., and Young, B.A., 2000. Greenhouses Gases and Animal Agriculture. University of Queensland, Brisbane. pp106.

Young, B.A., 2000. Greenhouse gases in Australia – Where are the animal industries? In: Greenhouses Gases and Animal Agriculture. Editors (J. Takahashi and B. A. Young). University of Queensland, Brisbane. 1-5.

Young, B.A., Takahashi, J. and Copland, R., 1999. Greenhouse Gases in Animal Husbandry. University of Queensland, Brisbane. pp57.

Greenhouse Gases and Animal Agriculture
J. Takahashi & B. A. Young (Editors)

Greenhouse gases and animal agriculture in Asia

J. B. Liang

Department of Animal Science, Faculty of Agriculture, Universiti Putra Malaysia, 43400 UPM Serdang, Selangor, Malaysia

Asia has been one of the most rapidly developing regions in the world over the last two decades. More than 55% of the 6 billion world inhabitants live in Asia and with many countries within Asia recording an average annual GDP three times higher than the world average of 2.2%. Intensification of livestock production to meet the rapid increased demand for animal protein is inevitably associated with environmental pollution including the emission of greenhouse gases. The transition from traditional livestock production to the intensive industrialized system is characterized by a shift from ruminant to monogastric livestock farming, the use of imported grain concentrates over the traditional roughage feeds and incapable to utilize the excessive nutrients from livestock manure. The above influence the emission of greenhouse gases and thus affecting global warming. The result of global climatic change is postulated to affect Asian farming systems through a rise in sea level, higher tropical surface temperatures, increased tropical cyclone frequency and severity, and changes in cloud cover and precipitation. This paper highlights some of the above issues with reference to Asia.

Keywords: animal agriculture, greenhouse gases, Asia

1. INTRODUCTION

Asia is the fastest developing region in the world. More than 55% of the current 6 billion world inhabitants live in Asia (not including the Middle East) and with an average annual population increase of between 0.2 in Japan (the lowest) to 2.6 in Laos People's Democratic Republic (the highest) (Table 1). Average annual GDP growth rates (1988 to 1997) ranged between 2.2% for Japan to 8.2% for China, with several countries in South East Asia, such as Indonesia (6.7%), Malaysia (7.5%), Singapore (7.3%), Thailand (6.8%) and Vietnam (6.9%) recorded more than 3 folds of the world average (2.2%) (Table1). To meet the sudden surge in the demand for animal proteins as the results of the increased population and affluence, livestock production had intensified and shifted from the traditional integrated farming to an intensive 'industrialized' system.

The most distinct changes in the shift from traditional to industrialized livestock production include (i) a shift of livestock species farming from ruminant to monogastric because of the apparent better feed conversion efficiency of the latter, (ii) an increasing dependency on imported cheap grain concentrates over grazing and or the use of locally available resources and (iii) incapable to recycle the excessive nutrients from livestock manure within the farm because of restricted available farmland. It is believed that the new industrialized livestock production systems will impact on the environment including the emission of greenhouse gases and thus on global warming. This paper discusses the above scenario with particular reference to the Asia.

Table 1: Human population, mean annual population growth rate and annual GDP growth rate of selected Asian countries as compared to other regions of the world (source: World Resource Institute; wttp://earthtrends.wri.org)

	Population (million)			Annual Population Growth rate (%)		Annual GDP growth rate (%)
Year	1950	2000	2025	1975-1980	1995-2000	1988-1997
WORLD	2,522	6,055	7,824	n.a.	1.3	2.2
ASIA	1,334	3,420	4,308	n.a.	n.a.	n.a.
China	555	1,278	1,480	1.5	0.9	8.2
India	358	1,014	1,330	2.1	1.6	5.0
Indonesia	79	212	273	2.1	1.4	6.7
Japan						
Lao PDR	2	5	10	1.2	2.6	6.4
Malaysia	6	22	31	2.3	2.0	7.5
Myanmar	18	46	58	1.7	1.2	n.a.
Philippines	21	76	108	2.3	2.1	3.0
Singapore	1	4	4	1.3	1.4	7.3
Thailand	20	61	73	2.4	0.9	6.8
Vietnam	30	80	108	2.2	1.6	6.9
EUROPE	547	728	702	0.5	0.0	n.a.
NORTH AMERICA	172	310	364	0.9	0.8	n.a.
OCEANIA	13	30	40	1.1	1.3	n.a.

n.a. = not available

2. INDUSTRIALIZED LIVESTOCK PRODUCTION

2.1. Ruminant versus Monogastric Animals

Modern monogastric livestock, particularly pigs and poultry are extremely efficient in converting feed nutrients into high value animal protein. With balanced nutrition and modern management technology, the feed conversion efficiency (kg feed/kg weight gain) for pigs and poultry is less than two whereas in the case of ruminant livestock such as cattle, buffaloes, sheep and goats the values are three times or more higher depending on quality of feed. This is the primary basis which had driven the rapid shift from ruminant to pig and poultry farming in many rapidly developing nations including those in Asia in the last two decades (Table 2).

The shift from ruminant to monogastric livestock production should theoretically help to slow down the increased production of greenhouse gases, particularly methane and, to a lesser extend, carbon dioxide associated with the livestock production. To illustrate the above point, depending on productivity and thus feed intake, a matured lactating cow produces between 400 to 500 liters of methane daily from its enteric fermentation (Kurihara, *et. al.*, 1997). Broadly speaking other ruminants such as beef cattle, buffaloes, sheep and goats fed similar diet also produce an equivalent amount of methane per unit body weight as those of a dairy cow, but production of the above greenhouse gases is much lower in pigs and negligible in poultry.

Table 2. Changes (%) in animal populations of selected Asian countries in the last two decades (source: FAOSTAT Agriculture Data, Http://www.fao.org)

Country	1980-1990				1990-2000			
	Cattle	Buffaloes	Chicken	Pig	Cattle	Buffaloes	Chicken	Pig
China	51.43	16.17	126.93	10.81	31.55	5.49	73.42	21.24
India	8.63	21.95	57.39	33.33	7.60	16.39	36.73	37.50
Indonesia	61.65	35.74	282.69	126.17	16.25	-14.27	40.07	31.07
Malaysia	21.51	-28.10	12.20	45.79	8.33	-24.34	110.53	-31.72
Myanmar	9.13	8.42	4.09	3.73	17.77	18.45	86.61	71.83
Philippine	-13.47	-3.67	55.99	0.71	56.69	9.15	72.54	30.14
Thailand	39.19	-9.85	91.92	57.62	11.28	-58.78	59.91	61.33
Vietnam	87.29	23.39	35.07	22.59	32.73	1.51	152.17	64.71

The presence of methane and carbon dioxide in the atmosphere results in the retention of solar radiation. Methane, although present at relatively lower concentration as compared to carbon dioxide, on the molecular basis, is more than 50 times more efficient in absorbing infrared radiation than the latter. And about 60% or more of the methane released has been directly or indirectly attributed to human activities. For these reasons methane is currently receiving considerable attention in the greenhouse gas debate (Young, 2000). Methane in the atmosphere may account for15% to the global greenhouse effect (Williams and Kelly, 1994). Global methane emission from livestock industry accounts for 10-30% of total emission, with some countries reaching as high as 70% (e.g. New Zealand). The above values for most Asian countries are about 25% with Japan on the high end of nearly 50% (Table 3).

Table 3: Inventories of national greenhouse gas emissions in selected Asian countries compared to countries in other regions, 1989-94. (Source: World Resource Institute; wttp://earthtrends.wri.org).

	Methane Emissions from Anthropogenic Sources (000 metric tons)						
	Year	Fossil Fuel Extraction	Fuel Combustion	Agriculture Livestock	Other	Waste	Total Methane
ASIA							
China	1990	5,650	50	8,940	18,400	790	33,830
Indonesia	1990	527	316	864	2,039	n.a.	3,746
Japan	1994	94	26	520	276	400	1,316
Philippines	1990	8	220	315	559	138	1,290
EUROPE							
France	1993	333	182	1,550	38	725	2,831
German	1993	1,460	130	1,688	n.a.	1,917	5,203
Netherlands	1994	159	28	478	n.a.	376	1,041
NORTH AMERICA							
Canada	1994	1,695	40	964	n.a.	815	3,514
United States	1994	7,630	945	8,558	638	10,400	28,171
OCEANIA							
Australia	1990	1,026	28	3,005	396	1,391	6,243
New Zealand	1994	23	8	1,436	n.a.	423	1,895

n.a. = not available

2.2. Roughage versus grains feeding

The association of increased roughage/fiber to ruminant diets with increased production of methane gas been identified almost half a century ago (Blaxter, 1967). More recently Kurihara, *et. al.*, (1997) analyzed 30 years data obtained from 115 energy balance trials with lactating cows in Japan had shown that methane production was strongly associated with digestible carbohydrate intake, particularly the component of crude fiber. The above authors estimated that due to an increase use of grain concentrates, 4% fat-corrected milk production increased 36% over a period of 13 years (1980 to 1993) while methane production increased only by 13% in the same period of time. Their results thus imply that increased feeding of concentrates (less crude fiber) reduces methane production. The production of methane (particularly resulting from maintenance metabolism) is not only energy inefficient for animals but also has a negative impact on the environment as discussed above. Industrialized livestock production system including feedlotting of beef cattle and intensive dairy production has moved away from the use of bulky roughage to the conveniently available high-density nutrient grain concentrates. Modern pigs and poultry productions have done away with most of the roughage/fiber in their rations, replacing them with grain concentrates instead.

The question to be addressed here is "whether the shift from roughage to grain concentrates in the livestock feeding reduces the emission of greenhouse gases"?

We have earlier pointed out that livestock fed grain concentrates emits much lower or negligible amounts of methane as compared with those fed high roughage diet. However, large scaled mechanized cropping of grains to produce animal feed requires energy derived from the burning of fossil fuels and chemical fertilizers. Extraction and combustion of fossil fuels for the above purposes emit huge quantity of greenhouse gases and these are the major contributor of the total emission (Table 3). The reduction in emission of greenhouse gases from grain-fed livestock is apparent on one hand, but has a hidden component in the use of fossil fuels for the production of grains.

An interesting phenomenon associated with industrialized livestock, particularly pigs and poultry productions in many Asia countries, is the heavy dependence on imported grains. This seems to speak well for the countries concern in term of emission of greenhouse gases *in situ* from fossil fuels for production of grains because they do not have to produce these grains locally. Nevertheless, the fact remains that the increasing quantum of livestock production in the Asian countries contributes to the increasing total global emission of greenhouse gases and thus global warming.

2.3. Waste treatments

The traditional integrated agriculture system depends on recycling of livestock manure for cropping. Such a system creates a balance between nutrient input and output in the soil and thus maintains an equilibrium environment in the farms. Economic pressures have led to intensification of livestock enterprises where animals are housed at high stocking densities producing a huge volume of waste, which is often beyond recycling capacity of the limited farmland (Koshino, 1988).

Livestock manure from industrialized farming units in most developing Asian countries are not returned to the soil due to shortage of land. The manure is often washed and flushed into nearby waterways directly or pre-treated depending on how stringent are the local environmental regulatory authorities. Treatment of livestock manure, particularly in the form of slurry from pig and cattle farms, is mainly based on temporary holding of the slurry in a series of ponds in the farms before discharging it into the surrounding waterways, often resulting in contamination of ground water. Anaerobic fermentation in the ponds emits greenhouse gases, widely referred to as biogas, which can serve as useful household fuels. Because of its potential as a household fuel, the biogas technology was thought to be an economically viable and environmentally friendly solution of livestock waste management, particularly for family farms in the rural Asia. Unfortunately the acceptance of this technology, which was extensively introduced to many rural Asia in the 60's and 70's was short-lived and thousands of biogas digesters constructed then are now derelicts.

3. IMPACTS OF CLIMATE CHANGE ON ASIAN AGRICULTURE

The overall effects of global warming due to anthropogenic (human activities) increases in greenhouse gases are well understood, but less so for its effects at regional levels and higher-order (e.g. hydrologic) (Bennett *et al.*, 1999). The possible resultant effects of climatic change could affect farming systems in Asia through rising sea level (estimated to be 0.5 to two meters between 1990 and 2100), higher tropical surface temperatures (0.028° to 0.074° C per decade), increased tropical cyclone frequency and severity, and changes in cloud cover and precipitation (Bennett *et al.*, 1999).

Increased ultraviolet radiation (UV-B) due to ozone layer depletion by chlorofluorohydrocarbons (CFCs) may also affect elements of farming systems. Tropical UV-B flux after ozone depletion could exceed that experienced anywhere on earth in recent geological history, since the highest naturally occurring UV-B fluxes are at the equator (Caldwell *et al.*, 1989)

According to Bennett, *et al.* (1999) rising sea level, in the absence of defensive measures, would gradually inundate coastal farmland, forcing shifts to more salt-tolerant activities like shrimp farming and, ultimately, coastal retreat. In Asia, low-lying countries such as Bangladesh, Indonesia and Thailand and many low-lying islands will be most affected. In quite similar manner, more and stronger cyclones would cause more crop damage. Frequency of tropical cyclones in the north Indian Ocean during the last 30 years was found to be inversely correlated with sea-surface temperature even though the energy source for the cyclones is the heat in the surface mixed layer (McBride, 1989). Changes in precipitation would reduce economic returns on existing water resources infra-structural investment in affected areas. Other more subtle effects such as changes in natural plant propagation rates and species mix and in plant disease patterns are also not ruled out.

4. CONCLUSION

Animal agriculture contributes substantial amount of greenhouse gases and thus global warming. Current statistics indicated that the majority of Asian countries emit lower greenhouse gases, both in total and on per capital basis, as compared with their counterparts in the western world. The above scenario, however, is expected to change in the near future as most Asian countries are developing their livestock industry very rapidly to meet the increased demands for animal protein. Therefore, it is necessary that Asian countries take cognizance of the consequences if the phenomenon is not checked. The issue is truly a global one and requires transnational co-operation. It is, however, heartening to know that many Asian countries are signatories to the 1997 Kyoto Protocol, indicating their concern and willing to work for a better world.

Table 4. Inventories of methane emissions (1989-94) in selected Asian countries compared to other regions (source: World Resource Institute; wttp://earthtrends.wri.org)

	Methane emission			
Country	Total*	Per capita**	Livestock*	Per capita**
ASIA				
China	33,830	26.5	8,940	7.0
Indonesia	3,746	17.7	864	4.1
Japan	1,316	10.4	520	4.1
Philippines	1,290	17.0	315	4.1
EUROPE				
France	2,831	67.7	1,550	37.1
Germany	5,203	76.1	1,688	24.7
Netherlands	1,041	65.9	478	30.3
NORTH AMERICA				
Canada	3,514	112.8	964	31.0
United states	28,171	101.2	8,558	30.7
OCEANIA				
Australia	6,243	330.6	3,005	159.1
New Zealand	1,895	490.7	1,436	371.8

* 000 metric tons; ** metric tons

REFERENCES

Bennett, S.L., Rahman, A. and Huq, S. 1999. Climate change and Asian farming systems. http://bicn.com/wei/resources/BennettRahmanHuq1990.htm

Blaxter, K.L. 1967. The efficiency of energy transformations in ruminants. Proceedings of 4th Symposium on Energy Metabolism of Farm Animals. (ed. K.L. Blaxter, J. Kielanowski & G. Thorbek). EAAP Publication No. 12. p. 21 – 28

Caldwell, M.M., A.H. Teramura and M. Tevini. 1989. The changing solar ultraviolet climate and the ecological consequences for higher plants. Tree, 4 (12): 363 -366

Koshino, M.1988. Recent Advances in the Application of Livestock manure to Farmland in Japan. Proceedings In International Seminar on the Utilization of Livestock and poultry manure. Los Banos, Phillipines. June 1988. p. 1-17

Kurihara, M., F. Terada, M. Shibata and T. Nishida. 1997. Methane emission from lactating cows in Japan during past 30 years. Proceedings of the 8th International Conference on Anaerobic Digestion. May 25-29, 1997. Sendai International Center, Sendai, Japan. Vol. 3: 329-332

McBride, J.L. 1989. The effect of greenhouse warming on global tropical cyclone activity. In: The greenhouse effect and coastal area of Bangladesh – proceedings of an international conference, Dhaka, 9th March 1989. (Ed. H.J. Moudud, H. Rashid, A. Rahman, M. Hossain). P. 56-58 (cited Bennett et al., 1999)

Williams, P.E.V. and J.M. Kelly. 1994. Animal agriculture and pollution problems. In: Livestock Production in the 21st Century Priorities and Research Needs. (Ed. P.A. Thacker). p.159-186

Young, B.A. 2000. Where are the animal industries in the greenhouse debate? Malaysian Journal of animal Science. Vol. 6 (1/2): 39-44

Greenhouse Gases and Animal Agriculture
J. Takahashi & B. A. Young (Editors)

Global warming and animal agriculture in Japan

F. Terada

National Institute of Livestock and Grassland Science, Tsukuba Norin-kenkyu-danchi, Ibaraki, 305-0901 Japan

The estimated greenhouse gases (GHG) emitted from Japan was 1,335,537Gg/year (CO_2 equivalent). Methane (CH_2) and nitrogen oxide (N_2O) contributed 2.14 % and 1.49 % of the total, respectively. Animal agriculture produced 32% of total CH_4, especially, enteric fermentation emitted a huge amount of CH_4 (328 Gg/year). Much research was conducted to reduce the CH_4 from this sub-sector. The estimated CH_4 and N_2O emission from manure management were 105.7 and 4.3 Gg/year, respectively, but variation in the emission factors were very high.

The demand of animal products has reached a plateau and total number of animals to meet the demand for animal protein will decrease gradually due to improvements of productivity. Consequently, GHG emission will be reduced in Japan. However, this subject must be solved on a global scale and we do not have enough information concerning the effects of global warming on productivity of animal agriculture and effects of animal agriculture on global warming. To deal with this problem, there is a need to improve the accuracy of GHG inventory, generating technologies to reduce GHG emission from animal agriculture sources and to clarify the effects of global warming on the productivity of animal agriculture components.

Keywords: methane, nitrous oxide, enteric fermentation, manure management

1. INTRODUCTION

The environmental crisis from global warming or the commonly called ‘greenhouse effect’ is a major issue in these days. Methane and N_2O are potent greenhouse gases produced from animal production system. Methane emission from ruminants accounts for about 16% of total global CH_4 emission (Fig.1), and N_2O emission from animal wastes is 6% of total N_2O emission (IPCC, 1995). A huge amount of GHG are produced in Japan, but these gases are mainly derived from industrial sectors. However, the agriculture sector contributes 57 % of CH_4 and 13 % of N_2O emission in Japan (Ministry of Environment, 1998).

The major objectives of this paper are 1) to present the amount of GHG emitted from animal agriculture sector in Japan, and, 2) to review the obtained results of the research activities regarding the estimation and suppression of GHG emission, especially CH_4, from animal agriculture in Japan.

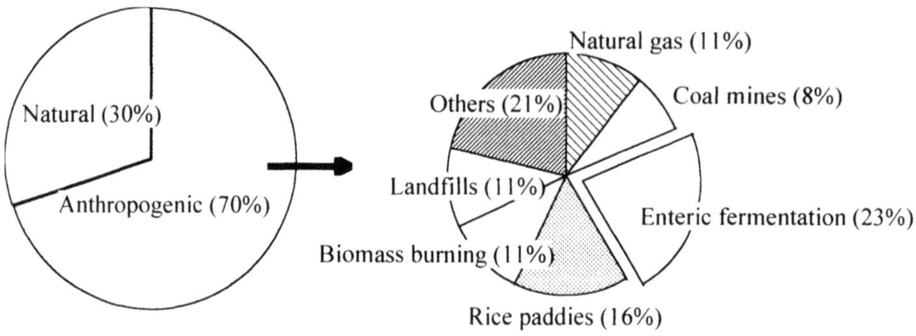

Fig. 1 Estimated sources of methane in Tg/yr (IPCC, 1995)

2. DEVELOPMENT OF ANIMAL AGRICULTURE IN JAPAN

Historical documents showed that animal agriculture in Japan has expanded rapidly for the last few decades, with increasing demand for animal products. The expansion of animal agriculture is characterized by a noticeable increase in the number of animals per farm. For example, presently, the average number of dairy cattle per farm is 51 head which is more than 10 times higher than that was in 1960. The situation for beef production is similar, and the number of swine per farm has also increased 86 times. The total dairy cattle population was 1.82 million in 1999 while beef cattle was 2.84 million and the numbers are almost steady. A dairy cow produces on an average 7,600kg milk/year. Beef cattle (Wagyu) are fed on concentrate rich diets as finishing diets and is slaughtered at about 700 kg body weight at 30 months old. The population of swine, layer and broiler chickens are 9.88, 143.1 and 107.4 million, respectively, but there is a slight decreasing trend. The population of horse, sheep and goats are not so important in Japan.

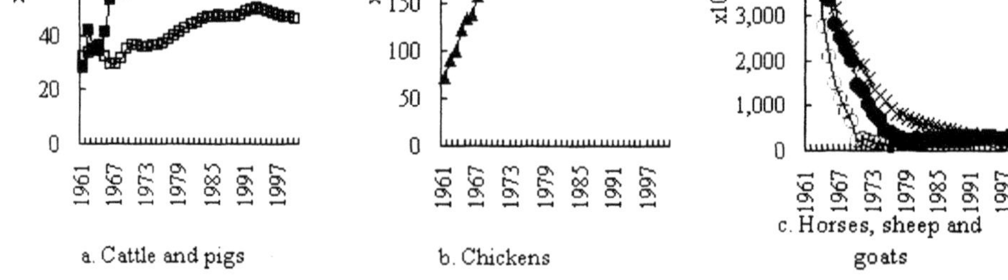

Fig. 2 Livestocks population in Japan (1961-1999)

The estimated amount of animal waste excretion was 95 million tons per year, suggesting that 11.2 million tons organic matter and 717 thousand tons total nitrogen are excreted from animals per year (Haga, 2001).

3. AMOUNTS OF GREENHOUSE GASES EMITTED FROM JAPAN

According to Ministry of Environment, 1,363.3 Gg CH_4 was emitted from Japan in 1998. Emission from enteric fermentation was 327.8 Gg, while that from manure management was 105.7Gg. Emission of CH_4 from animal agriculture was equivalent to 31.8 % of total. Emission of N_2O was 4.3 Gg from manure management, which was 6.7 % of total. Since total GHG emission from Japan was 1,335,536 Gg (CO_2 equivalent), both the CH_4 and N_2O in animal agriculture contributed 0.78 % of total GHG emission in Japan.

Table 1. Methane and Nitrous Oxide Emissions in Japan (1998)

	CH_4	N_2O
	-- Gg/year --	
Energy	177.6	24.3
Industrial Processes	47.3	24.8
Agriculture	771.4	8.1
Enteric Fermentation	327.8	-
Manure Management	105.7	4.3
Rice Cultivation	333.5	-
Agricultural Soils	-	3.1
Field Burning of Agricultural Residues	4.5	0.7
Waste	367.0	6.0
Solvent & Other Products Use	-	1.2
Total	1363.3	64.4

4. METHANE EMISSION FROM ANIMAL AGRICULTURE

4.1 Methane emission from enteric fermentation in Japan

4.1.1. Cattle

Methane emissions from Holstein and Japanese black cattle measured in Japan (Shibata et al., 1993) are shown in Fig.3 Lactating cows, which had a dry matter intake (DMI) of 3 to 4 % of body weight (BW), produced 464 L/day CH_4. Meanwhile, the other type of cattle produced 200 to 300 L/day. Expressing the data as CH_4 per kg DMI, cows had approximately 30 L/kgDMI with a small variation (27-36 L/kgDMI). In Tier 1 method of the Guidelines for National Greenhouse Gas Inventories, IPCC (1996), it was recommended that the dairy cattle population is estimated separately from other cattle. According to results in Fig.3, this is highly agreeable. In Tier 2, methane conversion rate (MCR, percentage of feed energy converted to methane energy), for feedlot cattle is lower than the other cattle, because CH_4 production is suppressed under the high concentrate based diet conditions. Also, according to our observation, cattle fed more than 20 % roughage diets produce a normal volume of CH_4 (linearity between CH_4 emission and DMI is very high), but cattle fed high concentrate diets produce a more variable volume of CH_4 (Fig.4).

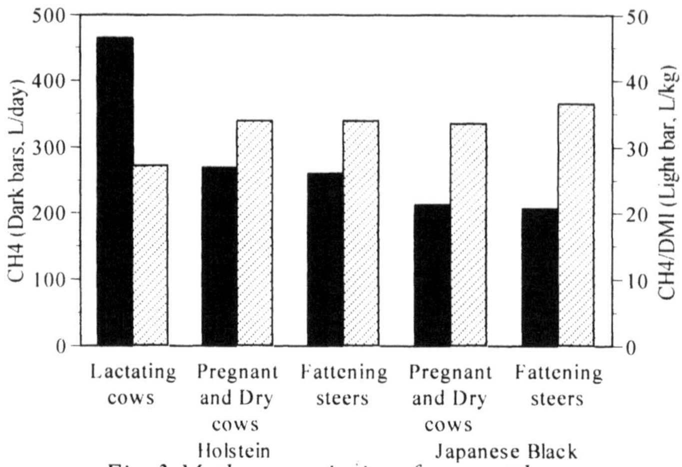

Fig.3 Methane emission from cattle

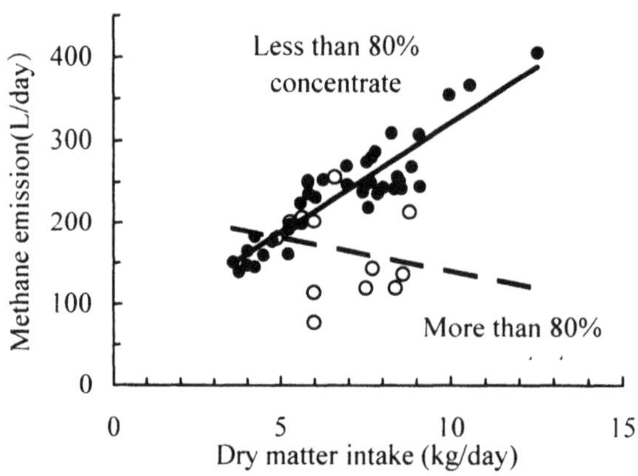

Fig.4 Relationship between dry matter intake and methane emissin in fattenig cattle

4.1.2. Sheep

The average MCR of sheep in Japan is 6.6 % (n = 21). Based on the data, the calculated CH_4 emission is 6.4 kg/year/head assuming DM intake 0.8 kg/day. This value is within the IPCC default value of 5-8 kg/head (IPCC, 1996).

4.1.3. Goats

Goats are an important species in Asia but a small population in Japan. The average MCR value of 52 diets measured in our laboratory was 6.4 % and the CH_4 emission was 4.3 kg/head/year. This value was similar to the default value of 5 kg/head/year (IPCC, 1996).

4.1.4. Swine

Saitoh (1988) reported that the MCR value was 0.65 % and 0.5 % for fattening pigs and pregnant sows, respectively under thermo-neutral conditions but the value was higher on high fiber diets.

4.2. Prediction of methane emission from enteric fermentation in Japan

The method of Shibata et al (1993) was used to estimate the enteric fermentation CH_4 in Japan. This method is based on the relationship between CH_4 emission and DMI (Fig.5). According to the report, 96 % of total CH_4 emitted from enteric fermentation in Japan was by dairy and beef cattle. Kurihara et al (1997) developed an estimation equation of CH_4 emission from lactating cattle based on fat corrected milk (FCM) yield and reported that CH_4 emission per kg FCM from lactating cattle were decreased by increasing the FCM yield. Estimated CH4 emission from lactating cows by Kurihara's method is similar to that of Shibata's.

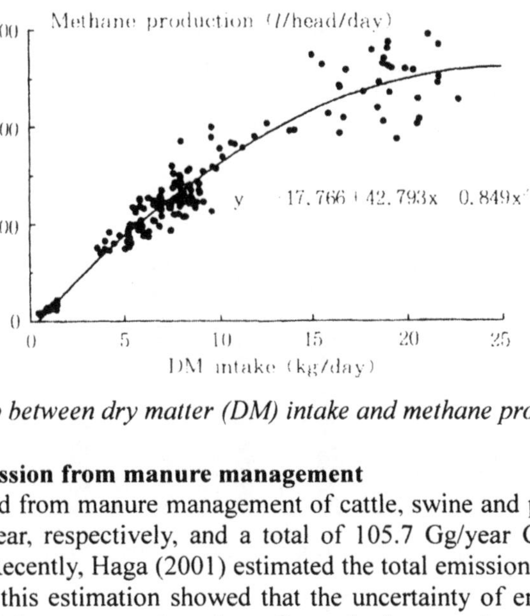

Fig.5 Relationship between dry matter (DM) intake and methane production

4.3. Methane emission from manure management

Methane emitted from manure management of cattle, swine and poultry was 25.2 Gg, 78.1 Gg and 2.4 Gg/year, respectively, and a total of 105.7 Gg/year CH_4 in 1998 (Ministry of Environment). Recently, Haga (2001) estimated the total emission to be 32.4 Gg/year. The large variation in this estimation showed that the uncertainty of emission factor for manure management is higher compared to that of enteric fermentation. In Japan, a composting system is followed to manage livestock manure. To suppress the CH_4 emitted from the system, it is recommended to use a forced aeration system and maintain aerobic conditions (Osada, K., et al., , 2000).

5. NITROUS OXIDE EMISSION FROM ANIMAL AGRICULTURE

Nitrous oxide is emitted from manure management, especially from the wastewater treatment process using activated sludge.. A value of 4.3 Gg was reported from this sub-sector in 1998 (Ministry of Environment), but the value reported by Haga (2001) was 8.8 Gg. Similar to CH_4, the N_2O emission factor from manure management also has large variability.

6. AIMING TO REDUCE 6 % OF GHG EMISSION FROM ANIMAL AGRICULTURE

According to Kyoto Pact on 1997, Japan has to reduce 6 % of it's GHG emission before 2008 to 2012.

Presently, animal agriculture contributes 40 % of total protein needed for the nation and animal products supply 33.8g protein per person per day, which meets the requirement. The human population in Japan will reach a peak by the year 2007 and then will reduce gradually. The demand for animal products has reached a plateau (Fig.6), and the total number of animals to meet the demand for animal protein will decrease gradually due to improvement in productivity. Consequently, GHG emission from this sub-sector will be reduced in Japan. Improvements in the efficiency of animal production is a way to reduce CH4 emission and to supply sufficient foods for humans.

However, the information concerning the effects of global warming on productivity of animal agriculture is scarce. Moreover, it is important to develop the technology needed to estimate CH_4 and N_2O emission accurately from animal agriculture and to reduce the amount of the gases emitted from animal production. Global environmental pollution must be solved globally, therefore collaborative studies are needed among countries to achieve the technologies, and that should begin as soon as possible.

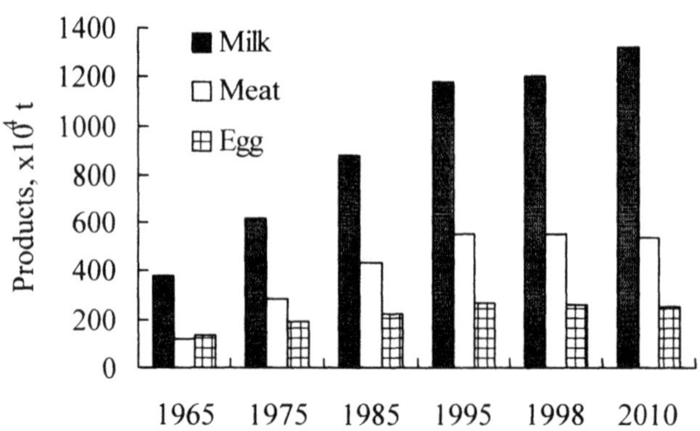

Fig. 6 Demands for Animal Products in Japan

REFERENCES

Haga, K. (2001) Generation and regulation of methane and nitrous oxide from livestock waste. In Emission control of greenhouse-gas in livestock agriculture No.6, Japan Livestock Technology Association, pp70-92.

IPCC, 1995. Climate change 1994: Radiative forcing of climate change and an evaluation of IPCC IS92 emission scenarios. J.T. Houghton et al., eds. Camb. Univ. Press. UK, pp73-126

IPCC (1996). Greenhouse gas inventory. Revised methodology. Guidelines for National Greenhouse Gas Inventories. Vol. 3. OECD, Paris, France.

Kurihara, M., F. Terada, M. Shibata, & T. Nishida (1997). Methane emission from lactating cows in Japan during past 30 years. Proc. 8th Int. Conf. on Anaerobic Digest..Vol.1 329-332.

Osada, K., Kuroda K. & Yonaga M .(2000) Determination of nitrous oxide, methane, and anmonia emissions from swine waste composting process. J. Mater. Cycles Waste Manag, 2: 51-56.

Saitoh, M. (1988). Methane excretion in fattening pigs and pregnant sows. Jpn. J. Zootech. Sci. 59:773-778.

Shibata, M., T. Terada, M. Kurihara, T. Nishida and K. Iwasaki (1993). Estimation of methane production in ruminants. Anim. Sci. Technol. (Jpn) 64:790-796.

Greenhouse Gases and Animal Agriculture
J. Takahashi & B. A. Young (Editors)

Appropriate effluent management of intensive animal industries

T.A. Streeten

Intensive Livestock Environmental Management Services, Department of Primary Industries, PO Box 102, Toowoomba, Qld 4350, Australia

When deciding what is appropriate effluent management for intensive animal industries, we have to ask ourselves what environmental outcomes do we want to achieve. In broad terms this should be protection of the soils, surface and groundwater and community amenity eg. air quality, odour, noise, dust etc.

We have seen the ground rules continually changing in response to the state of the existing environment and its long term sustainability. Communities in general are more discerning about environmental issues and recognise that our environment is not a continually renewable resource. Government response to community pressure through State, National and International environmental legislation and agreements impact on intensive animal industries

There is also a recognition that there needs to be a balance between social, economic and environmental impacts and that focusing purely on economic/production issues is no longer acceptable.

Appropriate effluent management should no longer include the direct discharges of effluent from animal housing into nearby water courses. The use of effluent ponds, whilst the most common form of effluent treatment and storage, should be critically reviewed to ensure that the environmental outcomes mentioned above are being achieved. The traditional "waste products" should and are being viewed as a resource containing valuable plant nutrients and organic matter and used accordingly.

If this is not occurring, what changes need to be made and what alternatives are there?

Keywords: effluent management, manure, feedlot, piggery, dairy, Australia

1. INTRODUCTION

This paper discusses the existing systems and what the future is likely to hold.

For the purpose of this paper, intensive animal industries include cattle feedlotting, piggeries, dairies and poultry. Other intensive industries to which environmental management principles apply and have attracted attention by either local councils or the Environmental Protection Agency are sheep feedlots, emu and ostrich farming, deer farming and livestock holding facilities such as saleyards and cattle export holding facilities.

Effluent and manure are used interchangeably to describe slurries of liquid and associated solid waste produced by animals and birds housed and fed intensively.

Intensive animal industries achieve high product output per unit area via high inputs and animal throughput and as such have been compared to industrial manufacturing of meat, eggs and milk.

With the high input of feedstuffs (grain and concentrates) and water and the resultant large animal throughput, management of the manure produced is of paramount importance to ensure that these industries are environmentally sustainable.

Manure has been described as a resource in the wrong place. In the past its potential as a fertiliser source and a soil conditioner was not generally realised. It was basically considered as a waste which had to be disposed of.

How much manure is produced and what is its fertiliser value? A 100 sow farrow to finish piggery produces in the order of 1000 tonnes of manure per year, which equates to approximately 8 tonnes of nitrogen, 2.5 tonnes of phosphorus and 4.5 tonnes of potassium. Kruger *et al.*, (1995) In terms of nitrogen this equates to around $8000 worth of urea. For a 1000 head feedlot, this equates to around 10 tonnes of nitrogen and 4 tonnes of phosphorus from the manure stockpile. Watts *et al.*, (1994)

The structure and health of the soil is also improved through the high organic matter content of manure.

As demonstrated above, a lot of manure is produced. What are the likely adverse impacts if not managed appropriately? To phrase this in another way what are the environmental outcomes we would want to achieve when operating a piggery, feedlot, poultry farm or dairy.

The environmental outcomes we have listed in the Environmental Code of Practice for Queensland Piggeries are:

- Piggeries are to be managed so that the generation of waste should be prevented or minimised, and utilised according to the principles of ecologically sustainable development.
- Piggeries are to be managed so that there is no degradation of the productive characteristics and qualities of the soil
- Piggeries are to be sited, designed, constructed and operated so that no adverse impacts occur on the environmental value of the surface and groundwaters.
- Piggeries are to be sited, designed, constructed and operated to avoid unreasonable interference with the comfortable enjoyment of life, and property, both on and off site.

The issues to consider when addressing community amenity are odour, noise, dust, public health, transport, and visual impacts. Streeten and McGahan (2000)

Whilst the management of manure differs according to the type of enterprise eg. feedlot, piggery dairy or poultry, some basic principles apply. These are collection and transport, treatment and storage, and utilisation.

2. COLLECTION

Effluent (liquid manure) needs to be contained within the system so no runoff outside designated controlled drainage areas occurs. Well designed and constructed pits under sheds, drains, feedlot pens, dairy sheds etc. will ensure containment which also prevents potential leaching of nutrients through the soil. In some cases a screen or separator is used to further reduce the solids requiring treatment in the pond, thus reducing the rate of sludge accumulation and reduce pond sizes.

As part of a general move worldwide to develop alternative systems to anaerobic treatment and storage ponds, dry manure systems and more efficient separation technologies are being implemented. The challenge is to develop processes that are low maintenance and cost effective. There is considerable expansion of straw based ecoshelters for weaner and grower pigs. Improved solids separation to the existing 20% removal rates using run down screens is being achieved by screwpress separators and chemical flocculation, the latter still in the development stage in Australia.

The move to solid based systems is also been driven by the availability and cost of fresh water. Recycling of effluent for shed flushing needs to be carefully managed to prevent the build up of struvite (magnesium ammonium phosphate) and salinity levels injurious to pond function and plant survival.

3. TREATMENT AND STORAGE

Unless utilised directly onto agricultural land, effluent is stored and treated in anaerobic lagoons where breakdown of organic matter occurs leaving a sludge at the bottom of the lagoon. Correct sizing of the lagoons is critical to ensure adequate treatment, sludge storage, reduced odour emissions and to prevent overtopping during rainfall periods.

As up to 80% of odours from piggeries are emitted from anaerobic ponds considerable research is being undertaken in Australia to better understand factors influencing these emissions including volatile solids loading rates, spacial variation in emissions and the presence of bacterial populations.

The use of straw pond covers which are commonly used in Canada and USA is being trialed under Australian conditions. Whilst research has shown effective odour emission reduction, the practicality and cost of applying such covers to large lagoons is being questioned. Hudson (2001)

The Government's response to greenhouse gas emissions under the Kyoto Agreement may have an impact on intensive animal industries, which contribute 2% emissions. As anaerobic lagoons are a primary source of methane production compared to other waste management systems there could be increased pressure to cover ponds with permeable or impermeable covers to address air quality concerns. Farron *et al.*, (1997)

There are virtually no small to medium sized biodigesters in operation in Australia. The high initial capital cost and ongoing management expertise required has prevented the adoption of this technology.

In many states of USA, the construction of new earthen lagoons for waste storage and treatment is no longer permitted or is under serious investigation. The primary concern is accessions to groundwater through seepage, however catastrophic failure, overtopping and atmospheric emissions are also significant concerns. Casey (2000)

More stringent environmental regulation is being applied to Australian intensive animal industries to ensure the correct construction and sizing of ponds prevents contamination from occurring. More research is needed into management systems which include effective solids separation, reduced odour emissions and pond designs which enables regular cleaning and efficient use of stored nutrients.

4. UTILISATION

There is a marked change of culture brought about by increased community concern, awareness by industry and more stringent regulation which has seen a shift from waste disposal to effluent utilisation.

Conventionally manure and effluent has long been applied to crops and pasture owned or controlled by the feedlot, piggery or dairy as a source of nutrients, organic matter and water. However, as enterprises have increased in size, not all the nutrients produced may be able to be sustainably utilised on the property.

Manure and effluent do not provide a balanced source of nutrients (as a fertiliser) and careful management of their application is required through balancing the amount applied and time of application with crop removal and soil storage. Thus to gain the greatest benefit from manure application to land, the addition of inorganic fertilisers may be required to match the nutrients requirements of the crop. By carefully managing the application of manure through estimation of application rates via mass balance of nutrients, measurement and monitoring, its potential negative effects from over-application of a particular nutrient can be avoided.

To ensure that there is no degradation of the soil, surface and groundwater it is important that the maximum total application of a nutrient = amount removed in the crop + safe soil storage + allowable losses.

Soil monitoring is an integral part of any manure management plan and in a number of instances inappropriate utilisation practices have revealed potential environmental degradation either through high nitrate levels at depth or the level of phosphorus being stored in the soil reaching saturation point.

Spreadsheets and models such as MEDLI (Model for effluent disposal using land irrigation), PIGBAL, BEEFBAL and DAIRYBAL have been developed to calculate mass balances for nutrients and water and predict likely impacts from manure and effluent applications from various sized enterprises on different soil types and areas, under differing climatic conditions.

A number of commercial enterprises are contracting to remove solids and sludge from ponds and value adding the product through vermiculture, composting and pelletting. This is viewed as a viable alternative to intensive livestock enterprises which have insufficient land of their own to sustainably apply manure.

The other major area of environmental management of intensive livestock facilities is carcase disposal. Whilst the main options used in the past have been burial and burning, there is increasing interest in carcase composting using other carbonaceous waste material such as sawdust and grain husks. This is enabling the reuse of the composted product and avoiding potential leaching of nutrients and odour nuisances.

The challenge for the future is for researchers and industry to develop and adopt technology which meet the three key criteria of ecologically sustainable development. ie. technology which is economical for producers to use, ensures environmental protection and satisfies community expectations.

REFERENCES

Casey, K, 2000; Final Travel Report to PRDC.

Farron, I., Maul, C. and Charles, S. 1997; Methane Emissions Program Final Report, Pig Research and Development Corporation and Department of Primary Industries and Energy.

Hudson, N., Casey K., Melvin, S. and Nicholas, P. 2001; Laboratory Investigation of Supported, Permeable, Organic Covers for the Management of Odour Emissions from Anaerobic, Piggery Waste Ponds.

Kruger, I., Taylor, G. and Ferrie, M., 1995; Effluent at Work, NSW Agriculture, RMB 944, Calala Lane, Tamworth NSW 2340, Australia.

Streeten, T., and McGahan, E. - Editors 2000; Environmental Code of Practice for Queensland Piggeries, Department of Primary Industries, Queensland.

Watts, P.J., Tucker, R.W., Gardner, E.A., Casey, K.D. and Lott, S.C., 1994; Characteristics of Feedlot Wastes, Chapter 7.1, Designing Better Feedlots (Ed. Watts P. and Tucker R.), Department of Primary Industries, Queensland.

Greenhouse Gases and Animal Agriculture
J. Takahashi & B. A. Young (Editors)

Greenhouse gas emission from livestock in Brazil

M. A. Lima & C. Campanhola

Researcher Scientists of the Brazilian Agricultural Research Corporation – Embrapa
Rodovia SP-340, Km 127,5, Jaguariúna, SP, 13820-000, Brazil

Agribusiness represents more than 32% of the Gross Domestic Product in Brazil, employs more than 40% of the labor force and is responsible for about 38% of the country exports. On the other hand, agriculture activities have released an average amount of 9.9 Tg year^{-1} of CH_4 and 0,5 Tg year^{-1} of N_2O to the atmosphere, in the period 1989-1995. From the total figure of methane emitted, 96% is assigned to livestock, mainly cattle raising in extensive systems, with a cattle herd of about 165 million head. Following are the flooded rice fields and biomass burning with, respectively, 3% and 1% of the total methane emission by agriculture. The N_2O emissions from animal sources are also significant. Emissions from manure applied to the soil correspond to 10.3% of the direct N_2O emissions from agricultural soils in Brazil, and are estimated at 8.24 Gg of N-N_2O, in 1994. N_2O emissions from animal production systems (grazing animals) accounted to 141.8 Gg of N-N_2O in the same year. Indirect N_2O emissions derived from animal sources accounted to 72.3 Gg of N-N_2O, or 87.8% of the total indirect N_2O emissions from agricultural soils. Due to the relevant amounts of methane and nitrous oxide emissions from livestock sources in Brazil, this paper presents estimates of methane and nitrous oxide emissions from livestock and potential measures that could be used for gas emission mitigation.

Keywords: Brazil, livestock, methane, nitrous oxide, mitigation

1. INTRODUCTION

According to official statistics (IBGE, 1998), Brazilian agricultural areas occupy nearly 228 million hectares and pasturelands cover the most part of the agricultural lands compared with the cropped areas. In 1995, 15% of the total cultivated area corresponded to permanent crops and 85% to temporary crops. However, the area occupied with agriculture is very variable from region to region. Only 1.7% of the North region is planted with crops and 6.3% is covered with pastures, whereas in the South, Southeast and Northeast regions the cropped area accounts to 24.5%, 14.0% and 14.8% in each region, respectively. In addition, pasture areas are more significant in the Southeast, Western-Central and South regions, with 40.7%, 38.9% and 35.8% representation in each region, respectively.

Brazil has currently one of the largest extensive domestic ruminant herds in the world, with 159.4 million head of bovines, 22.7 million head of goats and sheep, above 31.3 million head of swine and 3.25 billion head of broiler chicken (FNP, 2001). The main characteristic of the bovine, buffalo, mule, asinine, goat and sheep production systems countrywide is the use of large grazing lands with a low animal load per area unit. Most part of them are characterized by deposition of wastes in the soil or even in water bodies, the latter being very common in most of the swine production systems.

Bovines are the most numerous herd in all the Brazilian regions with beef cattle representing around 87% of the total bovine herd. In this production system, the Zebu cattle represents 80% of breeds raised for slaughter, and predominates in the Southeastern, Western-Central, Northern and Northeastern regions. However, in the South region, European breeds outweigh the others. In average, Brazilian beef cattle average indexes are: birth rates of 60%, death rates from birth to weaning of 8%, interval between parturitions of 21 months, slaughter age of 48 months and 34 kg of carcass per hectare. The low support capacity of pastures shows a low digestibility rate of the material consumed, which varies from 50% to 55%. Currently, the bovine herd fattening by means of the confinement management corresponds only to 1.2% of the total herd, with the highest concentration in São Paulo State (30% of the total herd confined). In the same way, the pasture-supplemented management represents only 1.4% of the total herd.

The most common variety of dairy cattle in Brazil is a crossbreeding between Zebu and Holstein breeds. These herds are generally not specialized, presenting mixed aptitude (meat and milk). Females present low birth rates (50 to 60%), on average, with intervals of parturition varying from 19 to 24 months. Hay is used mainly for lactating cows during the dry season (Lima, 2000).

This paper presents estimates of methane and nitrous oxide emissions from livestock in Brazil, considering the differences among geographical regions. There is also a brief discussion on the potential measures for gas emission mitigation.

2. METHODOLOGY

Methane and nitrous oxide emissions were estimated using methods described in the IPCC/OECD Guidelines for Greenhouse Gas Emissions (IPCC/OECD, 1996). Regional data were obtained from official production statistics and literature. When the information was not available in these two sources, data were complemented by consultation with institutions, breeders associations and experts. Parallel statistics surveys (Anualpec, 2001) were also accessed to obtain more recent estimates for livestock herds. Step 2 of the calculation of methane emissions proposed by IPCC was used for beef and dairy cattle as well as for swine (Table 1). Data used in the calculation of CH_4 and N_2O emissions for different management systems of animal wastes are shown in Table 2.

3. RESULTS AND DISCUSSION

Brazilian agricultural activities have contributed with significant amounts of greenhouse gas emissions to the atmosphere. In 1994, the contribution was estimated at 10.18 Tg of methane, 0.34 Tg of nitrous oxide, 0.24 Tg of oxides of nitrogen, and 2.79 Tg of carbon monoxide.

3.1. Methane emissions by livestock

In 1994, the methane emissions from livestock in Brazil were estimated at 9.77 Tg, being 9.38 Tg caused by enteric fermentation and 0.39 Tg by animal wastes management systems (Table 3). The category of beef cattle was responsible for 80.9% of methane emissions from cattle rising. The dairy cattle category contributed with 13.5% and the remaining livestock categories with 5.6% of emissions.

Table 1 – Livestock parameters used for estimating methane emission by enteric fermentation.

Technical parameters	*Value range for different regions*
Dairy cattle	
Weight (kg)	*400- 414*
Digestibility rate (%)	*55-60*
Pregnancy index (%)	*55-76*
Food Consumption (kg)	*7.9-8.3*
Milk production (kg)	*1.08-3.3*
Beef cattle	
Weight (kg)	
- Females	*380-400*
- Males	*450*
- Young	*230*
Digestibility rate (%)	*50-56*
Pregnancy index (%)	*58-67*
Food Consumption (kg)	
- Females	*7.6 – 8.0*
- Males	*9.0*
- Young	*5.75*
Milk production (kg)	*1.1*

Table 2 – Fraction of N in each waste management system (AWMS), according to type of animal and region of the country.

Management System	*Region*	*Dairy cattle*	*Beef cattle*	*Swine*	*Sheep*	*Poultry*	*Others*
				Fraction (AWMS)			
Pasture/Rangeland	*South*	*0.75*	*0.75*	*0*	*1.00*	*0*	*0.99*
	Others	*0.45*	*0.97*	*0*	*1.00*	*0*	*0.99*
Solid storage/ Dry lot	*South*	*0*	*0*	*0*	*0*	*0*	*0*
	Others	*0.20*	*0.030*	*00*	*0*	*0.20*	*0*
Liquid / slurry	*South*	*0*	*0*	*0*	*0*	*0*	*0*
	Others	*0.03*	*0*	*0*	*0*	*0*	*0*
Anaerobic lagoon	*South*	*0*	*0*	*0*	*0*	*0*	*0*
	Others	*0.01*	*0*	*0*	*0*	*0*	*0*
Daily spread	*South, São Paulo State*	*0.20*	*0.20*	*0.10*	*0*	*0.80*	*0*
	Southeast, W-Central	*0.20*	*0*	*0.05*	*0*	*0.80*	*0*
	Others	*0.20*	*0*	*0*	*0*	*0.80*	*0*
Other systems	*South*	*0.05*	*0.05*	*0.90**	*0*	*0*	*0.01*
	Southeast, W-Central	*0.11*	*0*	*0.95*	*0*	*0*	*0.01*
	Others	*0.11*	*0*	*100*	*0*	*0*	*0.01*

In the same year, methane emissions from management systems of animal wastes were estimated at 395.2 Gg (average of 396.80 ± 10.34 Gg in the period 1993-1995), being 50%

attributed to beef cattle rising, 15.5% to dairy cattle, 15.5% to poultry and 13.3% to swine. The total emissions from livestock represented nearly 95.9% of the total emissions of methane from agricultural activities (flooded rice fields – 2.8%, agricultural residues burning – 2.8%).

3.2.Nitrous oxide emissions by livestock

In 1994, direct N_2O emissions from manure applied to the fields (daily spread) were estimated at 8.24 Gg of N-N_2O, which represents 10.3% of the total direct N_2O emissions from agricultural soils (Table 4). N_2O emissions from animal production (grazing animal) were estimated at 141.8 Gg of N-N_2O. Beef cattle and dairy cattle accounted together to emissions of 119.87 Gg (84.5% of the total). Nitrogen leaching and runoff constituted the major sources of indirect N_2O emission in the country, contributing with 72.3 Gg, which is 87.8% of total indirect emissions.

Table 3. Methane an nitrous oxide emissions from livestock in Brazil, 1994 (Embrapa, 1999a,b).

Sources	*CH_4 (Tg)*	*N-N_2O (Tg)*
Enteric fermentation	*9.38*	-
Management Systems of Animal Wastes	*0.39*	-
Daily spread of animal wastes in cultivated soils	-	*0.008*
Wastes of grazing animals	-	*0.142*
Wastes as indirect source (deposition)	-	*0.015*
Wastes as indirect source (leaching)	-	*0.057*
Total	*9.77*	*0.222*

3.3. Considerations on methane and nitrous oxide mitigation measures for Brazilian livestock

These preliminary GHG estimates provide useful data to identify some mitigation measures for the Brazilian livestock sector. The diffusion of technologies for the improvement of production efficiency, the increase of technical and managerial capabilities, and a better adaptation of the animal production systems at the farm level are potential tools for the implementation of these measures. Incentives to the adoption of adequate technologies, supported by a set of policies oriented to improving regional animal production chains, are also important aspects to be considered. Based upon these premises, some feasible opportunities to reduce CH_4 and N_2O emissions from livestock sources are presented below.

3.3.1. Enteric fermentation:

Indications for methane mitigation in the domestic livestock sector are associated to the improvement of animal productivity (U.S.EPA, 1990a, 1990b, 1991). This can be achieved by food supplementation, control of diseases and genetic improvement. In Brazil, the most part of methane emissions come from areas of extensive pastures, and the food supplementation for grazing cattle is often a limited practice. Pasture supplemented systems have been shown a slight increase in the Southeast, South and Western-Central regions and there is a set of products for diet supplementation in local markets, such as corn and sorghum silage, soybean bran, sugar cane bagasse, etc. Recent studies carried out in beef cattle grazing systems with *Brachiaria spp.*, a very frequent forage in Brazil, combined with supplementation along the year have shown very promising results on animal productivity.

Moreover, the carrying capacity of forages increases as the animals have diet supplementation, especially during critical periods of the year. Finally, there is a need for more studies on improvement of microbial process efficiency, in order to optimize fibbers digestion in the rumen, on microbial synthesis, and on design of bioengineering strategies.

3.3.2. Animal wastes:

CH_4: The mitigation of methane emissions from animal wastes in confinement systems of bovine and swine can be obtained by energy recuperation processes, using anaerobic digesters. But, in relative terms, this methane source is not critical in Brazil, due to the extensive character of livestock. In the case of swine, efforts have been made in implementing the building of dunghills and bio-dunghills to treat the wastes.

N_2O: In the South region, approximately 80% and 10% of poultry and swine manure, respectively, are applied to the soil as nitrogen source. In these cases, formulation of diets with a better amino acid balance has a good potential to both reduce excretion of urinary nitrogen and promote a better digestion of food. For grazing animals, studies should be carried out to find the best balance of C and N in soils.

REFERENCES

Embrapa, 1999a. *Inventory of greenhouse gas emissions from agricultural activities in Brazil: methane emissions from livestock (revised report).* Jaguariúna, SP, Brasil.

Embrapa, 1999b. *Inventory of greenhouse gas emissions from agricultural activities in Brazil: nitrous oxide emissions from agricultural soils.* Jaguariúna, SP, Brasil.

FNP, 2001. *Anualpec 2001 - Anuário da pecuária brasileira.* São Paulo: FNP.

IBGE, 1998. *Censo Agropecuário* 1995-1996, n. 1, Brasil. Rio de Janeiro. p. 1-366.

IPCC, 1996. IPCC Guidelines for National Greenhouse Gas Inventories. Bracknell: IPCC.

Lima, M.A., 2000. Methane emissions from livestock in Brazil: an overview of the potential mitigation options. Second International Conference on Methane Mitigation, 18-23jun, 1999, Novosibirsk, RU, Proceedings... U.S. EPA, Siberian Branch of Russian Academy of Sciences. p.161-168.

U.S. ENVIROMENTAL PROTECTION AGENCY. 1990a. Greenhouse gas emissions from agricultural systems, v. 1, Summary Report. *Proceedings of the Workshop on Greenhouse Gas Emissions from Agriculture.* Washington: U.S. EPA.

U.S. ENVIROMENTAL PROTECTION AGENCY. 1990b. Methane emissions and opportunities for control. *Proceedings of the Workshop results of Intergovernmental Panel on Climate Change.*

U.S. ENVIROMENTAL PROTECTION AGENCY. 1991. Improving Ruminant Production and Reducing Methane Emissions From Ruminants by strategic Supplementation. *Proceedings of the Workshop results of Intergovernmental Panel on Climate Change.*

Greenhouse Gases and Animal Agriculture
J. Takahashi & B. A. Young (Editors)

Emission of greenhouse gas from anaerobic pig and dairy cattle wastewater treatment system in Taiwan

J. J. Su, B. Y. Liu & Y. C. Chang

Department of Production Management, Pig Research Institute Taiwan, P.O. Box 23, Chunan, Miaoli, Taiwan, 35099, R.O.C.

Greenhouse gas can lower livestock production. Manure management systems in Taiwan vary from the IPCC model for the Far East and Asia region. This study centers on three pig farms and dairy farms in northern, central, and southern Taiwan. Analyzed GHG samples from *in situ* anaerobic wastewater treatment systems of pig farms revealed emission factors of GHG were CH_4 (1.39308855, 0.46201221, and 0.44907150 kg/head/yr), CO_2 (1.27948507, 0.44290146, and 0.420810940 kg/head/yr), and N_2O (0.00248757, 0.00113081, and 0.00135421 kg/head/yr) in three temperature periods (cool, temperate, and warm, respectively). Analysis results indicated that emissions factors of CH_4 from pig farms were lower than those of IPCC (1996b) under both temperate and warm temperature periods. Analyzed GHG samples from *in situ* anaerobic wastewater treatment systems of dairy farms revealed CH_4 emission from dairy farms were lower than those of IPCC (1996b).

Keywords: livestock wastewater, methane, anaerobic treatment, greenhouse gas, nitrous oxide

1. INTRODUCTION

The gases that cause the greenhouse effect including carbon dioxide (mainly produced from petroleum combustion), methane, and nitrous oxide are called greenhouse gases (GHG). The pig is the most important livestock animal in Taiwan. Thus, most studies of the consequences of high temperatures on livestock production have all focused on pig physiology (Hsia & Lu, 1987; Cheng & Wung, 1974; Hsia, 1989).

Biogas analysis from the samples of anaerobic piggery wastewater treatment systems (i.e. Three-step Piggery Wastewater Treatment System, TPWT) (Su *et al.*, 1997; Su, 1999) revealed the following ratio of major GHG in sequence: methane (60~80%) > carbon dioxide (20%) > nitrous oxide (<1%). Hong (1998) estimated emission (0.25 m^3/head/d) of biogas from anaerobic fermentation of pig manure by laboratory-scale reactors. Establishing the data of GHG emission to compromise with the IPCC on the contribution of GHG in Taiwan is extremely important. This study therefore attempts to establish the emission level of GHG from on-site anaerobic livestock wastewater treatment systems in Taiwan.

2. METHODS AND MATERIALS

2.1. Livestock farms and composting houses

Three integrated pig farms and three dairy farms were selected from three separate counties: Miaoli, Changhua, and Tainan.

2.2. Collection of gas samples

A 500-ml gas collector and 1-L Tedlar® gas sample bags were used to collect gas samples from the biogas outlets of covered anaerobic wastewater treatment basins.

2.3. Measurement of flow rate of the biogas from the anaerobic treatment basins of livestock farms on sites

A mass flow meter (Sierra, 780S Series, Flat-Trak Mass Flow Meter, USA) directly measured the flow rate of biogas. The instrument manufacturer in the USA initially calibrated this mass flow meter with a gas mixture containing 60% CH_4 and 40% CO_2.

2.4. Analysis of greenhouse gases

Gas samples were analyzed for both methane and carbon dioxide using gas chromatography (GC) with a thermal conductivity detector (TCD) and a stainless steel "Porapak Q" column (1/8" × 3m). The carrier gas was helium. In addition, gas samples were also analyzed using GC with an electron capture detector (ECD) and a stainless steel "Porapak Q" column (1/8" × 4m) to reveal nitrous oxide levels. The carrier gas was P-10 containing 90% Argon and 10% methane.

2.5. Calculation of greenhouse gas emission

Percentages of methane (R_{CH4}), carbon dioxide (R_{CO2}), and nitrous oxide (R_{N2O}) in biogas samples were calculated based on the measurement of gas samples by gas chromatography. The calculations were listed as follows:

(1)Daily production of CH_4, CO_2, and N_2O in livestock farms (m^3/day) = Daily production of biogas × (R_{CH4}, R_{CO2}, and R_{N2O})..[A_1]

(2)Daily production of CH_4, CO_2, and N_2O in livestock farms per livestock animal (m^3/head/day) = A_1 ÷ (the number of raising animals)................................[B_1]

(3)Based on the Ideal Gas Law (PV=nRT), daily production of CH_4, CO_2, and N_2O by weight can be calculated by equation [C_1] as follows:

Daily production of CH_4, CO_2, and N_2O per livestock by weight (kg/head/day) = 〔B_1/1000×1/(R×T)× (MW_{CH4}, MW_{CO2} or MW_{N2O})÷1000〕........................[C_1]

Here, P = pressure (=1 atm), V = volume (liter), n = number of moles (moles), R = 0.0821 *l*·atm/(mole·°K), and T = (273 + 25°C) = 298°K. Molecular weight of CH_4 (MW_{CH4}), CO_2 (MW_{Co2}), and N_2O (MW_{N2O}) are equal to 16, 44, and 44 g/mole, respectively.

(4)Annual production of CH_4, CO_2, and N_2O per livestock (kg/head/yr) = C_1 × 365 d/yr

3. RESULTS

Total annual emissions of CH_4, CO_2, and N_2O from anaerobic wastewater treatment of pig and cattle farms in Taiwan are listed in Table 1 to 4. Production of biogas from anaerobic piggery wastewater treatment was estimated at 0.1~0.4 m^3/head/d, although it had been defined as 0.25 m^3/head/d (Hong, 1998) since 1983 (Koh, M.T., unpublished data). Also, production of biogas in summer was supposed to be exceeding that in winter. However, this study measured with mass flow meters the average biogas production on sites from the anaerobic wastewater treatment systems of pig farms (0.0048 m^3/head/d) and dairy farms (0.0298 m^3/head/d). Measured biogas production is much lower than what Hong estimated.

In addition, production of biogas in winter exceeds that in summer. This observation contradicts the former estimation, which was based on some laboratory data. Actually, production of biogas closely corresponds to the hydraulic retention time of anaerobic treatment sections and organic concentrations in wastewater. Increasing temperature can only accelerate the efficiency of biogas production with a certain concentration of organic matter and produce the same amount of biogas as at a lower temperature. Analyzing gas samples from full-scale wastewater treatment facilities and composting houses with gas chromatography and mass flow meters yielded the results related to the emission factors of GHG for livestock industry in this study. According to our results, methane emissions from anaerobic wastewater treatment of the chosen pig farms in both temperate and warm periods were lower than that established by IPCC (1996ab).

Table 1. Emission factors of GHG from the anaerobic wastewater treatment systems of the chosen pig farms in all regions of Taiwan

GHG	Monthly average temperature (°C)		
	<20	20~25	26~30
CH_4 (kg/head/year)	1.39308855	0.46201221	0.44907150
CO_2 (kg/head/year)	1.27948507	0.44290146	0.42081094
N_2O (kg/head/year)	0.00248757	0.00113081	0.00135421

Table 2. Emission factors of GHG from the anaerobic wastewater treatment systems of the chosen dairy farms in all regions of Taiwan

GHG	Monthly average temperature (°C)		
	<20	20~25	26~30
CH_4 (kg/head/yr)	5.39817150	4.80173256	4.49468423
CO_2 (kg/head/yr)	4.22198468	4.25285745	4.115933664
N_2O (kg/head/yr)	0.00763958	0.01027064	0.01584225

Table 3. Total emission of GHG from anaerobic wastewater treatment system of pig farms in Taiwan based on the number of pigs on farms in 1998 (Mt = 1000 kg)

	Emission factor (kg/head/yr)	Total emission (*Mt*/yr)	Total number of pigs on farms in 1998 (head)
CH_4	0.76805742	5022.017	
CO_2	0.71439916	4671.167	6538596
N_2O	0.00165753	10.838	

Table 4. Total emission of GHG from anaerobic wastewater treatment system of dairy farms in Taiwan based on the number of dairy cattle on farms in 1998

	Emission factor (kg/head/yr)	Total emission (*Mt*/yr)	Total number of dairy cattle on farms in 1998 (head)
CH_4	4.89819610	810.157	
CO_2	4.19692527	694.167	165399
N_2O	0.01125082	1.861	

REFERENCES

Cheng, S.P. & Wung, S.C., 1974. A study of seasonal changes in the boar semen characteristics. Annual Research Report of Animal Industry Research Institute. Taiwan Sugar Corporation 62/63: 89-103. [in Chinese]

Hsia, L.C., 1989. Effect of cool and warm temperature on pig growth in Taiwan. Journal of Chinese Society of Animal Science 18: 3-4. [in Chinese]

Hsia, L.C. & Lu, G.H., 1987. The effect of high environmental temperature on the protein and energy requirements of growing-finishing pigs I. The effect of dietary levels of protein and energy on the performance of pigs under Ad Lib feeding. Journal of Chinese Society of Animal Science 16(1-2): 1-10. [in Chinese]

Hong, C.M., 1998. Technology of swine waste resource treatment. Technical Manual of Taiwan Livestock Research Institute Vol.60, pp. 27-57. [in Chinese]

IPCC, 1996a. Chapter 4. In: Revised 1996 IPCC Guidelines for National Greenhouse Gas Inventories, Vol. 2: Greenhouse gas inventory workbook, J.T. Houghton, L.G. Meira Filho, B. Lim, K. Tréanton, I. Mamaty, Y. Bonduki, D.J. Griggs, B.A. Callander.

IPCC, 1996b. Chapter 4. In: Revised 1996 IPCC Guidelines for National Greenhouse Gas Inventories, Vol. 3: Greenhouse gas inventory reference manual. J.T. Houghton, L.G. Meira Filho, B. Lim, K. Tréanton, I. Mamaty, Y. Bonduki, D.J. Griggs, B.A. Callander.

Su, J.J., Liu, Y.L., Shu, F.J. & Wu, J.F., 1997. Treatment of piggery wastewater treatment by contact aeration treatment in coordination with the anaerobic fermentation of three-step piggery wastewater treatment (TPWT) process in Taiwan. Journal of Environmental Science Health. Part A, 32(1):55-73.

Su, J.J., 1999. Experience of operating a full-scale sequencing batch reactor system for treating piggery wastewater in Asia--the Taiwan experience. Proceeding of The 3rd Inter-national Symposium of the Environmental Center for Livestock Waste Management--Innovative Livestock Waste Management Technologies for Today and the Future, pp.167-185.

Greenhouse Gases and Animal Agriculture
J. Takahashi & B. A. Young (Editors)

Emissions inventories and their implications for intensive livestock production

K. D. Casey

Intensive Livestock Environmental Management Services, Department of Primary Industries, PO Box 102, Toowoomba, Queensland 4350, Australia

Australia as a party to the United Nations Framework Convention on Climate Change (UNFCCC), is required to produce and regularly update National Greenhouse Gas Inventories (NGGI). Separately, Australia has implemented the National Pollutant Inventory (NPI) which is an internet database designed to provide community, industry and government with information on the types and amounts of certain substances being emitted to the air, land and water.

Emissions of methane and nitrous oxide from intensive livestock facilities are calculated as part of the National Greenhouse Gas Inventories. While emissions of ammonia, particulates (PM10) and possibly total nitrogen and total phosphorus must be reported by individual intensive piggeries and feedlots exceeding the respective threshold levels in the National Pollutant Inventory.

This paper discusses some of the potential effects of changing management by intensive livestock industries on emissions reported under the national inventories and the potential for community scrutiny of industry and individual facility emissions.

The accuracy of the estimation methods used in some of the NPI Industry Handbooks has been questioned. Over estimates of these emissions that result may expose the industry to unfair criticism. This all now occurs in an environment of community scrutiny via the publicly available national inventories.

The increasing use of straw bedded housing by the intensive pig industry to reduce facility costs and odour emissions, potentially reduces odour and methane emissions however increases nitrous oxide emissions. Decisions made by producers related to housing style and waste management system for new and expanding facilities often do not take potential greenhouse gas emissions into consideration. Currently, these increased emissions are not being reflected in the greenhouse gas inventory.

Keywords: intensive livestock, emissions inventory, methane, ammonia, particulates

1. INTRODUCTION

Australia as a party to the United Nations Framework Convention on Climate Change (UNFCCC), is required to produce and regularly update National Greenhouse Gas Inventories (NGGI). These inventories provide a baseline from which it is possible to monitor and review response action and develop projections of greenhouse gas emissions. The inventory is compiled using the *Australian Methodology for the Estimation of Greenhouse Gas Emissions and Sinks* (NGGIC, 1996) which has been developed from Intergovernmental Panel on Climate Change (IPCC) guidelines to better reflect Australian production systems and conditions. Input data for intensive livestock is sourced from the Australian Bureau of Statistics, Australian Lot Feeders Association, Meat and Livestock Australia, Wool International, and Australian Dairy Corporation (NGGIC, 2000).

The Australian National Greenhouse Gas Inventory covers the major greenhouse gases – carbon dioxide (CO_2), methane (CH_4), nitrous oxide (N_2O), perfluorocarbons (PFCs), hydrofluorocarbons (HFCs), and sulphur hexafluoride (SF_6), and the indirect greenhouse gases – carbon monoxide (CO), oxides of nitrogen (NO_x), non-methane volatile organic compounds (NMVOC) and sulphur dioxide (SO_2). The only emissions estimated for livestock is methane while emissions of methane and nitrous oxide are estimated for livestock manure management systems. Carbon dioxide produced by livestock during respiration is not estimated in the inventory, in line with IPCC convention.

Separately, Australia has implemented the National Pollutant Inventory (NPI) which is an internet database designed to provide community, industry and government with information on the types and amounts of certain substances being emitted to the air, land and water (NPI, 2001). The NPI has been developed by the National Environment Protection Council (NPEC) whose two primary functions are; to make National Environmental Protection Measures (NEPM) and to assess and report on their implementation and effectiveness in participating jurisdictions. The objectives of National Environmental Protection Measures are to ensure that; the people of Australia enjoy the benefit of equivalent protection from air, water and soil pollution and from noise, wherever they live, and decisions by businesses are not distorted and markets are not fragmented by variations between jurisdictions in relation to the adoption or implementation of major environment protection measures. The main objectives of the NPI are to: provide information to industry and government to assist in environmental planning and management; satisfy community demand for accessible information on pollutant emissions to the atmosphere; and promote waste minimisation, cleaner production, and efficient energy and resource use (NPI, 2001).

An Industry Handbook has been developed for each industry required to report its emissions to aid the industry in meeting the requirements of the NPI. Currently, 80 Industry Handbooks have been developed. To date, the only Industry Handbooks developed for intensive livestock industries are for lot feeding (Intensive Livestock (Beef Cattle)) (NPI, 1999a) and pig production (Intensive Livestock (Pig Farming)) (NPI, 1999b) although it is reported that an Industry Handbook for the Poultry Industry is being developed following concerted community pressure (Petrov, 2001). Emissions from industries for which no Handbook exists and for facilities that are below the reporting threshold are estimated and reported by the relevant state environmental agency. Twenty-one (21) Aggregated Data Manuals have been developed to assist state governments in estimating these emissions.

The NPI substance reporting list for up to and including 2000/2001 lists 36 substances. From 2000/2001, a further 54 (90 total) substances are added to the list. A facility is required to estimate and report emissions if it uses more than 10 tonne of a Category 1 substance, consumes more than a specified amount of fuel or electrical power (Category 2) or emits more than a specified amount of total nitrogen or total phosphorus to water (Category 3). Emissions of ammonia, particulates (PM10) and possibly total nitrogen and total phosphorus must be reported by individual intensive piggeries and feedlots exceeding the respective threshold levels in the National Pollutant Inventory.

Reporting for industries for which handbooks have been developed has been phased in over a number of years. The emissions from piggeries and feedlots in the year March 1st, 2000 to February 28th, 2001 will be the first reported year for these industries. It is likely that many piggeries and feedlots will exceed the Category 1 threshold of 10 tonnes for ammonia (NH_3). A large feedlot with a feedmill or abattoir on site may exceed Category 2 fuel burning threshold requiring it to report all substances listed in that Category (2a or 2b). A feedlot or piggery that has suffered a spill or overtopping event from its waste treatment and storage facilities into a water body of greater than the Category 3 threshold for either total phosphorus (3 tonne) or total nitrogen (15 tonne) will be required to report those emissions.

2. DISCUSSION

2.1. National pollutant inventory thresholds

The substance that is most likely to be emitted by an intensive livestock facility at greater than its Category 1 threshold is ammonia. Ammonia emissions contribute to atmospheric deposition of ammonia on both land and water potentially contributing to eutrophication of lakes and rivers. Atmospheric deposition of ammonia on land will in turn contribute to the emission of nitrous oxide, a greenhouse gas from the land. Emissions of ammonia also potentially contribute to the level of fine particles in the air (PM2.5) through the formation of ammonium salts (McCulloch *et al.*, 1998).

The Industry Handbooks for both lot feeding and pig production present methodologies for estimating emissions of ammonia from each component of the facility. The handbooks also include a calculation of the minimum facility size to trigger the Category 1 threshold for a facility of industry standard configuration. The minimum size feedlot to trigger the Category 1, ammonia threshold of 10 tonne/year is 122 standard cattle units (NPI, 1999a). It can clearly be seen from the calculation that almost all feedlots will be required to report their emissions of ammonia under the NPI. The minimum size of intensive piggery to trigger the Category 1, ammonia threshold of 10 tonne/year is 1535 standard pig units (spu) for a farrow to finish unit and 2227 spu for a grower unit (NPI, 1999b). Again, many pig production facilities will exceed this threshold requiring them to estimate and report their emissions under the NPI.

Distance to neighbours is often the most limiting factor in applications for new and expanding pig production units. In an attempt to minimise the required buffer distances, proponents with their consultants will specify a deeper than normal anaerobic lagoon, minimising the surface area for the required volume. This reduces the odour emission rate used during modelling as total emission is assumed to by controlled by surface area. These changes in design of anaerobic lagoon system design aimed at limiting odour emissions will also alter ammonia emissions, potentially impacting on nutrient balance of both the lagoons and waste utilisation areas.

A large feedlot that exceeds a Category 2 threshold for fuel burning will be required to estimate and report all the substances listed under the relevant Category 2a/2b including their PM10 particulate emissions. The Industry Handbook for lot feeding suggests that emissions can either be measured, which would be a very expensive and difficult task or calculated using the conservative USEPA emission factor. Researchers have shown that this factor may over estimate emissions for US conditions (Sweeten *et al.*, 1988) which would imply that this factor would substantially over estimate emissions under Australian conditions.

2.2. Greenhouse gas emissions and pig industry changes

Estimates of methane emissions from manure management systems (NGGIC, 1996) have been calculated using an assessment of the proportion of different manure management systems used in the Australian pig industry based on Kruger *et al.* (1995). This assessment was based on a survey of 20% of the Australian industry undertaken in the early 1990s. Since that time, the pig industry has undergone considerable change both in the production systems used, the style of new housing constructed and manure management system used. The use of straw-based housing systems, often for growing pigs by the intensive pig industry to reduce capital costs has greatly increased. This change in housing style potentially reduces odour emissions and eliminates the need for anaerobic lagoons as manure is handled as a solid, thus reducing methane emission, however increasing nitrous oxide emissions. Farran *et al.*, (1995) showed in a series of case studies of estimated greenhouse gas from different housing system and waste management system combinations, that a straw-based system had the highest greenhouse gas emissions. The changes in housing style and waste management system and thus estimated greenhouse gas emissions are not currently being reflected in the inventory.

3. CONCLUSION

Emissions reporting under the NPI will expose the intensive pig and lot feeding industries to a level of public scrutiny not previously experienced. Even relatively small facilities will exceed the ammonia emissions threshold. These reported emissions will show up in non-urban, non-industrial areas for which there will be no other reported emissions.

The accuracy of estimates of PM10 emissions from feedlots made using NPI Industry Handbook is questionable. Over estimates of these emissions that result may expose the industry to unfair criticism. This all now occurs in an environment of community scrutiny via the publicly available national inventories.

Decisions made by producers related to housing style and waste management system for new and expanding facilities often do not take potential greenhouse gas emissions into consideration. Currently, these increased emissions are not being reflected in the greenhouse gas inventory.

REFERENCES

Farran, I., C.R. Maul, & S.M.W. Charles, 1997. The Methane Emissions Project – Final Report, Report No DPI 3/1156, Pig Research and Development Corporation, Kingston, ACT. pp62.

Kruger, I., G. Taylor & M. Ferrier, 1995. Effluent at Work. NSW Agriculture, Tamworth, NSW. pp211.

McCulloch, R.B., S. Few, G.C. Murray & V.P. Aneja, 1998. Analysis of ammonia, ammonium, aerosols and acid gases in the atmosphere of a commercial hog farm in eastern North Carolina, USA. Environmental Pollution. 102: 263-268.

McGahan, E.J. & K.D. Casey, 1997. "A new unit for defining the size of a pig enterprise for environmental purposes". In: Proceedings of Manipulating Pig Production VI, P.D Cranwell, Australian Pig Science Association, Werribee, VIC. 298.

NGGIC, 1996. Australian Methodology for the Estimation of Greenhouse Gas Emissions and Sinks. National Greenhouse Gas Inventory Committee, Canberra, ACT.

NGGIC, 2000. National Greenhouse Gas Inventory 1998, National Greenhouse Gas Inventory Committee, Canberra, ACT. pp274.

NPI, 1999a. Emission Estimation Technique Manual for Intensive Livestock – Beef Cattle, National Pollutant Inventory, Environment Australia, Kingston, ACT. pp34.

NPI, 1999b. Emission Estimation Technique Manual for Intensive Livestock – Pig Farming, National Pollutant Inventory, Environment Australia, Kingston, ACT. pp32.

NPI, 2001. National Pollutant Inventory Guide, Version 2.4 – 26 March 2001, Environment Australia, Kingston, ACT, pp31.

Petrov, R., 2001. National Pollutant Inventory – Estimating your emissions and working through your return. In: Proceedings of 2001 Production and Environmental Monitoring Workshop, UNE, Armidale, NSW, Paper No PEM 14, pp9.

Sweeten, J.M., C.B. Parnell, R.S. Etheredge & D. Osborne, 1988. Cattle Feedlot Dust Emissions. ASAE Paper No 88-4047. American Society of Agricultural Engineering, St Joseph, MI. p29.

Greenhouse Gases and Animal Agriculture
J. Takahashi & B. A. Young (Editors)

Perspectives for realizing agricultural production systems with material circulation

S. Hoshiba

Dept. of Dairy Science, Faculty of Dairy Science, Rakuno Gakuen University, 582 Midorimachi, Ebetsu, Hokkaido, 069-8501, Japan

In this paper the reasons of the lack of material circulation in agricultural production systems will be summarized into three points :

(1)Biased evaluation of agricultural production systems only from economic efficiency for long time,

(2)Insufficient understanding of the differences in production systems between agriculture and industry and

(3)Almighty science or dreams of a perpetual motion machine,

and concepts for realizing agriculture with material circulation will be discussed on two topics :

(1)Evaluation of agricultural production systems by multiple criteria and

(2)Re-construction of agricultural production systems based on material circulation.

Keywords: material circulation, fossil energy, animal welfare, nitrogen load

1. INTRODUCTION

Agriculture is the production system based on the material circulation (Fig. 1) as it was presented by Kurosawa(1971). However, present agricultural production systems in Japan can not keep this basic theory in most of the cases. In this paper the reasons of the lack of material circulation will be summarized into three points and concepts for realizing agriculture with material circulation will be discussed on two topics (Hoshiba, 1996).

2. REASONS OF ENVIRONMENTAL PROBLEMS IN AGRICULTURE

2.1. Biased evaluation only from economic efficiency

Fig. 2 is the schematic diagram showing the efforts to improve economic efficiency and the problems as the results in Japanese agriculture. Agricultural production systems have been constructed to obtain maximum economic profit. Three methods have mainly been used to realize the maximum economic profit :

(1) improvement of labor productivity,

(2) improvement of land productivity and

(3) raising value-added.

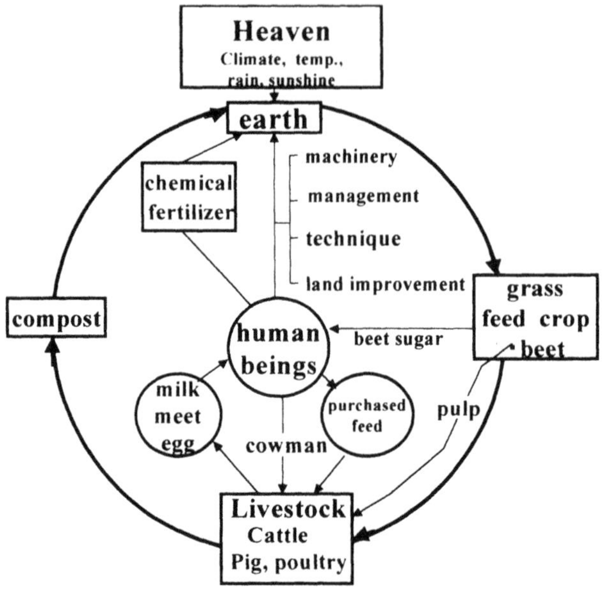

Fig. 1 Chart of circulation agriculture by Kurosawa.

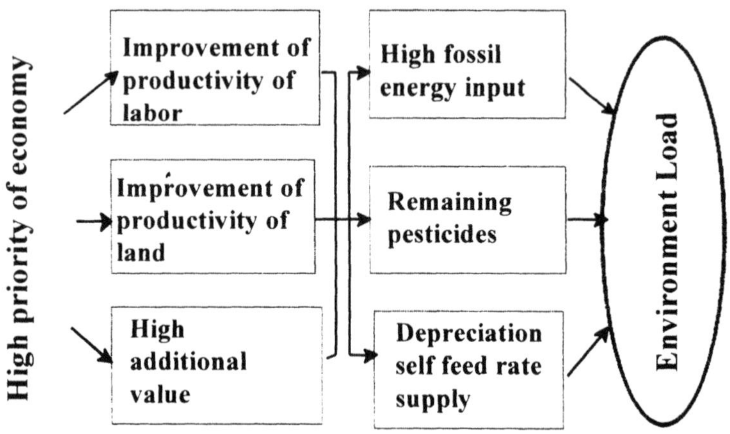

Fig. 2 Schematic diagram showing the efforts to improve economic efficiency and the problems as the results in Japanese agriculture.

To improve labor productivity mechanization has been contributed in all production processes such as tractors and machinery for cultivation, seeding and harvesting. To improve land productivity large amount of chemical fertilizer and pesticide have been used. To raise the value-added of agricultural products green houses or glass houses with supplemental heating systems have been used to produce some agricultural products such as strawberry in the period when the production had not been made in natural condition.

Those three methods had succeeded well until a certain point of time. As a matter of fact the economic profit had increased dramatically. However, after the certain point of the time various problems such as high energy consumption, remaining of pesticide in the soil and decreasing self sufficiency of food started to appear. These problems consequently led to the environmental problems. When is "the certain point of the time" will be dicussed later.

2.2. Lack of understanding the differences in production systems between agriculture and industry

It is the fact that Japanese agriculture has been suddenly developed by utilizing industrial techniques. However, it is thought that there are obvious differences between agriculture and industry. Industry has the production systems of a straight line which means that industry utilizes raw materials and produces products but throws away the rest of the raw materials or even the products when they lose their original functions.

On the other hand agriculture originally has the production systems of a circle. In a milk production system, for example, dairy cows intake feed plants and produce milk and manure. Manure is returned to the soil and feed plants are produced again by absorbing nutrients included in the manure. When the circulation is well managed almost no wastes can appear.

However, at present in Japan, milk is produced by feeding large amount of imported grains. Only about 30% of the nutrient of the imported grain can be utilized by milking cows and the rest, which is manure (about 70%), can not be utilized sufficiently. This production system is similar to that of the industry, but worse in the sense of volume because the volume of the waste from cows is enormous.

It is acceptable to use industrial techniques for improving productivity in agriculture but it should be limited within sustaining material circulation. "The certain point of the time" in the previous section is when the material circulation can not exist.

2.3. Almighty science or dreams of a perpetual motion machine

Science has been contributing to solve most of the problems and improve our everyday life. Most of the people believe that the development in scientific technology keeps contributing on and insuring high quality of our life. However, we have to know that it is not always true that science can solve all the problems. Although we hope high productivity in agriculture and the harmony with the environment at the same time we have started to notice that the simultaneous existence is not always possible, especially in these era, the beginning of the 21 century when the scientific technology is matured. At present pursuing higher productivity conflicts with environmental conservation.

Thinking back to hundreds of years ago, many scientists tried to make a perpetual motion machine which is the system running without energy input from outside or the system performing the same amount of work as energy input. Finally they found that there was no such a system, and this understanding leads to the second law of thermodynamics. We must have already known the limitation of the scientific technology. However, it seems people are trying to make perpetual motion machines now again.

3. HOW TO SOLVE THE ENVIRONMENTAL PROBLEMS IN AGRICULTURE

3.1. Evaluation of agricultural production systems by multiple criteria

3.1.1. Definitions of five evaluation criteria and two combined indexes

Problems led by the biased evaluation only from economic profit have already been mentioned above. The proposal of the multiple evaluation criteria and the examples of the evaluation will be discussed in this section.

The multiples evaluation criteria composed of not only (a)economic efficiency but also other four criteria : (b)energy, (c)environmental load, (d)animal welfare and (e)human welfare are proposed by Kawakami et al. (2000) to evaluate agricultural production systems totally. The definitions and methods of calculation are as fellows.

(a) Economic efficiency : Net agricultural income (NAI) was used as a criterion for economic efficiency, and were calculated as follows :
NAI [1,000 yen]= Gross Agricultural Income - Agricultural Production Costs

(b) Energy : Fossil energy input (FEI) was used as a criterion for enrgy. Energy input is the total fossil energy input used for feed production and transportation, chemical fertilizer and various dairy head management.

(c) Environmental load : Nitrogen load (NLD) was used as a criterion for environmental load. Nitrogen load used in this paper is defined as the difference between input and output nitrogen. Nitrogen inputs are nitrogen included in purchased feed, chemical fertilizer and beddings etc.. Nitrogen outputs are calculated from milk production and number of animals sold. Fig. 3 shows the definition of nitrogen load and nitrogen flow.

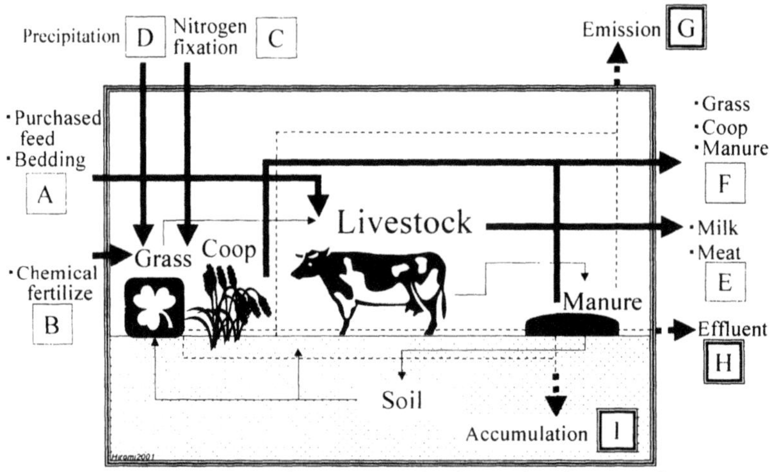

Fig. 3 The definition of the nitrogen load and nitrogen flow.

(d)Animal Welfare : Animal health condition (AHC) was used as a criterion for animal welfare and was expressed by medication fee [yen].

(e) Human Welfare : Human satisfaction (HSF) was used as a criterion for human welfare and was investigated by questionnaire to dairy farmers and expressed by five steps according to the degree of satisfaction: Level 1 (extremely unsatisfied), Level 2 (unsatisfied), Level 3 (common), Level 4 (satisfied), Level 5 (extremely satisfied). The contents of the questionnaire were composed of 25 questions related to satisfaction and dissatisfaction.

Two combined indexes composed of three evaluation criteria (NAI, FEI and NLD) were also used to evaluate agricultural production systems. First combined index was [FEI /NAI] ratio (GJ/1,000 yen) and indicates how much of fossil energy is consumed for milk production to earn 1,000 yen of net agricultural income. Second index was [NLD /NAI] ratio (kgN/1,000 yen) and indicates how much of nitrogen is spread to the circumstances to earn 1,000 yen of net agricultural income.

3.1.2. Example 1. - Evaluation using three criteria

Evaluation using three criteria : NAI, FEI and NLD was made on 197 farms in H-district, East part of Hokkaido, Japan. Fig. 4 shows the relationship between NLD and NAI and indicates that there are large differences in NLD among farms having the same NAI. Farms having better evaluation in NAI sometimes have worse evaluation in NLD. Fig. 5 shows the relationship between FEI and NAI and indicates similar results to Fig. 4.

Combined indexes are the inclinations of lines draw through the origin and each point of farms in both Fig. 4 and Fig. 5. In Fig. 4 farms were divided into four groups: Group A, B, C and D depending on [NLD/NAI] ratio. Farms in Group A have the highest [NLD/NAI] ratio and farms in Group D have the lowest.

3.1.3. Example 2 - Evaluation using five criteria

A radar chart was used for expressing the results of total evaluation using five criteria. The characteristics of a radar chart are to express a pattern of numerical values of five criteria by single graph. In this case four indexes excluding human satisfaction were expressed by the numerical values per unit hectare [ha] of operating scale of land, so that these four indexes are not influenced by scale of farms. For example, economic balance was expressed as net agricultural income obtained by operating scale per unit area of land [ha] . In general outer value in each axis in a radar chart shows a certain value judgment, in this case more favorable conditions. Therefore, the reciprocal numbers were used in three axes indicating fossil energy input, nitrogen load and animal health. For example, the unit of an axis for nitrogen load was [kgN/ha] showing how much of nitrogen load [kgN] is given to the environment.

Fig. 6 shows the comparison between farms in Group A, B, C, and D by a radar chart. The tendency can be seen that farms in Group D having the lowest [NLD/NAI] ratio obtain better evaluations than those in Group A in all five criteria.

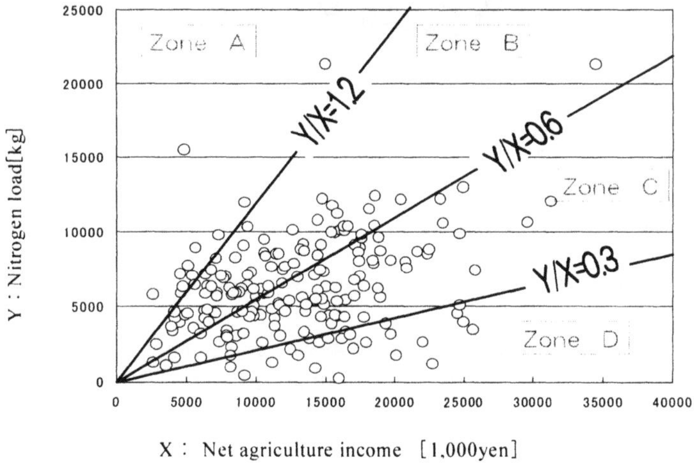

Fig. 4 Relationship between nitrogen load and net agricultural income.

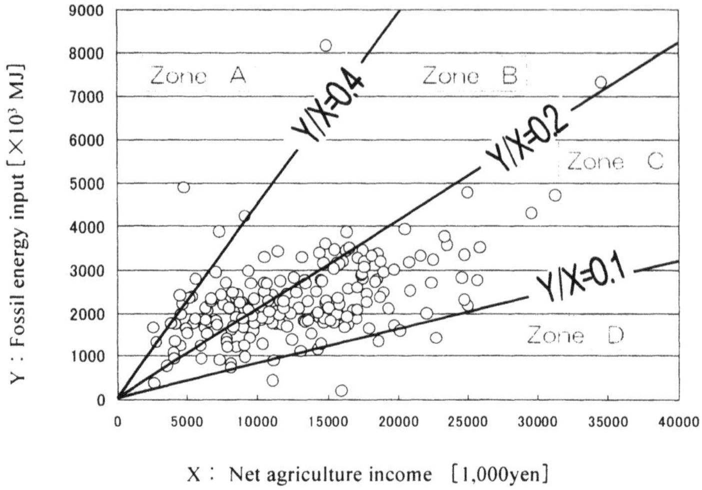

Fig. 5 Relationship between fossil energy input and net agricultural income.

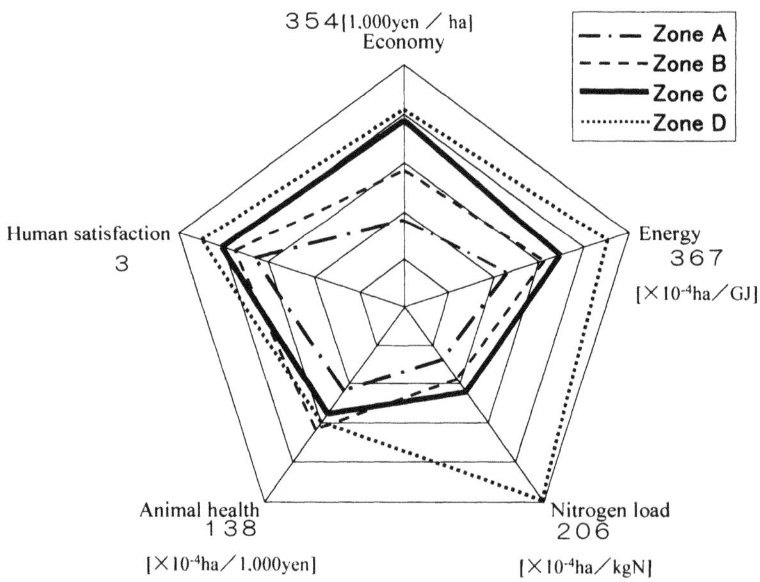

Fig. 6 Total evaluation by radar chart.

3.2. Re-construction of agricultural production systems based on material circulation

*3.2.1. **Understanding of inexistence of a perpetual motion machine***

It might be said to be natural that we always expect the simultaneous existence of both high productivity and environmental conservation when we build the agricultural production systems. At present when all the technology has matured, however, no such a system can exist. There must be some conflicts. Problems in livestock manure handling systems such as bad smell and pollution of rivers are their good examples, which can be seen often at large scale farms. We need to construct the agricultural production systems with understanding the inexistence of a perpetual motion machine.

3.2.2. ***How to line up agricultural production systems to decrease environmental load - Entropy theory***

Recently the biogas plant system is attracting a great deal of attention as a good livestock manure handling system. We tend to think we can solve the livestock manure handling problems if the biogas plant can be built. However, it is not always true. Although the biogas plant system is a good system there must be premises to give full play of the systems ability.

The most important premise is the land space where the digested manure can be spread. Discussion on this subject was presented by Hishinuma (2001) at GGAA Conference. Therefore the biogas plant system should not be built where the digested manure can not be spread to the field. It is even further out of the question that the wastewater treatment is built after anaerobic digestion for livestock manure, because of extremely high inputs in both energy and economy.

When the utilization of food wastes is considered, adding those to a biogas plant for livestock manure is sometimes said to be favorable method. It is sure that this method is better than the waste disposal treatment. However, the most effective method is to utilize all eatable food wastes as feed to livestock animals first, and then, to take livestock manure to biogas plant according to the entropy theory (Tsuchida, 1992). The entropy theory suggests that to line up systems according to the increase of entropy is the most effective.

3.2.3. What is the highest technology in the agricultural production systems ?

Agricultural production systems are being developed by using high technology and advanced biotechnology. However, we need to notice that those advanced technology sometimes breaks the balance of the total system of agricultural production. For example, high milk producing dairy cows were bred by using biotechnology. However, they need large amount of concentrates which can not be produced inside Japan and have to be imported from outside. As a result the circulation of materials in producing milk was broken and led to various environmental problems. Therefore, the highest technology nowadays is the technology which can establish the circulation of materials as shown in Fig. 1.

REFERENCES

Hishinuma, T., S. Hoshiba, S. Morita, T. Amano & Y. Tsukada, 2001. Estimation of biogas plant from the energetic and economic point of view. GGAA Conference.

Hoshiba, S., 1996. Premises for sustainable dairy production. Manure manual '96 (Edited by Hoshiba & Ichikawa), Extension Center, Rakuno Gakuen University, Ebetsu, Japan. 13-21. (in Japanese)

Kawakami, H., S. Hoshiba, Y. Tamura, S. Morita, T. Noda & A. Ikeguchi, 2000. Total evaluation of sustainable dairy production systems using five indexes including animal health and human satisfaction. CIGR World Congress 2000, Abstracts. p218.

Kurosawa, T., 1971. Significance of agriculture and desired agricultural technology. Agricultural Co-op. Management Research Center. (in Japanese)

Tsuchida, A., 1992. Introduction to theory of heat, Asakura Shoten, Tokyo, Japan. pp197. (in Japanese)

Part 2

Measurement and manipulation of rumen methanogenesis

Greenhouse Gases and Animal Agriculture
J. Takahashi & B. A. Young (Editors)

Strategies for mitigating methane emissions from livestock – Australian options and opportunities

R. S. Hegarty

NSW Agriculture Beef Industry Centre, University of New England, Armidale NSW 2351, Australia

Methane produced by enteric fermentation within the digestive tract of sheep and cattle is the largest source of agricultural methane emissions in Australia and contributes 13.2 % of total national greenhouse gas emissions. Reducing emissions from this sector by nutritional management, animal selection and rumen modification are possible, but require a long-term research and development commitment. While rumen modification and animal selection offer direct tools to reduce daily methane emissions from domestic ruminants, feed management does not offer such a simple solution to achieve reduced annual methane emissions from the livestock sector. A continual focus on the improved production efficiency resulting from improved diet digestibility (g methane/g product) is considered unhelpful in achieving reduced emissions from the sector unless incentives to decrease livestock numbers are also implemented. An array of opportunities for animal selection and rumen modification are discussed and their development is recommended to create 'environment-friendly' livestock.

Keywords: selection, efficiency, rumen, diet, ruminant

1. INTRODUCTION

Agriculture contributes 20.5% of Australia's total greenhouse gas emissions and 64% of this emission is due to enteric fermentation in the digestive tract of sheep and cattle (NGGI, 2001). Cattle are the largest single source of agricultural methane, contributing 9.6% of national emissions while sheep generate 3.5% and rice paddies only 0.1%. The majority of Australian livestock graze year round and are never housed. Of a national flock of 116 million sheep, only 0.01% are fed total mixed rations, all other sheep graze pasture or fodders and are generally only supplemented during drought and/or late pregnancy. While 39% of Australian slaughter cattle are finished in feedlots, this is only for the final 70 – 300 days of life (Graham, 2001). Pasture consumed by Australian ruminants is of lower quality that that in temperate countries, with average digestibilites of 68% and 63% for sheep and cattle over the annual pasture cycle (NGGI, 1994). This low pasture digestibility, together with the extensive grazing areas and minimal human contact with stock, significantly constrains the opportunities to reduce emissions from Australian ruminants. This paper describes some of the possible strategies to reduce enteric methane emissions from Australian ruminants.

2. ABATEMENT OPPORTUNITIES

2.1. Altering feed quality

Improving the digestibility of the diet is often considered a simple means of decreasing methane emissions from ruminants. For forage-fed ruminants however, simply increasing the digestibility of available feed is unlikely to achieve reduced annual methane emissions from the sector. Pelchen and Peters' (1998) review demonstrates that increasing ration digestibility or energy density of Australian pastures from their current mean levels to 10.5 MJ/kg DM, would increase rather than decrease daily methane emissions per animal (Figure 1). This result from a large increase in the voluntary feed intake associated with improving forage digestibility. Importantly however, the greater digestible nutrient intake achieved with higher digestibility diets improves livestock performance disproportionately, so methane emission per unit animal product is decreased (Table 1).

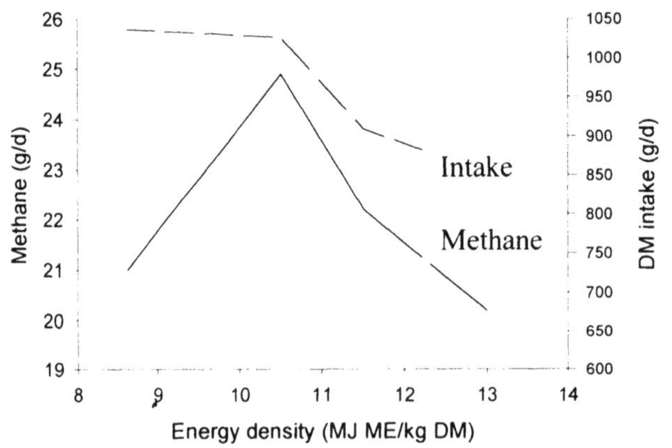

Figure 1. Changes in dry matter intake and daily methane emission by sheep consuming diets of increasing metabolisable energy (ME) content (after Pelchen and Peters, 1998).

Table 1. Feed intake, methane emission and growth rate of a 30 kg Merino wether when given free access to forages differing in digestibility but of equal protein content (15% CP). Simulations based on GrazFeed version 4.1.

Sheep parameter	Digestibility		
	55%	65%	75%
Dry matter intake (kg DM/d)	1.09	1.42	1.74
Gross energy intake (MJ GE/d)	20.1	26.1	32.0
Digestible energy intake (MJ DE/d)	11.0	17.0	24.0
Methane emission (g methane/d)	24.5	32.4	35.4
Methane emission (% of DE)	12.0	10.3	8.0
Growth Rate (g liveweight/d)	5	57	162
Grams methane emitted/g liveweight gained	4.9	0.6	0.2

Table 1 demonstrates that simply improving forage digestibility without mechanisms in place to either constrain feed intake or decrease livestock numbers, will increase rather than decrease the daily (and therefore annual), methane emissions from the sector. In contrast, increasing the digestibility of diets already providing more than 10.5 MJ/kg DM will decrease daily methane emissions (Figure 1). In Australia, use of high-energy diets is limited to small feedlot beef finishing and dairy cattle sectors, and their use is impractical for the breeding herd and flock that presently consume more than three quarters of total feed eaten.

2.2. Animal characteristics

Considerable scope exists to reduce not only animal numbers, but also to breed animals which consume less feed or which produce less methane per unit feed, through genetic selection.

2.2.1. Selection for net feed efficiency

Natural variation in the quantity of feed eaten per unit liveweight gain can be exploited to breed animals that consume less feed than the unselected population while achieving a desired rate of growth. This trait, referred to as the net feed efficiency (NFE) of the animal is heritable ($h^2 = 0.39$) and is independent of the rate of gain (Arthur et al., 2001). Lines of cattle in Australia divergently selected for NFE and low NFE have achieved a divergence of 1.2 kg in daily feed intake after only two generations of selection, without altering liveweight gain. Okine et al., (2001) calculated annual methane emissions from Canadian high NFE steers to be 21% lower than for low NFE steers.

2.2.2. Selection for digesta turn-over

The mean retention time (MRT) of digesta in the reticulorumen is one of the key determinants of fermentation pattern and enteric methane emission. MRT is a highly repeatable trait (Smuts et al., 1995) and a short MRT has been shown to be one of the principal physiological traits of sheep that emit less than expected levels of methane (Pinares-Pattino et al., 2000). These facts imply there is scope for selecting animals for MRT in order to reduce methane emissions per unit food eaten. Since MRT also affects the propionate and microbial protein yield from rumen fermentation, this selection procedure may also have major production advantages.

2.3. Rumen modification to reduce daily methane emissions

The ideal rumen manipulation will achieve a stable ecological change supporting low methane emissions but a high yield of absorbable nutrients, and will require only a single treatment to establish. Potential rumen manipulations are described below.

2.3.1. Feed additives to reduce methanogen activity

Compounds inhibitory to methanogen metabolism has been extensively screened but their generally short period of action in the rumen and loss of efficacy over time limits their use (Van Nevel & Demeyer, 1996). Longer acting chemical agents (McCrabb, 2000) may offer improved efficacy but their use will still be restricted to fully-fed or regularly supplemented animals. Inclusion of other dietary additives such as medium chain fatty acids (Machmuller et al., 2001), bacteriocins (Callaway et al., 1997) or yeast cultures, are similarly constrained by the need for regular intervention in the animal's diet and limits their Australian application.

2.3.2. Enacting stable long-term ecological change in the rumen

Achieving a stable low methane fermentation in the rumen may potentially be achieved through either direct suppression of methanogenesis or by modifying the balance of rumen organisms. Direct suppressions of methanogens may be possible through immunisation of the animal (Baker, 1999), by stimulating either natural or introduced ruminal organisms to produce bacteriocins or by use of Archaeal viruses as a means of biological control (Klieve & Hegarty, 1999). While an immunisation strategy is been evaluated, biological control and bacteriocin options for control have not been thoroughly explored.

Reducing methane production by targeting other rumen organisms to achieve ecological control is also possible, and may be more easily achieved than direct suppression of methanogens. Elimination of rumen protozoa has consistently resulted in reduced methane production and increased microbial protein supply from the rumen and the simplicity and efficacy of this change makes it worthy of attention (Kreuzer, 1986). An alternate ecological change is the stimulation of reductive acetogenesis in the rumen but discovery of acetogens with lower hydrogen thresholds will be required.

3. CONCLUSION

The fundamental responsibility of the livestock industries in regard to global warming is to reduce the total annual emissions of methane from the sector. While this can be achieved by improving pasture digestibility, the area of pasture and the number of livestock must be decreased accordingly. Improving pasture digestibility without these associated management changes will increase annual sector emissions. Production of ‘environment-friendly’ ruminants that generate less methane per day can potentially be achieved by animal selection and by initiating stable ecological changes in the rumen. These strategies require developing.

REFERENCES

Arthur, P.F., Archer, J.A., Johnston, D.J., Herd, R.M., Richardson, E.C. & P.F. Parnell, 2001. Genetic and phenotypic variance and covariance components for feed intake, feed efficiency, and other postweaning traits in Angus cattle. J. Anim. Sci. *In press.*

Baker, S.K., 1999. Biology of rumen methanogens and stimulation of animal immunity. In. “Methane from Livestock - Understanding emissions and mitigation options. Proceedings. Bureau of Resource Sciences, Canberra Nov. 4 1998. pp29-37.

Callaway, T.R., Carneiro De Melo, A.M.S. & J.B. Russell, 1997. The effect of nisin and monensin on ruminal fermentations in vitro. Curr. Micro. 35: 90-96.

Graham, J., 2001. The significance of the feedlot sector of the beef industry. *In.* Proc. NSW Agric. Beef Prod. Conf. Dubbo, May 2001. pp145-146.

Klieve, A., & R.S. Hegarty, 1999. Opportunities for biological control of methanogensis. Austr. J. Agric. Res 50: 1315-1319.

Kreuzer, M., 1986. Methodik und Anwendung der Defaunierung bein wachsenden Wiederkauer. *J. Vet. Med. (A)* 33: 721-45.

Machmuller, A., Dohme, F., Soliva, C.R., Wanner, M. & M. Kreuzer, 2001. Diet composition affects the level of ruminal methane suppression by medium chain fatty acids. Aust. J. Agric. Res. 52: 713-722.

McCrabb, G., 2000. The relationship between methane inhibition, feed digestibility and animal production in ruminants. 2nd Int. methane mitigation conference proc., Novosibirsk, Russia. pp125 – 132.

NGGI, 1994. Agriculture : Workbook 6.0 for Livestock. Department of the Environment, Sport and Territories, Canberra.

NGGI, 2001. National Greenhouse Gas Inventory 1999 with methodology supplements. The Australian Greenhouse Office, Canberra.

Okine, E.K., Basarab, J.A., Baron, V. & M.A. Price, 2001. Net feed efficiency in young growing cattle: III. Relationship to methane and manure production. Proc. Can. J. Anim. Sci Conf. *In press*.

Pelchen, A.& K. Peters, 1998. Methane emissions from sheep. Small Rum. Res. 27: 137-150.

Pinares-Pattino, C., Ulyatt, M.J., Holmes, C.W., Barry, T.N. & K.R. Lassey, 2000. Some rumen digestion characteristics and methane emission in sheep. Proc. 15th Symp. Energy Metabolism in Animals, Elsinore, Denmark pp 117-120.

Smuts, M., Meissner, H. & P. Cronje (1995). Retention time of digesta in the rumen: its repeatability and relationship with wool production of Merino rams J. Anim. Sci. 73: 206-210.

Van Nevel, C. & D.I Demeyer, 1996. Control of rumen methanogenesis. Env. Monitor. Asses. 42: 73-97.

Greenhouse Gases and Animal Agriculture
J. Takahashi & B. A. Young (Editors)

Environmental control of methane production by ruminants

Angela R. Moss

ADAS Nutritional Sciences Research Unit, Drayton Manor Drive, Alcester Road, Stratford-upon Avon, CV37 9RQ, UK.

Methane is an important greenhouse gas, contributing approximately 18% to the overall radiative forcing that is currently occurring. The "Kyoto Protocol" agreed a decrease of greenhouse gas emissions by an average of 5.2 % below 1990 level during the period 2008-2012 in industrialised countries. The level of allowed emissions during this period varies according to the countries: e.g., +8 % for Australia; -8 % for EU; -6 % for Japan. The agreement is applied only to six greenhouse gases: Carbon dioxide, methane, nitrous oxide, two fluorocarbons and sulphur hexafluoride. Enteric fermentation in commercial ruminant livestock represents one of the major agricultural sources of methane on a world wide basis and in the UK, it is estimated (CONAIR) that it contributes some 26% of the national emission to the atmosphere. The information on which this estimate is based, has been derived from measurements with cattle or sheep in open-circuit respiration chambers at a standard temperature. The ambient temperature at which methane measurements are made may have a significant impact on the estimate for the national emission of methane from ruminants. The objective of the current study was to determine the impact of ambient temperature (5 and 24 °C constant over the measurement period and 8 and 15 °C with diurnal variation) on methane production and the rumen environment (temperature and pH). Dehydrated lucerne (*Medicago sativa)* was offered at maintenance to four mature wether sheep. Methane production was measured in open-circuit respiration chambers for five consecutive 24h periods at a mean temperature of 5, 8.1, 15.3 and 24 °C. Three periods of data were collected at each temperature (two periods for the diurnal study). Apparent digestibility of organic matter (OM) was measured over the five days and 24h *in situ* OM degradability was determined on each of the five days. Rumen temperature and pH was measured at 10 min intervals throughout the methane measurement periods. In Study 1 methane production ($l\ d^{1}$) was significantly higher at the lower environmental temperature and this higher production may be partly explained by the increased feed degradability ($P<0.10$) in the rumen. There was no significant effect of ambient temperature on apparent digestibility or on mean rumen temperature, although rumen pH was reduced at the high temperature. In Study 2 methane production ($l\ d^{-1}$) was significantly higher at the higher environmental temperature and this higher production may be partly explained by the increased degradability ($P<0.17$) in the rumen. There was no significant effect of ambient temperature on apparent digestibility, but mean rumentemperature increased and as in Study 1, mean rumen pH decreased. The work shows that both mean ambient temperature and the diurnal variation in ambient temperature can influence the amount of methane produced from a particular diet and this should be considered when using data from different laboratories. More work is required to investigate interactions between environment and diet.

Key words: methane, ambient temperature, in vivo, sheep, rumen temperature.

1. INTRODUCTION

Methane is an important greenhouse gas, contributing approximately 18% to the overall radiative forcing that is currently occurring. The "Kyoto Protocol" agreed a decrease of greenhouse gas emissions by an average of 5.2 % below 1990 level during the period 2008-2012 in industrialised countries. The level of allowed emissions during this period varies according to the countries e.g.,: +8 % for Australia; -8 % for EU; -6 % for Japan. The agreement is applied only to six greenhouse gases, namely, carbon dioxide, methane, nitrous oxide, two fluorocarbons and sulphur hexafluoride.

Enteric fermentation in commercial ruminant livestock represents one of the major agricultural sources of methane on a world wide basis and in the UK, it is estimated (CONAIR) that it contributes some 26% of the national emission to the atmosphere (see Moss *et al.*, 2000, for a review). The information on which this estimate is based has been derived from measurements with cattle or sheep in open-circuit respiration chambers at a standard temperature. Other systems for determining methane production from ruminants are being developed which allow the measurements to be made at the natural ambient temperatures prevailing, for example the poly-tunnel system (Murray *et al.*, 1999) and the sulphur hexafluoride (SF_6) tracer technique (Johnson *et al.*, 1994) for determining methane production by grazing ruminants. Murray *et al.* (1999) compared the poly-tunnel system with open-circuit respiration chambers using sheep fed either zero-grazed grass or dried grass pellets and reported greater methane output (15% increase) from sheep, when determined in open-circuit respiration chambers compared with the tunnel system. This work implied thatthat ambient conditions may influence methane production.

It is important to know if the ambient temperature at which methane measurements are made has a significant impact on methane emissions and hence on the estimate for the national emission of methane from ruminants. The objective of the current study was to determine the impact of ambient temperature (5, 8.5, 15.3 and 24 °C) on the rumen environment and methane production by sheep.

2. MATERIALS AND METHODS

Dehydrated lucerne (*Medicago sativa)* was offered at maintenance to four mature wether sheep (Lleyn, liveweight 75kg). In addition all animals received a mineral/vitamin supplement (7g d^{-1}) and had free access to water. The diet was offered for a period of 56d, with 14d acclimatisation to the diet, followed by five consecutive 24h periods when methane production was measured in four individual open-circuit respiration chambers at 15 min intervals at 5 and 24 °C (Study 1) and was repeated with the same animals one year later at 8.1 and 15.3 °C (Study 2). Methane output volumes were corrected to STP. Apparent digestibility of organic matter (OM) was measured over the five days and *in situ* DM degradability was determined after 24h for the dehydrated lucerne using polyester fibre bags (pore size 43 μm) on each of the five days. Rumen temperature and rumen pH were measured at 10 min intervals throughout the methane measurement periods. The above was repeated to provide a total of three periods of five days at 5 and 24 °C (Study 1) and two periods of five days at mean daytime temperatures of 10.2 and 19.3 and mean night-time temperatures of 7.2 and 13.9 °C for cold and hot respectively to determine the effects of diurnal temperature variation (Study 2) (Figure 1). During the periods of measurement the sheep had a full fleece for the cold environment and were shorn for the hot environment. Dry matter and total ash contents of the diets were measured.

3. RESULTS

The results for DM intake, methane production, apparent digestibility, *in situ* DM degradability at 24h and mean rumen temperature and pH are shown in Table 1 for Study 1. Methane production (l d^{-1}) was significantly higher at the lower environmental temperature and this higher production may be partly explained by the increased degradability ($P<0.10$) in the rumen. There was a significant ($P<0.001$) time x environmental temperature interaction for rate of methane production (Figure 2), with methane production rate being lower during the first 8h in the hot environment and then similar for the two temperatures later on. There was no significant effect of temperature on apparent digestibilitynor on mean rumen temperature, but the mean rumen pH was significantly lower for the hot environment.

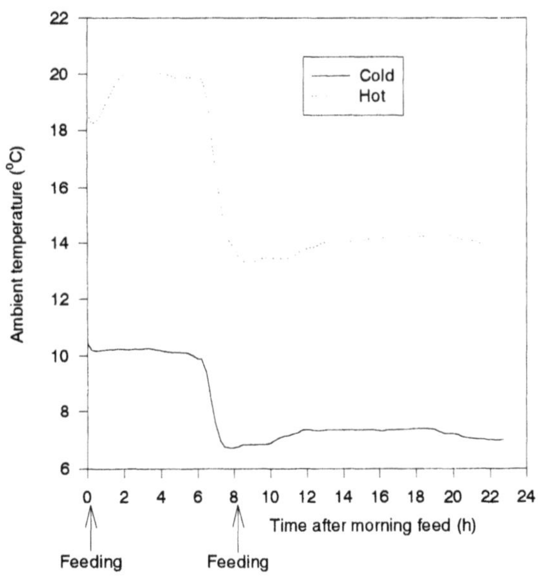

Figure 1. Pattern of ambient temperature in open-circuit respiration chambers for the cold and hot environment in Study 2.

Table 1 Dry matter intake, methane production, apparent digestibility, in situ dry matter degradability and mean rumen temperature and pH determined in sheep at target ambient temperatures of 5 and 24 °C (Study 1).

Parameter	Ambient temperature (°C) 5	24	SED
Ambient temperature (°C)	4.9	21.8	0.92***
Dry matter intake (g d^{-1})	1212	1212	
Methane (l d^{-1})	39.59	37.55	0.542***
Methane (l kg^{-1} DMI)	32.89	31.00	0.430***
Apparent DM digestibility	0.694	0.701.	0.008
Apparent OM digestibility	0.709	0.713	0.009
In situ 24h DM degradability (g kg^{-1})	693	679	8.13
Efficiency of methane production (l kg^{-1}DM degraded)	47.5	45.6	1.17
Rumen temperature (°C)	39.2	39.1	0.011
Rumen pH	6.51	6.38	0.006***

*** P<0.001

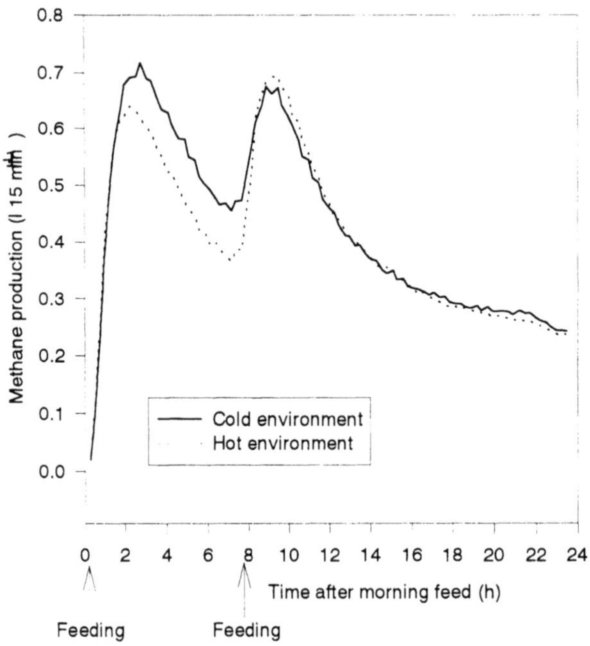

Figure 2 Diurnal variation in rate of methane production at 4.9 and 21.8 °C (Study 1).

There was a significant ($P<0.001$) time x temperature interaction for both rumen temperature and pH regardless of the environmental temperature, with rumen temperature being raised immediately post-feeding and rumen pH being depressed (Figure 3). Rumen temperature and pH were lower for the hot environment compared with the cold particularly between the morning and afternoon feeds for temperature and post the afternoon feed for pH. Rumen temperature (Y) and pH (X) were negatively related ($Y=47.7-1.3X$, $R^2=83.8\%$; $Y=45.4-0.99X$, $R^2=71.5\%$ for cold and hot environments respectively).

The results for DM intake, methane production, apparent digestibility, *in situ* DM degradability at 24h and mean rumen temperature and pH are shown in Table 2 for Study 2.

Methane production ($l\ d^{-1}$) was significantly higher at the higher environmental temperature and this higher production may be partly explained by the increased degradability ($P<0.17$) in the rumen. There was a significant ($P<0.001$) time x environmental temperature interaction for methane production (Figure 4), with methane production being similar for the two ambient temperatures during the first 8h and then higher in the hot environment for 8 to 16h. There was no significant effect of ambient temperature on apparent digestibility. There was a significant increase ($P<0.001$) in mean rumen temperature anda significant depression in mean rumen pH for the hot environment.

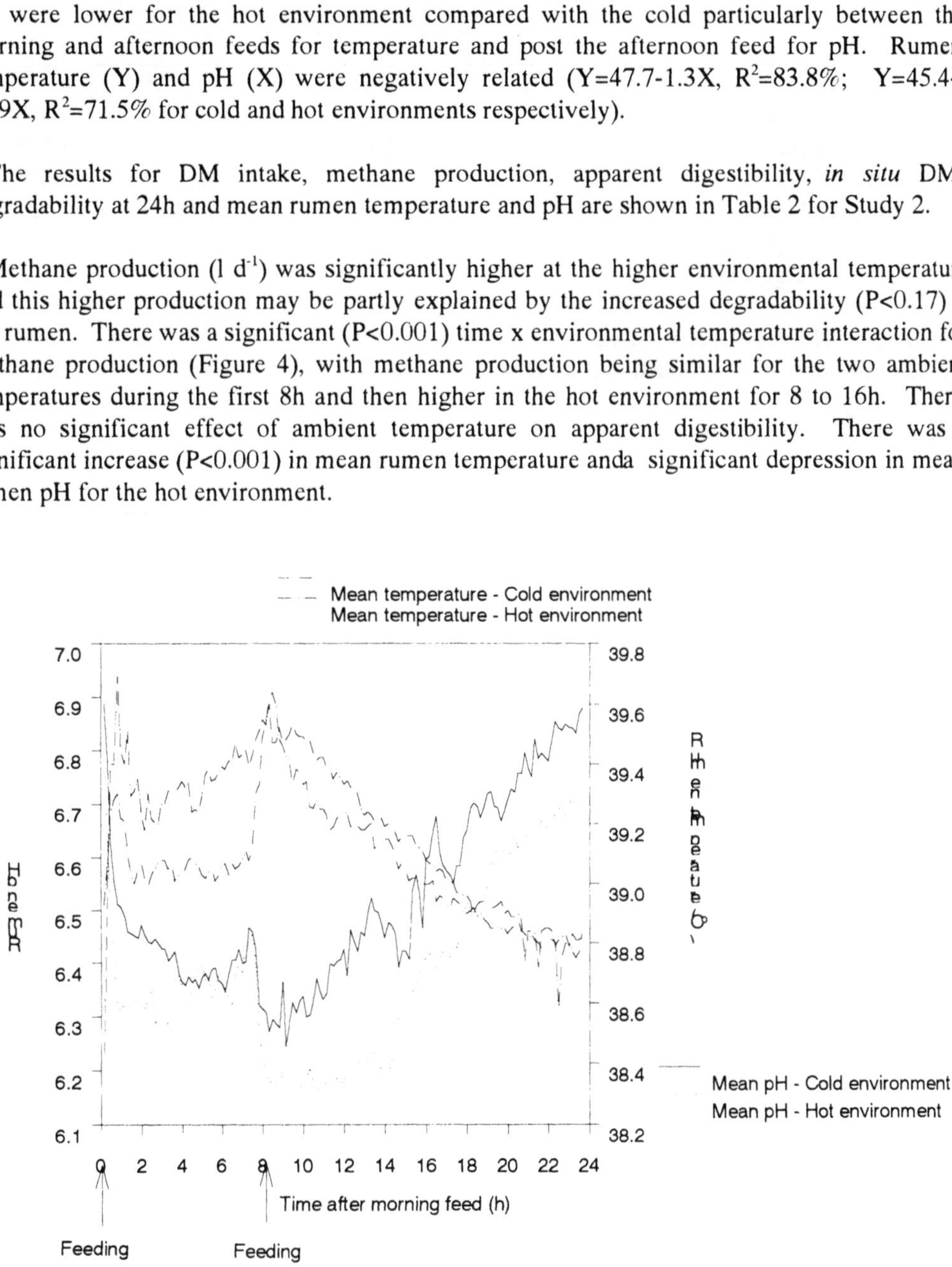

Figure 3 Diurnal variation of rumen temperature and pH at 4.9 and 21.8 °C (Study 1).

Table 2 Dry matter intake, methane production, apparent digestibility, in situ dry matter degradability and mean rumen temperature and pH determined in sheep at target ambient temperatures of 8 and 16 °C (Study 2).

Parameter	Ambient temperature (°C) 8	16	SED
Ambient temperature (°C)	8.04	15.26	0.497***
Dry matter intake (g d^{-1})	1135	1180	17.1*
Methane (l d^{-1})	36.2	38.8	0.68***
Methane (l kg^{-1} DMI)	31.9	32.9	0.52*
Apparent DM digestibility	0.680	0.698	0.0094
Apparent OM digestibility	0.697	0.713	0.0101
In situ 24h DM degradability (g kg^{-1})	685	696	8.5
Efficiency of methane production (l kg^{-1}DM degraded)	46.1	47.5	1.11
Mean rumen temperature (°C)	39.00	39.17	0.013***
Mean rumen pH	6.50	6.43	0.008***

*** P<0.001

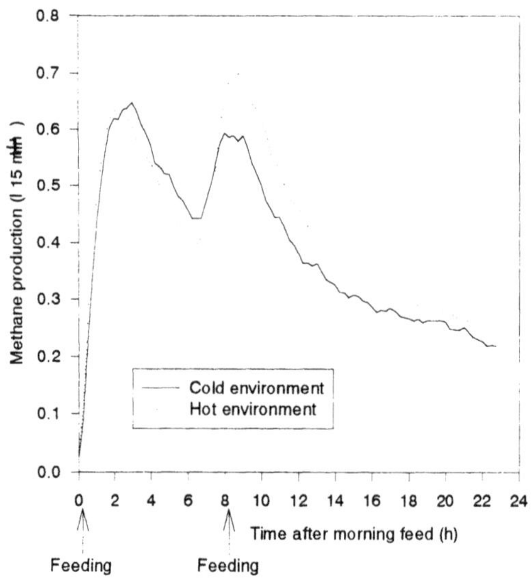

Figure 4 Diurnal variation in rate of methane production at 8.0 and 15.3 °C (Study 2).

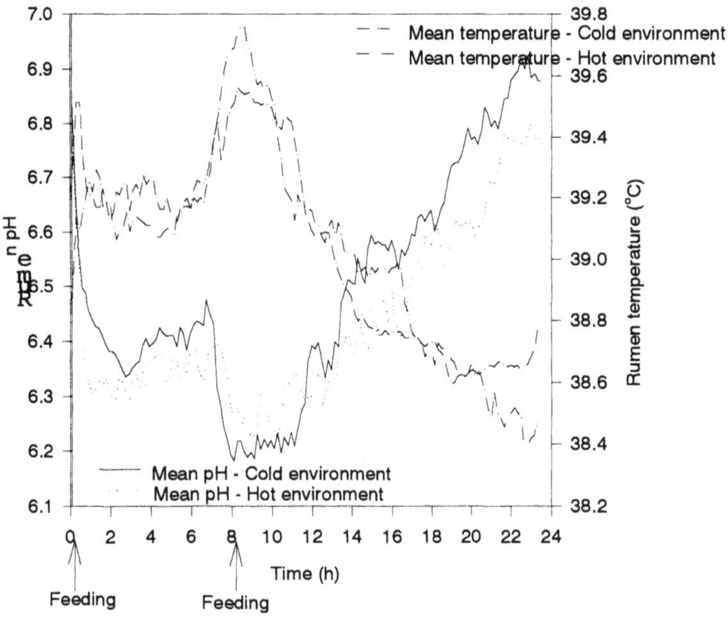

Figure 5 Diurnal variation of rumen temperature and pH at 8.0 and 15.3 °C (Study 2).

There was a significant ($P<0.001$) time x environment interaction for both rumen temperature and pH regardless of the ambient environmental conditions, with rumen temperature being raised immediately post-feeding and rumen pH being depressed (Figure 5). Rumen temperature was significantly higher for the hot environment compared with the cold after the afternoon feed. Rumen temperature (Y) and pH (X) were negatively related ($Y=46.4-1.13X$, $R^2=50.5\%$; $Y=47.6-1.32X$, $R^2=49.4\%$ for cold and hot environments respectively).

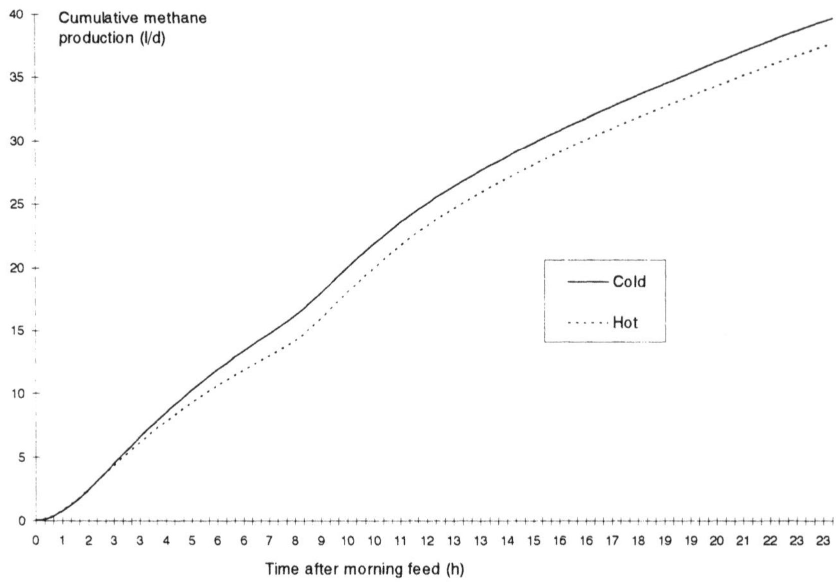

Figure 6 Cumulative methane production from sheep housed at cold and hot ambient temperatures with diurnal variation in open-circuit respiration chambers (Study 2).

4. DISCUSSION

In Study 1 methane production was significantly reduced when the ambient temperature was raised from approximately 5 to 22 °C and this was associated with a reduced rumen DM degradability ($P<0.10$). Other workers suggested that methane production would be reduced at cold ambient temperatures (Kennedy and Milligan, 1978) and that this was associated with increased passage rate. The opposite was observed here. Rumen temperature and pH were raised in the cold environment which is associated with enhanced microbial andcellulolytic activity respectively in the rumen and corresponds to the increased level of methane produced per kg DM intake. Abdel-Baki *et al.* (1995) noted raised rumen pH for sheep maintained at 18.2 °C compared with 35.8 °C.

The DM degradability at 24h (Study 1) was significantly ($P<0.10$) lower for the hot environment compared with the cold. Yousri *et al.* (1977) suggested that elevated ambient temperatures depressed rumen cellulolytic activity and that the negative effect of such a depression on diet degradability might overcome the positive effects of the decline in rumen passage rate with increased ambient temperature. Diet apparent digestibility of DM was unaltered by environment in Study 1 but tended to increase with the hot environment in Study 2 and this has been reported by others (Christopherson and Kennedy, 1983). Abdel-Baki *et al.* (1995) observed that Egyptian sheep were more adapted to summer and winter conditions than the transitional seasons (autumn and spring) as they observed lower apparent digestibility of nutrients and lower rate of fibre degradation compared with summer and winter. The range in ambient temperature was 13 °C in the spring and only 8 °C in the summer. The range in ambient temperature in the present study was < 8 °C for all the measurement periods.

Mean rumen temperature was significantly raised ($P<0.05$) in the hot environment compared with the cold for Study 2. There have been few studies determining the effect of environment on rumen temperature. Abdel-Baki *et al.* (1995) showed no effect of seasonal variation on either rumen temperature or rectal temperature, whereas Johnson (1980) and Bernabucci *et al.* (1999) reported that sudden exposure of heifers to a hot environment resulted in a rapid increase in rectal temperature followed by a gradual decline when the heifers were exposed to a hot environment for a long period. Brod *et al.* (1982) reported that rumen temperature in sheep was depressed by the addition of 2 l of water of varying temperature at feeding. Rumen temperature was depressed by 4.62 °C for water at 10 °C and by 4.01 °C for water at 20 °C and took approximately 90 min. to return to the pre-feeding rumen temperature. Water temperature had no significant effect on rumen pH, but pH was maximally depressed for a longer duration after feeding with increasing water temperature. Rumen pH in the current study was depressed longer at the higher temperature where water temperature was high. Sudden depression in observed rumen temperature (Figure 5) may be associated with the sheep drinking water.

The depression in rumen pH observed post-feeding regardless of ambient temperature is associated with rapid fermentation to short chain fatty acids of the soluble carbohydrates ingested (Church, 1976). Van Kessel and Russell (1969) observed *in vitro*, using rumen fluid sampled from animals fed on roughage-based diets, thatruminal methanogens lose the ability to use H_2 at low pH, giving rise to free H_2 in the gas phase when the pH was less than 5.5. Thus on roughage diets, a low pH leads to a decrease in methanogenesis independent from propionate formation. In Study 1 the low rumen pH observed post-second feed for the hot environment was associated with lowered methane production during that period (Figure 6). Lana *et al.* (1998) reported that ruminal pH was positively correlated with methanogenesis *in vitro* and to acetate to propionate ratio *in vitro* and *in vivo*. It is therefore likely that the decline in rumen pH influenced methane production but the mechanism by which rumen pH is controlled in this situation is unclear.

The impact of the environment on diet digestibility, rumen parameters and methane production is dependent on many factors. When undertaking rumen studies long-term acclimatisation to environmental conditions is required and the impact of diurnal variation in ambient temperature requires further research. The use of the open-circuit respiration chambers under carefully controlled and monitored environmental conditions is essential for determining methane production from test diets, but it is necessary to report the conditions used. For inventory purposes the poly-tunnel/SF_6 tracer systems of measuring methane production may provide conditions more representative of those experienced by animals in the field, but would require a systematic approach to include a range of diets under a range of seasons. These techniques also require a reliable method to determine feed intake in the grazing situation and would have to take into consideration the variation in intake that would be experienced in the true grazing situation due to the effects of rain etc.

ACKNOWLEDGEMENTS

I wish to acknowledge the funding of this work from the UK Ministry of Agriculture, Fisheries and Food (MAFF).

REFERENCES

Abdel-Baki, S. M., M. Y. Mekkaway, E. M. Hassona, M. S. Nowar, S. M. Bassuny & S. A. Ibraheem, 1995. Effect of seasonal variations on some nutritional and physiological parameters of Egyptian sheep. Ind. J. Anim. Sci., 65: 556-564.

Bernabucci U., Bani P., Ronchi B., Lacetera N. and Nardone A., 1999. Influence of short- and long-term exposure to a hot environment on rumen passage rate and diet digestibility by Friesian heifers. J. DairySci., 82: 967-973.

Brod D L., Bolsen K K. and Brent B E., 1982. Effect of water temperature on rumen temperature, digestion and rumen fermentation in sheep. J.Anim. Sci., 54: 179-182.

Christopherson R J. and Kennedy P M., 1983. Effect of the thermal environment on digestion in ruminants. Can. J. Anim. Sci., 63: 477-496.

Church, D. C., 1976. Digestive Physiology and Nutrition of Ruminants.Vol. 1. Digestive Physiology. O & B Books, Corvallis, OR.

Johnson H D., 1980. Environmental management of cattle to minimise the stress of climate change. Int. J. Biometeorol., 24 (suppl. 7): 65-78.

Johnson K., Huyler M., Westberg H., Lamb B. and Zimmerman P., 1994. Measurement of methane emissions from ruminant livestock using a SF_6 tracer technique. Environ. Sci. Technol., 28: 359-362.

Kennedy, P. M. & L. P. Milligan, 1978. Effects of cold exposure on digestion, microbial synthesis and nitrogen transformations in sheep. Brit. J. Nutr. 39: 105-117.

Lana R P., Russell J B. and Van Amburgh M E., 1998. The role of pH in regulating ruminal methane and ammonia production. J. Anim. Sci., 76: 2190-2196.

Moss, A.R., Jouany, J.P. and Newbold, J. 2000. Methane production by ruminants: its contribution to global warming. Annales de Zootechnie. 49: 231-253.

Murray, P. J., Moss, A. R., Lockyer, D. R. and Jarvis, S. C. (1999). A comparison of systems for measuring methane emissions from sheep. J. Agric. Sci., Cambridge. 133: 439-444.

Van Kessel JS & Russell JB (1969) The effect of pH on ruminal methanogenesis. FEMS Microbiol. Ecol.. 20: 205-210.

Greenhouse Gases and Animal Agriculture
J. Takahashi & B. A. Young (Editors)

Methane, nitrous oxide and carbon dioxide emissions from ruminant livestock production systems

D. E. Johnson[1], H. W. Phetteplace[1] and A. F. Seidl[2]

[1]Dept. of Animal Science, [2]Dept. of Agricultural & Resource Economics, Colorado State University, Ft. Collins, CO, USA

System inputs with greenhouse gas (GHG) implications were summarized for typical U.S. beef production systems and for representative U.S. and N.Z. dairy systems. The GHG emissions are expressed in carbon dioxide equivalents (CO_2eq) and related to livestock system product output. The U.S. beef systems are estimated to produce from 13 to 16 kg of CO_2eq per kg of live weight sold and the dairy systems 1.3 kg of CO_2eq/kg of milk while the pasture based N.Z. dairies produced 1.6 kg/kg milk. Relative source strengths varied. Nitrous oxide was the largest CO_2eq source of beef system emissions, 52%, with 36 and 12% from methane (CH_4) and carbon dioxide (CO_2). The amount of methane per wt sold was quite uniform across location in the U.S., while nitrous oxide was variable ranging from 7 in WI to 9 kg CO_2eq/kg sold in TX. Methane was the largest CO_2eq source from CA dairies, while nitrous oxide (N_2O) was the largest in N.Z. dairies. High CH_4 in CA was due to extensive use of anaerobic lagoon disposal of manure while excess N content and flux on N.Z. pastures caused higher relative N_2O emissions. Utilization of best management practices and manure amendments to sequester carbon in the crop and forage land of ruminant livestock production systems have modest potential to offset their GHG emissions.

Keywords: greenhouse gas, cattle, production system

1. INTRODUCTION

The unique gastrointestinal tract of ruminants coupled with large populations and/or body size, and appetites, results in annual global atmospheric emissions of about 77 Tg of methane enterically and 7 Tg from manure handling systems. Inventories and mitigation proposals of GHG's release from livestock have usually emphasized the enteric source (e.g., RLEP, 1998). Additional investigations of overall production systems, however, show other important sources (Ward et al., 1993; Johnson et al., 1997). With the more complete source estimates of N_2O (Mosier et al., 1998) as recommended for inventory purposes by IPCC (2000), the livestock production system emissions as carbon dioxide equivalents (N_2O-CO_2eq) of some systems exceeded those for CH_4 (Johnson et al., 2000b). Robertson et al. (2000) found N_2O to be the single greatest source of CO_2eq emissions from nearly all agronomic systems in Midwest U.S. observations. This paper briefly reviews and updates prior summaries (Johnson et al., 2000a, 2001) including embodied CO_2 inputs and expands coverage of management and mitigation strategy effects on cattle production system variations and relative source strengths of GHG emissions. Details of systems spreadsheet assumptions can be found in a report to E.P.A. (Johnson et al., 2000a) or by query to the authors.

2. ESTIMATION PROCEDURES

Beef and dairy systems were divided into 6 to 16 classes based on age, physiological or production stage, e.g., replacement animals, lactation, etc. The systems are based on the maintenance of and output from 100 head of mature cows. Inputs and outputs for each class were described from national (National Agricultural Statistics Service, National Animal Health Monitoring System), state (Cooperative Extension Service) and industry (National Cattlemen's Beef Assoc., Cattle-Fax) sources. Comparable information from the Livestock Improvement Corporation (LIC) and the Meat and Wool Economic Service (MWB) were consulted for New Zealand. Stocker cattle (weaned calves prior to feedlot) from each location were fed out in one of two representative feedlots, Texas (TX) or Iowa (IA). Production characteristics for each system include mature weights, mortality, culling, calving, replacement and growth rates, and live weight of beef output at the farm gate (Table 1). Dietary requirements and enteric CH_4 emissions were derived per IPCC (2000) Good Practice Inventory guidelines, an adaptation of NRC (1989, 1996) net energy requirements. Key assumptions include, 6 % of diet energy loss as methane generally, except for 3.5 % for the feedlot phase. Nitrous oxide estimates were 1.25 % of crop N-fertilization, and up to 2.5 % from manure or leached N. Carbon dioxide emissions from fertilizer synthesis and fossil fuel use for cropping, transportation, insecticide and herbicide synthesis and application, are based on Pimentel et al. (1980) in the U.S. and Wells (1998) in N.Z.. Methane and nitrous oxide were converted to CO_2eq using the factors of 21 and 310 g/g (IPCC, 1996). Local sources were used to define typical feedstuff ingredients, crop production practices and yields, fertilizer, insecticide and herbicide applications, manure handling, fuel use for crops, irrigation, transport, feed grain processing, cost of production for feed and animal management, and revenue (U.S. only) from cattle or milk sold (see partial list in Table 1). Byproduct feeds are apportioned to cropping inputs relative to $ value of the crop product(s). The CO_2 emission estimates are based on U.S. practices (Pimentel, 1980) with diesel as the fuel of choice.

The U.S. dairy cows averaged 613 kg live weight, 37% replacement rates and 11 mo lactations, producing 7200 to 9000 kg of milk annually containing 3.5% fat and 3.3% protein. The U.S. systems are dry lot, or 'mixed' operations with pasture comprising 0 to 31 % of the herd diet (Table 1). Alfalfa hay, corn silage, cereal grains, soybean meal and byproducts constituted the principal diet ingredients, averaging 10.5 MJ ME/kg DM in the U.S.. Anaerobic lagoon disposal was used for 50% of the manure in California (CA) and very little in Wisconsin (WI). Fertilizer application rates ranged from 27 to 91 kg of N and 13 to 18 kg of P per ha.

The New Zealand pasture based dairy was characterized as a Holstein herd of cows weighing 511 kg producing 3444 kg of milk with 4.34% fat and 3.41% protein during a 270 day lactation annually (LIC, 1999) with a 15% replacement rate (MWBNZ, 1999). Mortality rates of 1.3% for cows (Holmes, 1999) and 4 to 5% for heifers and calves were assumed. The diet contained 10.1 MJ ME/kg DM, 80 to 85 % from pasture, with the balance being haylage and a small fraction of maize silage. Pasture yields were 12 T DM/ha with fertilizer applications of 63 and 78 kg on N and P/ha (Wilcock et al., 1999) and no irrigation. Manure was applied to pasture, primarily by grazing animals, with 3% disposed through anaerobic lagoons at milk sheds (Russell, 1999).

3. RESULTS AND DISCUSSION

The 100 cow beef systems, cow-calf through feedlot, produced 37 t of live weight sales annually, which includes 9 t from cull cows and bulls (Table 1). Their diet was composed of 55 % pasture and 20% hay. These totals include the feedlot phase that used no pasture and only 10 to 20% forage. Manure (10 t of manure N per herd) was disposed on pastures for the cow-calf and stocker phases and by drylot methods from the feedlots. Land requirements (ha) for the herds, stockers and feedlot cattle from these five U.S. counties with large numbers of beef cows averaged 315 ha (range, 122 to 745). Nitrogen fertilizer use averaged 2.2 t/herd with moderate variation by location.

Table 1. Characteristics of simulated U.S. beef cow-calf through feedlot, U.S. and N.Z. dairy production systems.

Characteristics (100-cow herds)	U.S. Beef Average (CV %)	U.S. Dairies CA	U.S. Dairies WI	New Zealand Dairy
Mature cow weight, kg	497 (2)	635	590	511
Cows, calves, replacements, hd	168 (1)	167	180	127
Replacement, %	16 (5)	33	40	15
Calving rate, %	92 (3)	93	93	96
Calf mortality, to wean, %	10 (42)	9.1	10.8	9.0
Adult mortality, %	1.4 (64)	5.0	3.8	1.3
Live weight sold, t	37 (4)	21.3	23.2	18.3
Milk sold, kg/cow	0	8982	7169	3444
Pasture, %	55 (8)	0	31	81
Other roughage, %	20 (45)	43	30	15
Total ha/herd	315 (85)	103	166	41
Nitrogen, syn., 10^3 kg	2.2 (35)	9.3	4.4	2.6
Fuel, herd, 10^3 l	10 (13)	34	30	4
Manure on pasture, %	89 (.4)	0	42	94
Manure, anaerobic lagoon, %	0	50	3	3

CV = coefficient of variation; CA = California; WI = Wisconsin

The GHG emissions, as CO_2eq, averaged 580 t (CV=7%) from the cow-calf through feedlot beef production units (Table 2). These U.S. beef systems averaged 15.5 kg CO_2eq/kg of live weight produced. The emissions are predominantly from the cow-calf phase (77%) with lesser amounts from the stocker (11%) and feedlot (12%) phases. Division by gas source, shows approximately 36% from CH_4, 52% from N_2O, and the balance from CO_2. Manure disposal accounted for a small fraction of the CH_4 emissions since none of the beef systems simulated used anaerobic lagoons, however, manure from grazing or use for fertilizer produced a large part of the N_2O emissions. Nitrous oxide emission equivalents were the most variable among locations (CV = 33 %).

Total CO_2eq emissions from the 100-cow U.S. dairy systems were about double those from the beef systems, and 80 to100% higher than 100-cow N.Z. dairies (Table 2). When expressed as CO_2eq/product, however, the N.Z. herd emissions of 1.62 kg/kg of milk were slightly greater. These dairy system estimates are higher than found by Cedarberg (1998) who reported 1.0 kg CO_2 equivalent per kg milk on a whole farm or 'life cycle' basis for conventional dairy farms in Sweden.

The CA dairy has the lowest CO_2eq/milk even with a larger amount of methane production from the 50% of manure disposal via anaerobic lagoons. Comparison of the dairies according to enteric CH_4 per unit of milk, shows even more advantage to the CA dairy, which produces 12% less than those in WI and 40% less than those in N.Z.. This comparison illustrates the strong effect of increasing productivity per cow resulting in reduced feed requirements and thus GHG output per unit of milk. Examination of the fractional source of these varying amounts of GHG's also shows considerable variation by production system.

The CH_4-CO_2eq was the single largest contributor, 45%, in the CA dairy system with N_2O-CO_2eq and CO_2 contributing similar proportions of the balance of GHG emissions. In WI, CO_2 emissions contributed the most CO_2eq at 40% of the total emissions while CH_4 and N_2O CO_2eq were approximately equal. The N.Z. system was different in that the largest single source was N_2O-CO_2eq at 47% with 40% contributed from CH_4-CO_2eq. The N.Z. system was also unique in that only 13% of the GHG's were from CO_2, a reflection of dependence on hydroelectric power sources.

Table 2. Source strengths of beef and dairy GHG emissions (annual CO_2eq).

Source of GHG	U.S. Beef Cow-Feedlot CO_2eq Mean (%CV)	CA Dairy CO_2eq	WI Dairy CO_2eq	N.Z. Dairy CO_2eq
Enteric CH_4:				
t per herd	206 (10)	320	292	207
kg/kg product	5.5 (0.3)	0.36	0.41	0.60
Manure CH_4:				
t per herd	5.2 (1.5)	185	18	16
kg/product	0.14 (0.4)	0.21	0.03	0.04
N_2O:				
t per herd	301 (33)	330	298	259
kg/product	8.1 (0.9)	0.37	0.42	0.76
CO_2 :				
t per herd	67 (11)	296	407	74
kg/product	1.8 (0.3)	0.33	0.57	0.22
Total GHG:				
t per herd	580 (6.8)	1130	1015	556
kg/product	15.5 (1.0)	1.26	1.38	1.62
t per ha	1.8 (1.5)	11.0	6.2	13.6

Nitrous oxide emissions are inherently highly variable (Mosier et al., 1998) and were estimated by IPCC (2000) methods. Nitrous oxide (CO_2eq) emissions ranged from 301 to 330 t per 100-cow beef or dairy herds (Table 3). From 16 to 54% of the N_2O emissions resulted from manure application or deposition during grazing. A second source, about 30% is from indirect emissions of leached, gaseous, or runoff-N. Other sources include legumes, crop residues and manure management. Nitrous oxide emissions are likely to be even more variable than indicated by this analysis because of variations by N application rates to soils, season, soil type, ambient temperature, soil moisture, etc.

Table 3. Nitrous oxide emission sources (t CO_2eq/100-cow herd) from U.S. cattle and N.Z. dairy production systems.

	United States			New Zealand
	Beef	**CA dairy**	**WI dairy**	**Dairy**
N_2O Sources	---------Annual t N_2O-CO_2eq---------			
Synthetic fertilizer	12	51	24	15
Manure[a]	45	67	37	4
Manure management	10	70	28	0
Legume	6.3	29	43	5
Crop residue[b]	42	6	34	13
Grazing	90	0	43	139
Volatilized	19	19	17	16
Leaching	76	88	72	66
Total N_2O, t/herd	301	330	298	260

[a]Includes animal waste and broiler litter.
[b]Includes waste grass.

Total annual N_2O emissions expressed per land range from 0.4 from the UT beef system to 6.4 t N_2O-CO_2eq/ha from the N.Z. dairy. The higher emissions are considerably above the 2.4 and 2.6 t N_2O- CO_2eq $ha^{-1}yr^{-1}$ noted previously for N.Z. grasslands (Sherlock et al., 1992) and Dutch dairy systems (Velthof and Oenema, 1997). The higher emissions likely reflect the increased emphasis of IPCC (2000) default estimates for leached nitrogen (30% of manure and fertilizer applied to soil) plus the inclusion of emission estimates from the N in grass residue.

4. MANAGEMENT AND/OR MITIGATION STRATEGY EFFECTS

The variations in resource inputs and outputs from the dairy systems provide interesting insights to causes of variations in CO_2eq/product. Such variations in production or management techniques may serve as mitigation options. The magnitude of variations in GHG sources per milk vary from 2-fold for nitrous oxide to 8-fold for manure methane by production system. The wide range in manure methane emissions reflects the heavy dependence on anaerobic lagoon disposal used in the CA systems, resulting in 16% of CO_2eq from this source as compared to 3 to 2% from N.Z. and WI systems, respectively. Not surprisingly, anaerobic lagoon use for manure disposal is the most obvious target for GHG mitigation because of the major increases in methane and some increases in nitrous oxide emissions.

The dairy systems comparison reveals what appears to be a major limitation of pasture-only dairy production systems, the reduced milk production per cow, resulting in a higher feed and methane cost associated with the higher proportion of feed going to maintenance. This limitation occurs in spite of the approximately equal concentration of available energy (Mcal of ME/kg diet DM) in the pastures of N.Z. as found in the concentrate containing diets in the

U.S.. Ruminant nutritionists feel that the bulky, high moisture nature of grass causes the limitation to daily energy consumption and thus milk production of these cows since the genetic potential to produce milk has been shown to be approximately equal in U.S. and N.Z. cows. If the CA cows produced only as much fat plus protein in their milk as the N.Z. cows/yr, and lagoon use kept about equal, the CA GHG emissions per unit milk would increase about 73%. Conversely, if N.Z. cows could eat enough grass and produce as much milk fat and protein annually as the CA cows, the GHG emissions per unit milk would fall 37%. Such factors can be illustrated with our model comparisons.

If the body weights are all equalized to the heaviest cows (e.g. 635 kg as in CA), the emissions/milk increase 3% in WI and 15% in N.Z., other things being equal. Thus increasing annual milk per cow or better yet, milk energy per cow has marked impacts on GHG emissions per unit milk. Additionally, if the culling rate is equalized to that in CA, numbers of young developing heifers would need to approximately double in N.Z., increasing the associated GHG emissions by about 12% for the herd with the same milk production. If all of the above - productivity, weight, manure, etc., factors are equalized across systems, the total CO_2eq/milk are about equal, but from somewhat different sources, a 14% lower carbon dioxide cost in N.Z. is largely offset by a larger nitrous oxide emission. The pasture systems, at least in the N.Z. clover-ryegrass swards contain about 70% more nitrogen than is required which results in more nitrous oxide losses through manure, legume fixation, crop residue and leached N sources

An additional factor not considered, is the number of calves available to produce beef from the dairy herds. The numbers will be considerably greater, with fewer cull cows from the N.Z. herd. Lower replacement rates leave more calves to produce meat without "maternal maintenance" requirements and concomitant GHG emissions. It is clear that these systems each have advantages and disadvantages.

Previous exercises (Johnson et al., 2001) with the cow-calf phase model showed emission responsiveness to modified herd characteristics or management criteria. Changes of 10% in calving or calf mortality rates have little effect on herd emissions but nearly linear effects of 8 to 12% on GHG emission per gain of the cow-calf phase. Changing the culling or cow replacement rate from 17 to 27 % increased total herd CO_2eq by about 3 %, but gave a somewhat unexpected result of decreasing the cow-calf phase CO_2eq/gain by 9%. Thus, an increased rate of slaughter of mature cows improves the emissions per product (culls + calves) ratio. With a 20% heavier mature cow size there was an increase of about 15% in each of the GHG, but when these emissions were expressed on a per kg gain basis there was a 2% reduction in GHG emissions. From an economic standpoint, increasing cow weights in this scenario improved profitability. Feed cost will probably determine whether the large cow scenario is profitable.

Approximately 76% of the total GHG emissions result from the cow-calf phase of beef production, 11% from the stocker and 13% from the feedlot phases. Presented as CO_2eq/wt gain during each phase, the feedlot phase becomes the most efficient and produces only 60% as much CO_2eq as the stocker phase and approximately one-third that of the cow-calf phase (Figure 1). Large emissions by the cow-calf phase reflect the greater proportion of feed used for maintenance. Lower feedlot phase emissions per gain reflect the opposite, high rates of gain and low proportions of feed for maintenance as well as the much lower methane emissions per unit of feed from cattle fed these 92% concentrate diets. Emissions from CO_2 and N_2O become dominate and about equal for the feedlot phase of U.S. beef production.

A scenario was examined in which all of the weaned beef calves went directly to the feedlot, bypassing the stocker phase. Compared to the conventional cow through feedlot system, where only 18% went directly to the feedlot, the cattle are in the system 80 less days and produce just slightly more gain. The total number of ha required is reduced by about 13%, with little effect on total fuel and fertilizer use. As would be expected, these cattle emitted less CH_4 per unit gain, while N_2O and CO_2 emissions were similar to the baseline scenario. Overall the total CO_2eq were 8% less than the baseline. Of the mitigation strategies tested, delivering the calves directly to the feedlot from an intensive grazing (IG) cow-calf system was indicated to be the most profitable. The combination of the reduced CO_2eq and increased profitability made this the mitigation strategy of choice.

Best management practices (BMP) for pasture or rangeland (Follet, et al., 2001) and/or cropland (Lal, et al., 1998) used in, or to produce feedstuffs for livestock operations present significant potentials to offset GHG emissions with increased soil-C sequestrations. Land use by the U.S. beef systems that we simulated, vary from 1.2 to 7.5 ha per cow unit (Table 4). Average C-sequestrations resulting from BMP's applied to these hectares suggest offsets ranging from 32 to 60% of the beef-herd CO_2eq emissions (assuming no change in ha use per herd or other input requirements). Recent measurements by colleagues, (Conant et al., 2001, unpublished) support annual C-sequestrations of approximately 0.4 Mg/ha in IG pasture in the southeastern U.S., although such increases in soil-C are not likely in areas with less than moderate rainfall. Other factors that must be considered are yield per ha or fertilizer inputs.

Table 4. An estimate of annual land use and C sequestration potential of U.S. beef production systems (100-cow herd basis).

Items		**Location in U.S.**					
		AL	**TX**	**UT**	**VA**	**WI**	**Mean**
Land Use, ha/herd:							
Managed pasture		143	102	129	104	94	114
Rangeland		0	239	587	0	0	165
Hay crops		13.1	22.3	21.0	15.2	20.6	18.5
Grain crops		13.7	42.7	6.8	12.4	7.3	16.3
Total, ha/herd		170	406	745	132	122	315
C-sequestration:	**t/ha**	**---------------- t carbon/herd ---------------**					
BMP on pasture	0.4	57	41	52	42	38	46
BMP on range	0.05	0	12	29	0	0	8.2
BMP on hay crop	0.2	2.6	4.5	4.2	3.0	4.1	3.7
BMP on grain crop	0.8	11	34	5.4	9.9	5.8	13
Total C-sequestration		71	92	91	55	48	71
C-sequestration CO_2eq		259	336	332	201	176	260
CO_2eq emissions/herd		605	638	555	554	548	580
C-seq offset potential, %		43	53	60	36	32	45

A simulation of likely yield increases (+ 50%), fertilizer inputs (+20%), forage composition changes, and animal responses to intensive rotational grazing in the U.S., for example, would increase N_2O, decrease CH_4 and total CO_2eq emissions/gain slightly, offset

13 to17% of emissions by soil-C increases, and decrease net GHG/gain by 14 to 20%. In any case, these projections of C-sequestration should be viewed as 'ballpark' because of the highly variable and difficult to measure productivity, animal harvest efficiency, and nutritive value of pasture-range. Additionally, long term IG and fertilization of pastures can yield C-sequestration losses, as reported recently by Lambert et al. (2000) following an 18-yr study in New Zealand.

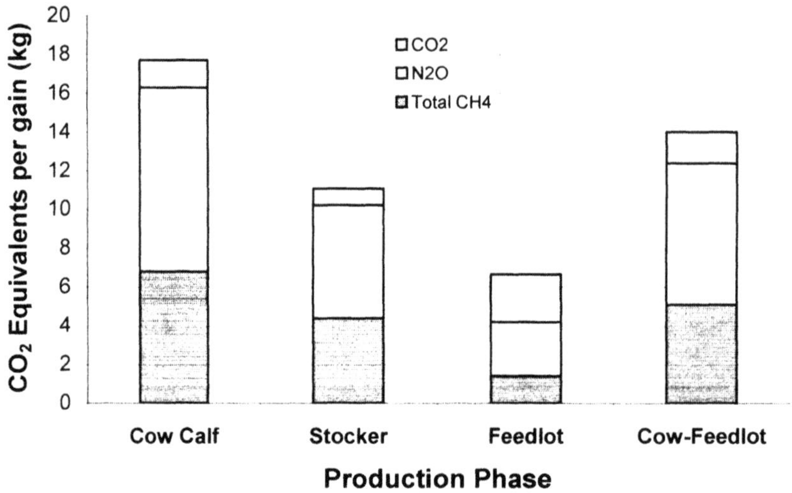

Figure 1. Relative source strengths of CO_2eq by beef production phase.

REFERENCES

Cedarberg, C., 1998. Life cycle assessment of milk production – a comparison of conventional and organic farming. Swedish Institute for Food and Biotechnology, Report Nr 643.

Conant, R.T., J. Six & K. Paustian, 2001. Land use effects on soil carbon fractions in the southeastern United States I: Management intensive versus extensive grazing (unpublished). Colorado State Univ., Nat. Res. Ecol. Lab.

Follett, R.F., J.M. Kimble, & R. Lal, 2001. The Potential of U.S. Grazing Lands to Sequester Carbon and Mitigate the Greenhouse Effect. Lewis Publishing, Wash., DC. 437 pp.

Holmes, C.W., 1999. An Outline of Dairy Production Systems in New Zealand, Massey Univ. (Unpublished Mimeo).

IPCC, 1996. Climate Change 95. Impacts, Adaptations and Mitigation of Climate Change: Scientific-Technical Analyses. Cambridge Univ. Press.

IPCC, 2000. Good Practice in Inventory Preparation for Agricultural Sources of Methane and Nitrous Oxide. (http://www.ipcc.ch/).

Johnson, D. E., A Seidl & H. Phetteplace. 2001. Methane, nitrous oxide and carbon dioxide emissions from U.S. beef production systems. In: Energy Metabolism in Animals. Eur. Ass. Anim. Prod. Publ. No. 103, pp. 161-164.

Johnson, D. E., A.F. Seidl, & H.W. Phetteplace, 2000a. Livestock system greenhouse gases: Effects of practices and policies on emissions and economic returns. Final report to US-EPA, Wash., D.C.

Johnson, D. E., H. Phetteplace & M. Ulyatt, 2000b. Variations in proportion of methane of total greenhouse gas emissions from U.S. and N.Z. dairy production systems. Proc. 2nd International Methane Mitigation Conference, Novosibirsk, Russia, USEPA, p 249.

Johnson, D.E., G.M. Ward & G. Bernal, 1997. Chapt. 32 Biotechnology mitigating the environmental effects of dairying: Greenhouse gas emissions. In: Milk Composition, Production and Biotechnology. Biotechnology in Agricultural Series, No. 18. Welch, R., et. al., Edit., CAB International, pp.497-511.

Lal, R., J.M. Kimble, R.F. Follett, & C.V. Cole, 1998. The Potential of U.S. Cropland to Sequester Carbon and Mitigate the Greenhouse Effect. Ann Arbor Press, Chelsea, MI. 128 pp.

Lambert, M.G., D.A. Clark, A.D. Mackay, & D.A Costall. 2000. Effects of fertiliser application on nutrient status and organic matter content of hill soils. N. Z. J. Agric. Res. 43: 127-139.

LIC, 1999. Dairy statistics, Livestock Improvement Corp., Hamilton, N.Z.

Mosier, A., C. Kroeze, C. Nevison, O. Oeneme, S. Seitzinger & O. van Cleemput, 1998. Closing the global N_2O budget:nitrous oxide emissions through the agricultural nitrogen cycle. Nutr. Cycling in Agroecosy. 52:225-248.

MWBNZ, 1999. Compendium N.Z. Farm Production Statistics, Meat and Wool Econ. Service of N.Z., Wellington, N.Z..

NRC, 1996. *Nutrient Requirements of Beef Cattle,* 7th ed. National Academy Press, Wash. DC.

NRC, 1989. Nutrient Requirements of Dairy Cattle, 6th ed. National Academy Press, Wash. DC

Pimentel, D., 1980. Handbook of Energy Utilization in Agriculture. Ed. D. Pimentel. CRC Press, Boca Raton, FL.

RLEP, 1998. Proc. Ruminant Livestock Efficiency Program. Edits. Conneely,D and M. Gibbs, US EPA.

Robertson, G.P., E. A. Paul, & R.R. Harwood, 2000. Greenhouse gases in intensive agriculture: contributions of individual gases to the radiative forcing of the atmosphere. Sci. 289:1922-5.

Russell, JM., 1999. Personal communication. N.Z. Dairy Research Inst., Palmerston North, N.Z..

Sherlock, R.R., C. Muller, J.M. Russell & RJ Haynes, 1992. Inventory information on nitrous oxide. Ministry of Environment, N.Z..

Velthof G.L. & O. Oenema, 1997. Nitrous oxide emission from dairy farming systems in the Netherlands. Netherlands J Agric Sci 45:347-360

Ward, G. M., K. G. Doxtader, W. C. Miller & D. E. Johnson, 1993. Effects of intensification of agricultural practices on emission of greenhouse gases. Chemosphere 26:87.

Wells, C.M., 1998. Total energy inputs as indicators of agricultural sustainability: Dairy industry case study. In: Mander, N.K. & R.E.H. Sims, Eds., Energy for Our Country - Opportunities for the 21st Century, Sustainable Energy Forum, Wellington, N.Z.. Pp.114-124.

Wilcock, R.J., J.W. Nagels, H.H.E. Rodda, M.B. O'Connor, B.S. Thorrold & J.W. Barnett, 1999. Water quality of a lowland stream in a New Zealand dairy farming catchment. N.Z. J Marine Freshwater Res. 33:683-696.

Greenhouse Gases and Animal Agriculture
J. Takahashi & B. A. Young (Editors)

Establishment of profitable dairy farming system on control of methane production in Hokkaido region

S. Kume

National Agricultural Research Center for Hokkaido Region (NARCH), Sapporo, 062-8555, Japan

Establishment of profitable large-scale dairy farming system is necessary not only to cope with competition from the imported products but also to control methane and wastes production in Hokkaido region. Because the annual increase in milk yield and composition is likely to continue in Hokkaido, the improvement in the feeding regimen for high producing cows, especially during periparturient and early lactation period, is a main strategy to reduce the methane production and enhance milk production. High quality alfalfa has been useful for the decreased methane production from cows due to the low NDF and high CP contents. Feeding high quality alfalfa and corn silage for high producing cows may decrease methane production by 10 to 20% in large-scale dairy farming system.

Keywords: alfalfa, dairy farming, methane production, milk production

1. INTRODUCTION

The Japanese dairy industry achieved dramatic growth in the 20th, because higher national income created an increased consumer demand for dairy products. However, dairy production has been affected by a strong trend towards trade liberalization and structural changes on dairy farms, while consumer needs in dairy products have become extremely diverse and environmental pollution from livestock wastes has a serious problem in Japanese dairy farming. To cope with competition from the imported products, the Japanese dairy industry must develop even more efficient milk production, which is however in harmony with the environment and rural society.

High producing cows require appropriate amounts of energy, protein, and minerals, because energy, protein, and minerals are mobilized from body stores to support high milk yield (NRC, 1988). The importance of high quality roughage has been recognized for maximal milk yield and maintenance of body weight in cows, especially during periparturient and early lactation period. Also, methane production was linearly related to dry matter intake (DMI) at intakes below 1.5 times of maintenance and methane production per unit of DMI decreased with the increased feeding level (Shibata et al., 1993). Therefore, high quality roughage may be useful for the reduction of methane production in cows due to the improved productivity and decreased dry matter intake of cows.

To build up a profitable large-scale dairy farming system, the high quality forage system introduced alfalfa in stead of conventional grass silage system has been performed in upland cropping region of Hokkaido (Itokawa, 1999). However, it is not clear whether high quality roughage such as alfalfa reduce methane production in large-scale dairy farming system. The present study was conducted to establish the profitable dairy farming system on the control of methane production in Hokkaido region.

2. MILK YIELD AND METHANE PRODUCTION OF HIGH PRODUCING COWS

Improving animal productivity will lead to the reductions in methane emissions per unit of product. According to the Japanese dairy herd improvement record, the average milk yield for 305 d increased from 5,826 kg in 1975 to 8,638 kg in 1999. Total numbers of dairy cows in Hokkaido were 866,900 head in 1999 and dairy cows produced 3,640,000 t of milk per year. In Hokkaido region, large-scale dairy farming has been developed utilizing vast land resources and the average number of cows per dairy farm and planted area in Hokkaido were 87 head and 40.9 ha in 2000.

Milk yield and milk composition of high producing cows varied by the parity and calving season in Hokkaido (Kume et al., 2001b). Milk yield increased with the increased parity, and milk yield of fourth and fifth lactation cows was approximately 10,000 kg for 305 d (Figure 1). The rapid increase of body weight during the first lactation improved high milk production during the second lactation. Kurihara et al. (1997) reported that methane production per unit of FCM hyperbolically decreased as the equation ($y = 8.19 + 300/FCM$), because methane production resulting from maintenance metabolism was regarded as non-productive. If the average milk yield for 305 d in large-scale dairy farming system increased from 8,500 kg to 10,000 kg, the methane production calculated from the equation decreased by 8.5 %.

Because the annual increase in milk yield and composition is likely to continue in Hokkaido, the improvement in the feeding regimen for high producing cows is a main strategy to reduce the methane production in large-scale dairy farming system. Especially, the decreased DMI by the increased parity or improved management in dairy farming system is beneficial to reduce the methane emissions per unit of product.

Heat stress represents sizable economic losses to dairy farms, because high environmental temperature is a primary factor that adversely affects the productivity of dairy cows. It is reported that the upper and lower critical temperatures for lactating cows are 25 and -25℃, and that relative humidity greater than 80% indirectly affects the upper critical temperatures. Summer of 1994 was very hot and humid in Japan, and approximately 4,600 dairy cows within 2.02 million head died due to heat stress (Kume et al., 1998). In the NARCH located in northern cool regions of Japan, mean temperature in August 1994 was 23.9 ℃, but mean temperatures during July/August from 1995 to 1997 were below 21℃. Additionally, about 300 dairy cows died in hot summer of 1999 in Hokkaido and the carryover effects of heat stress decreased milk production of cows in 2000.

In high producing cows, the summer environment around 20℃ decreased milk yield, milk fat and protein, and the decreased milk yield and milk composition may be partly due to the decreased feed intake affected by heat stress (Kume et al., 2001b). Methane production per DMI increased at 30℃, but methane production per DMI was not affected by high temperatures in dairy cattle fed high concentrate diets (Kurihara et al., 1995). Thus, the improved feeding regimen during summer is very important in Hokkaido to increase milk production and decrease methane emission of high producing cow.

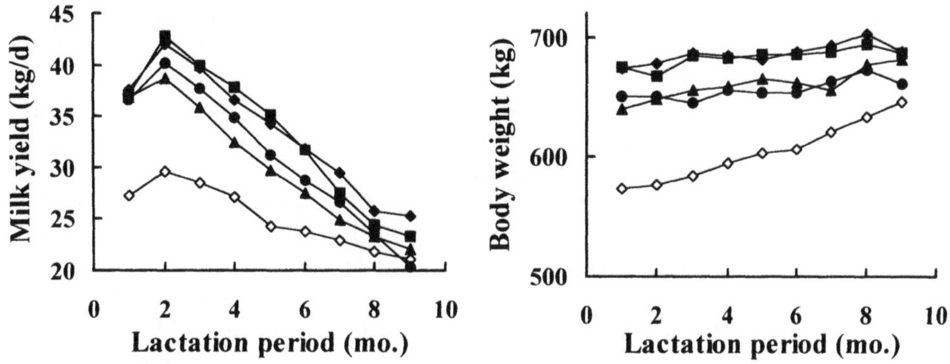

Figure1. Milk yield and body weight of first (◇), second (▲), third (●), fourth (■) and Fifth and more (◆) lactation cows (n=125) from 1994 to 1997 in NARCH (Kume et al., 2001b)

3. FEEDING ALFALFA FOR METHANE REDUCTION

Alfalfa is a suitable roughage for high producing cows because of the high protein content and high passage rate through the gut, and cows fed alfalfa silage generally produce more milk than those fed grass. The most important factor influencing forage quality is the growth stage, and a 1-week delay in harvesting of alfalfa will decrease CP by about 20g/kg and increase cell-wall concentration by approximately 30g/kg (Buxton, 1996).

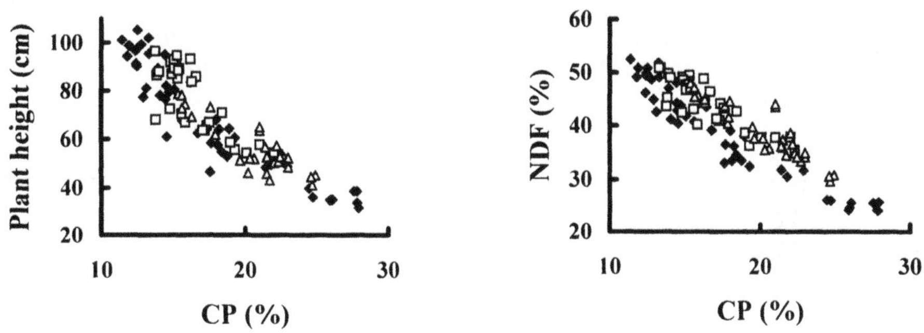

Figure 2. Relationships between CP and plant height or NDF of first (◆), second (□) and third (△) cutting alfalfa (Kume et al., 2001a)

In Hokkaido, "Makiwakaba" and "Hisawakaba", new cultivars of alfalfa, were developed by NARCH and recommended for cultivation. The CP content in alfalfa ranged from 11.4 to 27.9 % on DM during the growth period (Figure 2). The NDF content increased with the increased plant height, and NDF in the first cutting alfalfa was slightly lower. The slopes of CP for NDF during -1.58 and -1.67 in each cutting agreed with the slope of CP for cell-wall concentration of Buxton (1996). Methane production is closely related to cellulose digestion, and the utilization of mature alfalfa might lower feed intake of cows due to its higher NDF content. Also, methane production per DMI of cows decreased as CP and ME contents of diets increased (Kurihara et al. 1995). Belyea et al. (1999) reported that NE of alfalfa decreased with maturity, but that changes differed with year and cutting. The optimal CP content of alfalfa when fed in combination with corn silage should be 20% for high producing cows (Buxton, 1996), which showed 37% of NDF in Figure 2. Therefore, high quality alfalfa may be useful for the decreased methane production from cows due to the low NDF and high CP contents (Kume et al., 2001a).

Mechanical treatment of maceration could be a useful tool to produce high quality forage and increase the rate of drying relative to conventionally conditioned forage, such as tedder and wilted treatment. Also, maceration has been developed to increase forage digestibility, and maceration was estimated to increase NEl content of alfalfa silage by 5% (Broedrick et al., 1999). In the macerated treatment by NARCH, maceration increased CP contents in alfalfa and decreased NDF (Table 1), because maceration reduced the leaf loss of alfalfa (Nonaka et al., 2001). However, maceration had no effects on the digestibility and energy utilization of cows, and methane production per DMI was not affected by the macerated treatment (Table 2). Methane production per unit of DMI of cows fed alfalfa ranged from 32.5 to 37.3 l/kg (Mean: 35.1 l/kg) in 12 balance trials, and the methane production increased with the increased NDF digestibility of alfalfa.

Table 1. Digestibility of cows fed alfalfa silage (Nonaka et al., 2001).

	Macerated	Control
Chemical composition (% of DM)		
DM	48.7	51.1
CP	21.7^{a}	16.2^{b}
ADF	31.5^{b}	35.1^{a}
NDF	37.5^{b}	47.4^{a}
Digestibility (%)		
DM	68.8	69.8
CP	75.5	72.6
ADF	64.6	64.6
NDF	64.2	66.0

^{a,b} $P<0.05$.

Table 2. Methane production of cows fed alfalfa (Nonaka et al., 2001).

	Macerated	Control
n	2	2
BW, Kg	596	583
DMI, kg	7.6	8.6
DE/GE, %	68.9	69.7
ME/GE, %	53.9	56.0
CH_4, l/d	269.3^{b}	312.8^{a}
CH_4/DMI, l/kg	35.8	36.2

^{a,b} $P<0.05$.

Methane production and digestibility of cows fed typical forage in Hokkaido varied at intakes of maintenance level. The digestibility of DM, ADF and NDF of cows fed vegetative stage of timothy hay was higher than that of cows fed bloom stage, although the methane production of cows was not affected by the stage (Figure 3). Methane production of cows fed corn silage (44.3l/kgDM) was similar to the cows fed timothy hay, although energy utilization was higher. Methane production per DMI of cows fed grass-based mixed silage was higher than that of cows fed mixed silage plus alfalfa silage in the ratio of 50:50 (figure 4). Methane production of cows fed glass plus alfalfa was lower during 4 to 12 hours after feeding, and feeding 50% of alfalfa decreased methane production by 7%.

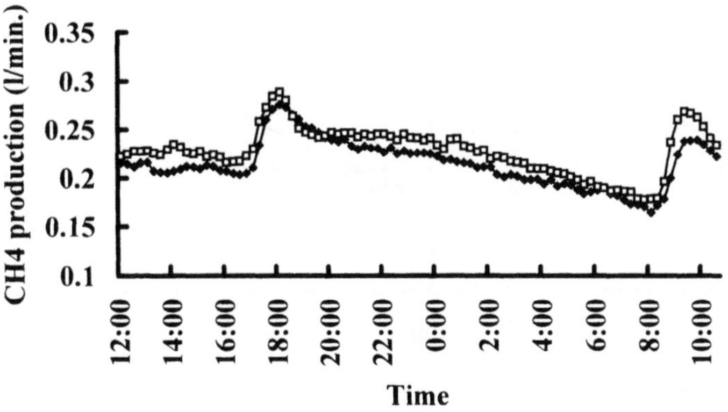

Figure 3. Methane production of cows fed bloom stage (45.0l/kgDM: ◆) and vegetative stage (49.0l/kgDM: □) of timothy hay. Feeding time was at 16:30 and 8:30 h.

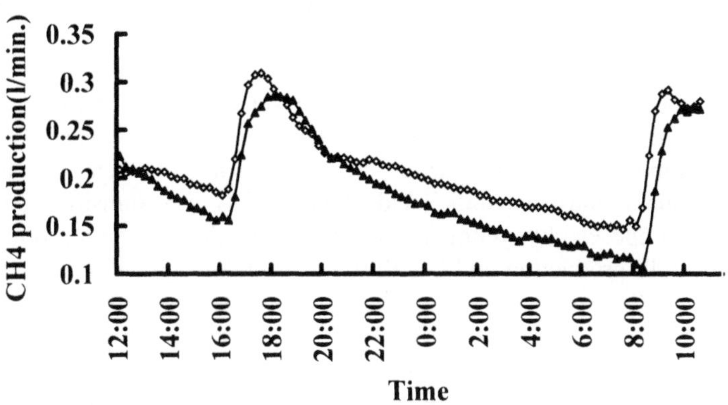

Figure 4. Methane production of cows fed mixed grass plus alfalfa silage in the ratio of 50:50 (32.9l/kgDM: ▲) and vegetative stage (35.6l/kgDM: ◇) of timothy hay. Feeding time was at 16:00 and 8:00 h.

Methane production of cows increased with the increased digestible OM or NDF, and methane production of cows fed alfalfa was lower (Figure 5). However, methane production of cows was not affected by dietary ME, although the range of ME in alfalfa and grass was similar. Compared to the typical forage in Hokkaido, methane production per unit of DMI of cows fed alfalfa decreased by 1 to 28%. Thus, the introduced alfalfa instead of grass and the macerated treatment of alfalfa may be beneficial method in the large-scale dairy farming system to reduce methane emission in Hokkaido.

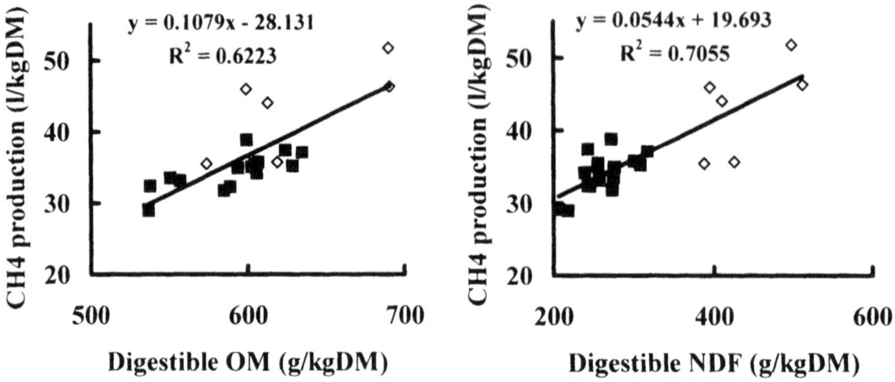

Figure 5. Relationships between digestible OM or NDF and methane production of cows fed grass (◇) and alfalfa (■).

4. METHANE REDUCTION OF HIGH PRODUCING COWS FED HIGH QUALITY ROUGHAGE

Most metabolic disorders occur around parturition and represent a failure of the cow to adjust to the rapid onset and stress of high milk production. The nutrient requirements of dairy cows increase with milk yield immediately after parturition, but cows in early lactation cannot consume sufficient DM to support maximal milk yield (NRC, 1988). The transition period from nonlactation to lactation imposes physiological and nutritional stress on high producing cows and may impair DMI, milk production and health (Grant and Albright, 1995; Kume et al., 1998).

The trend of DMI, milk yield, BW, and plasma NEFA of 4 periparturient cows was shown in Figure 6. No metabolic disorders occurred from 3 weeks prepartum to 1 week postpartum. Milk yield and DMI increased rapidly immediately after parturition, but BW decreased drastically from parturition to 2 weeks postpartum. Cows had negative N and mineral retention immediately after parturition (Kume et al., 2001a), and plasma NEFA increased drastically due to the rapid mobilization of body fat. On the other hand, positive retention of N, Ca, P, Mg, and K was observed in prepartum cows. Compared to high milk yield after parturition, the low DMI of cows immediately after parturition decreased BW and nutrient retention. Thus, the increased DMI after parturition and improved energy content of feed are necessary for high producing cows, because the low energy intake of cows adversely affects milk production and health status in large-scale dairy farming system.

Sixteen multiparous Holstein cows were assigned to an alfalfa silage (AS) diet or an orchardgrass silage (OS) diet from 3 weeks prepartum to 10 weeks postpartum. Cows in AS were offered 20 and 35% of alfalfa silage in prepartum period, and after parturition roughage and concentrates were given as a TMR in ratio of 50:50 with alfalfa or grass being 100% of the roughage. Milk fever occurred in 2 cows fed high K alfalfa immediately after parturition. Goff and Horst (1997) reported that the K content of forages like in alfalfa is more important than the Ca content in predisposing cows to milk fever, and the incidence of milk fever was increased simply by increasing the K concentration in the prepartum diet from 1.1 to 2.1%.

Thus, low K alfalfa must be utilized for periparturient cows to prevent the occurrence of milk fever. Additionally, the improvement of feeding regimen during periparturient and early lactation period is most important for high producing cows to increase milk production and prevent metabolic disorders.

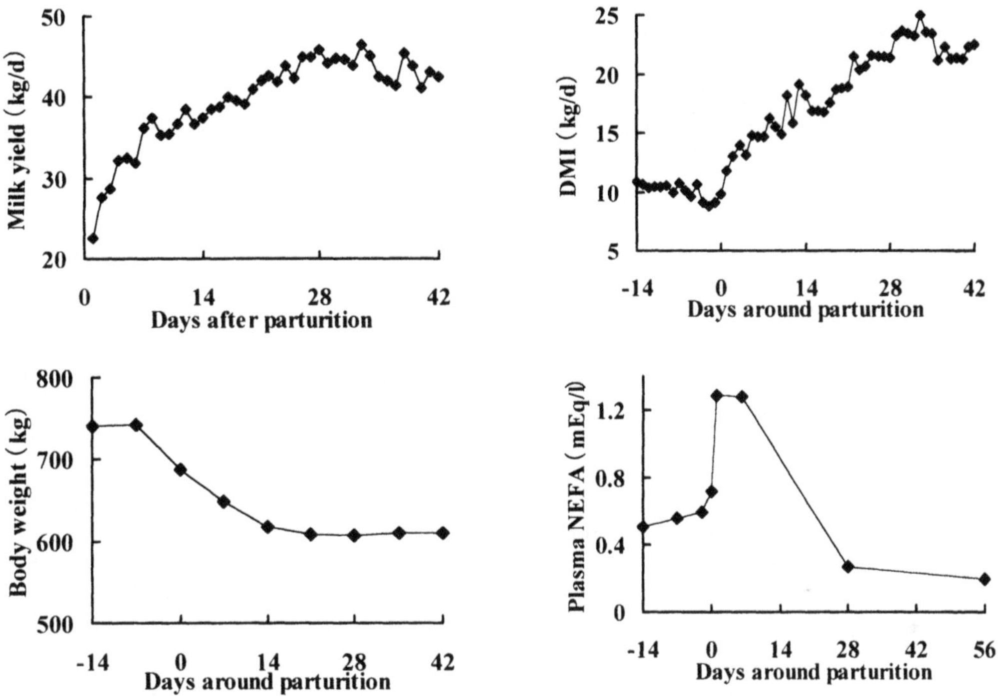

Figure 6. DMI, milk yield, BW, and plasma NEFA of 4 periparturient cows (Kume et al., 2001).

The digestibility of ADF and NDF of prepartum cows fed alfalfa was lower than that of cows fed grass. Energy utilization of the prepartum cows was not affected by feeding alfalfa, and the average methane production per unit of DMI was 36.9l/kgDM. Compared to the cows fed grass silage, the increased DMI and milk yield of cows fed alfalfa from 1 to 10 weeks postpartum were 3.2 and 2.6 kg/d, respectively. The increased DMI and the low digestibility of NDF may be due to the high passage rate of alfalfa through the rumen (Ueda et al., unpublished data). The tendency of methane production in cows fed alfalfa was not clear in the present study, but the low NDF contents and high passage rate of alfalfa may be profitable to reduce methane production of cow.

The yield of forage such as grass, legume or corn has not been changed in Hokkaido, but the import of concentrates and grain has increased. In Hokkaido, the high self-sufficiency rate of feed and high quality roughage has been a main target for the management of dairy farming system. According to the profitable large-scale dairy farming system, such as introduced alfalfa in stead of conventional grass silage system, feeding high quality alfalfa and corn silage for high producing cows may decrease methane production by 10 to 20% due to the increased milk yield and improved feed quality. Although, in large-scale dairy farming system, the target is to cope with competition from the imported products, the research forward increasing the efficiency of animal production by improving feed quality and nutritional management must be encouraged to increase the milk production and reduce the methane emissions.

REFERENCES

Belyea,R., R. Restrepo, F.Martz, and M. Ellersieck. 1999. Effect of year and cutting on equations for estimating net energy of alfalfa forage. J. Dairy Sci. 82;1943-1949.

Broderick, G.A., R.G.Koegel, M.J.C.Mauries, E.Schneeberger, and T.J.Kraus. 1999. Effect of feeding macerated alfalfa silage on nutrient digestibility and milk yield in lactating dairy cows. 82:2472-2485.

Buxton,D.R. 1996. Quality-related characteristics of forages as influenced by plant environment and agronomic factors. Anim. Feed Sci.Tech. 59;37-49.

Goff, J.P., and R.L.Horst. 1997. Physiological changes at parturition and their relationship to metabolic disorders. J.Dairy Sci. 80;1260-1268.

Grant, R.J. and J.L.Albright. 1995. Feeding behavior and management factors during the transition period in dairy cattle. J.Anim.Sci. 73;2791-2803.

Itokawa,N. 1999. Field machinery for forage production in Japan. Proc. International Workshop on Opening for Low-input Sustainable Forage Production and Use. 211-218.

Kume, S., T.Toharmat, and N.Kobayashi. 1998. Effect of restricted feed intake of dams and heat stress on mineral status of newborn calves. J. Dairy Sci. 81;1581-1590.

Kume,S., T.Toharmat, K.Nonaka, T.Oshita, T.Nakui, and J.H.Ternouth. 2001a. Relatioships between crude protein and mineral concentrations in alfalfa and value of alfalfa silage as a mineral source for periparturient cows. Anim. Feed Sci. Tech. 93:157-168.

Kume, S., T.Okamoto, O.Sasaki, K.Uetake, K.Yayou, and J.H.Ternouth. 2001b. Effect of calving season and parity on milk yield and milk composition of high producing cows in Hokkaido. Res.Bull.Hokkaido Natl. Agric. Exp. Stn. 173:37-45.

Kurihara, M., S.Kume, T.Aii, S.Takahashi, M.Shibata and T.Nishida. 1995. Feeding method for dairy cattle to cope with global warming–Technic assessment based on energy metabolism. Bull.Kyushu Natl. Agric. Exp.Stn. 29:21-107.

Kurihara, M., M.Shibata, T.Nishida, A.Purnomoadi, and F.Terada. 1997. Methane production and its dietary manipulation in ruminants. Rumen Microbes and Digestive Physiology in Ruminants. 199-208.

Livestock Improvement Association of Japan. 1998. Summary of Dairy Herd Improvement Records (1997). Livestock Improvement Association. Tokyo

National Research Council, 1988. Nutrient requirements of dairy cattle. 6th. rev. ed. p2-62. Natl.Acad.Sci., Washington,D.C.

Nonaka, K., S.Kume, and T.Oshita. 2001. Effect of maceration on *in sacco* degradability and Digestibility of low-moisture silage. Grassland Sci. 47:405-411.

Shibata, M., F.Terada, M.Kurihara, T.Nishida, and K.Iwasaki. 1993. Estimation of methane production in ruminants. Anim.Sci.Technol.(Jpn.), 64:790-796.

Greenhouse Gases and Animal Agriculture
J. Takahashi & B. A. Young (Editors)

Nutritional manipulation of methane emission from ruminants

J. Takahashi

Department of Animal Science, Obihiro University of Agriculture and Veterinary Medicine, Obihiro, Hokkaido 080-8555, Japan

The concentrations of greenhouse gases in the atmosphere are rising at a rapid rate which the human race has not experienced up to now. Methane which is the most important greenhouse gases after carbon dioxide is currently being paid the most attention. From recent assessments around the world, it is clear that the methane concentration in the atmosphere increased at 0.8-1.0% annually rate within the range of 1.7-1.8ppmv. The amount of emissions each year is presumed to be 410-660Tg, with an annual growth is 35-40Tg.

Rapid development of animal agriculture industry has resulted in a remarkable imbalance between the increased amount of animal effluents and cultivated land area to apply them. An absolute shortage of land area to produce animal feed and animal production highly dependent on imported ingredients of concentrates have amplified the imbalance. Excess amounts of the untreated effluent results in high nitrate contamination in underground water and pathogens such as criptosporidium in surface water. The complaints of pollution problems from animal agriculture industry have increased as have the drastic expansion of animal populations in limited areas.

In general, the environmentally toxic gases emitted from animal agriculture into the atmosphere does not become a problem as long as neither the odor nor the water pollution problem comes to light. Recognition of influence of methane emitted from animal agriculture on global warming as an environmental concern is still low because the gas is colorless, scentless, and does not pollute water. At last, grappling with the influence of methane emitted from enteric fermentation of farm animals on global warming has begun. The research and technical developments have sought natural manipulators and probiotics to control enteric methanogenesis instead of an anti-organism focus. Meanwhile, the annually accumulated animal effluent was estimated at 85,450,000 t in 1996, exceeding human effluent and organic wastes in Japan. The issues which should be resolved are the development of various technologies for animal effluent management to control offensive odors and wastewater, exploitation of effluent energy use through anaerobic methane fermentation and feeding management to reduce the excretion of nitrogen and phosphorus. Promoting environmentally conscious agriculture which does not rely on the chemical fertilizer can be realized by effective use of liquid manure and compost. Many research projects have been carried out to extract the waste material energy as methane by processes related to animal and organic effluent management. The fuel cell which uses energy derived from animal effluents and organic wastes is being considered as an epoch-making technology.

Keywords: methane, greenhouse gas, rumen, manipulator, effluent

1. INTRODUCTION

Extremely serious concerns for the global environment have been raised since the Kyoto protocol in COP 3 (at the 3rd session of the Conference of Parties to the United Nations Framework Convention on Climate Changes held in Kyoto in December 1997) is not yet ratified by thoroughly advanced nations which enjoyed prosperity in the 20th century while consuming large quantities of fossil energy. With the rapid expansion of human populations and their activities, global warming will progress drastically in near future. The increase in CO_2 originates largely from the combustion of fossil fuels and is closely linked to the development of the ore industries in advanced nations after the Industrial Revolution. Most publicity has focused on the contribution of CO_2 to global warming (Houghton, 1994). However, the importance of CH_4 as a greenhouse gas may turn out equally or even more important than that of CO_2. Although CH_4 concentration in the atmosphere is negligible compared to CO_2 concentration, it is a highly potent greenhouse gas due to its high absorption ability of infrared radiation from sun (IPCC, 1994). As for CH_4, the relative molecular absorptivity of the far-infrared radiation is about 21 times that for CO_2, 58 times on a weight basis. The absorptivity zone for CH_4 is not saturated as for CO_2. Most of absorption zone for CH_4 is in the 8-13 μm band and does not overlap the absorption zones of water vapor and CO_2. Therefore, even a small increase in CH_4 concentration in atmosphere exerts an extremely strong effect on global warming.

Methane is produced by microbial activities in wetlands, rice paddies, salt water, and enteric fermentation of animals. All living things on the earth have been composed of three domains, *Bacteria, Archaea* and *Eucarya*. Methanogens belong to *Archaea* distinct from *Bacteria* in genealogical evolution. So far, 59 species in 19 genera of methanogens are identified in wide range of ecosystems in spite of strict anaerobes. Rumen methanogenesis mainly coincides with CO_2 reducing reaction by hydrogenotrophic methanogens which use H_2 from several sources, e.g.; HCOOH, saturated fatty acids more than C3 or alcohol more than C2. Methyltrophic methanogens may be minor group in the rumen methanogens ecosystem due to absence of methyl donors. Aceticlastic methanogens were found the dominant group in anaerobic fermentation of animal effluent. Approximate 70% of the CH_4 formed in the anaerobic digester is derived through this pathway (Hill et al., 1987). Therefore, in the control and utilization of CH_4 attributed to animal agriculture, aceticlastic methanogens should be proliferated to generate the recycled biogas energy, whereas hydrogenotrophic bacteria in enteric fermentation must be suppressed to reduce atmospheric emission of CH_4.

The present paper deals with the comprehensive evaluation and possible manipulations of rumen methanogenesis and anaerobic methanogenesis in biogas systems.

2. MANIPULATION OF RUMEN METHANOGENESIS

In general, the amount of CH_4 produced in the rumen is large in feeds high in fiber content. The main substrate to yield H2 and subsequently, methane is digestible structural carbohydrates (cell wall constituents) such as hemicellulose. Ironically therefore, it is forages rich in digestible fibers which enhance performance of ruminant animals that result in large emission of methane.

Fig. 1 and 2 show profiles of *in vitro* CH_4 production rate and cumulative CH_4 production from fermentation of substrates with different fiber contents (Table 1). Digestible fiber contents of substrates may positively affect cumulative methane production, whereas crude protein contents could amplify CH_4 production rate. The potential CH_4 production of orchardgrass estimated from the extrapolation (parameter a) was twice as high as that of lucerne (Table3).

Table 1. Chemical composition of substrates used for incubations

	Orchardgrass silage	Alfalfa silage
Dry matter (%)	30.5	28.2
Crude protein (%)[1]	10.9	16.0
Soluble carbohydrate (%)[1]	1.2	1.3
Hemicellulose (%)[1]	20.5	12.3
Cellulose(%)[1]	30.5	22.2
Gross energy (MJ kg^{-1})[1]	17.70	16.07

1. Dry matter basis

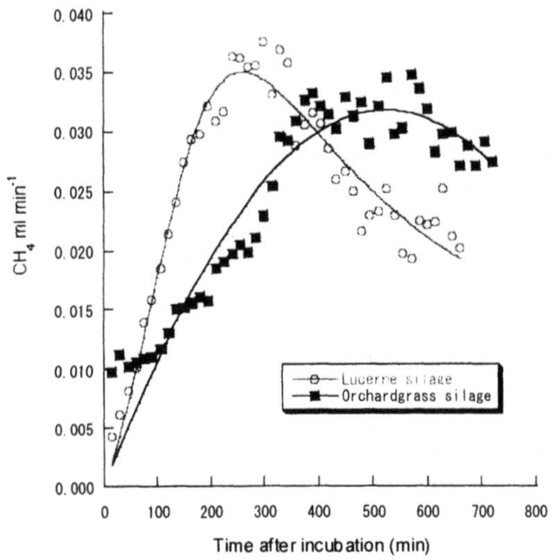

Fig.1.The in vitro methane production rates from fermentation of orchardgrass silage and lucerne silage.

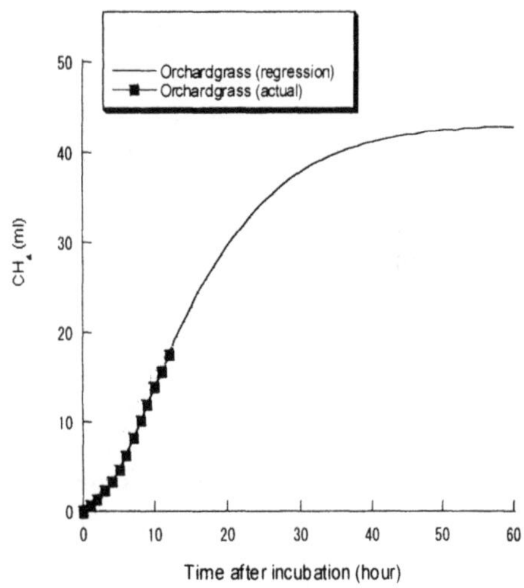

Fig.2. Cumulative methane production from fermentation of orchardgrass silage and lucerne silage.

Table 2. Parameters of inverse quadratic model for methane production rates [$a+(bt-a)/(1-ct+dt^2)$, t (min.)]

Silage	a	b	c	d	r
Orchardgrass	-0.83131	1.998×10^{-5}	−0.00012	1.338×10^{-7}	0.942
Lucerne	0.00190	1.416×10^{-4}	−0.00386	1.567×10^{-5}	0.972

Table 3. Parameters of Bertalanffy model for cumulative methane production $[a(1\text{-}be^{-ct})^3$, t (min.)]

Silage	a	b	c	r
Orchardgrass	43.116	0.8504	0.001658	0.999
Lucerne	22.584	0.8970	0.003435	0.999

While methane contains 892.6 kJ combustible energy per molecule at 25°C and 1013 hPa, it does not contribute to the total supply of usable energy to ruminants. Methanogenesis, however, plays an important role in rumen fermentation as a hydrogen sink. An alternative way to consume metabolic hydrogen (H_2) generated from the degradation of structural carbohydrate by fiber digestive bacteria may be effective to suppress rumen methanogenesis.

As for the removal of H_2 by a strong reduction reaction, the effect is highest in the CO_2 reduction obstruction. Alternative hydrogen sinks in the rumen, such as nitrate, markedly suppress methanogenesis. Fig. 3 shows the suppression of methane emission during the successive reduction of nitrate in the rumen. The free energy (G°) of nitrate reduction, $4H_2 + NO_3 + 2H^+ \rightarrow NH_4^+ + 3H_2O$, indicates -149kJ mol^{-1} H_2. As G° of hydrogen utilizing methanogen, $4H_2 + HCO_3 + H^+ \rightarrow CH_3 + 3H_2O$, is -33.9kJ mol^{-1} H_2, nitrate reduction acts as a highly competitive hydrogen sink against rumen methanogenesis. Ruminal reduction of nitrate contained in forage plants shows a remarkable effect on control of ruminal CH_4 generation (Takahashi, et al., 1983; Takahashi & Young, 1991). However, toxic nitrite will be accumulated in the rumen fluid due to relatively higher reduction of nitrate to nitrite than nitrite to ammonia. Excess intake of nitrate causes acute and sub-clinical poisoning of nitrate due to nitrite formation (Takahashi & Young, 1992). If the reduction rate of nitrate to nitrite is able to reduce to the rate of nitrite reduction, methanogenesis should decline without nitrite accumulation. Prophylactic effect of L-cysteine on nitrate-nitrite poisoning has been demonstrated *in vitro* and *in vivo* (Takahashi et al., 1989; Takahashi & Young, 1994; Takahashi et al., 1998). Sulphide generated form enzymic degradation of L-cysteine can bind molybdenum to form an insoluble inorganic compounds. Consequently, the activity of nitrate reductase will be suppressed depending on the ratio of L-cysteine to nitrate present. Thus, an appropriate combination of nitrate and L-cysteine makes it possible to suppress rumen methanogenesis without toxicity.

Since 20% of rumen methanogenic bacteria are parasitic on the surface of protozoans (Stumm, et al., 1982; Finlay, 1994; Newbold et al., 1995), it has been reported that CH_4 generation decrease by defaunation (removal of protozoa). Tremendous numbers of filamentous structures from bacteria attached to the surface of protozoa can be observed (Photo. 1. Right). These cocobacilli are almost methanogens depriving H_2 from protozoa. Excess feeding concentrate mixture such as cereals can suppress protozoa. In consequence, methanogenesis may decline. This is a reality with modern high performance milking cow and beef cattle feeding. However, for current consolidated dairy operations which pursue superior milk and meat production, a negative side of frequent incidence of metabolic diseases such as lactic acid acidosis and displaced abomasum etc. has come to light. Moreover, the adverse effect of H_2 is not removed in this case. A number of inhibitors of methanogenesis have been developed to improve feed conversion efficiency of ruminant feeds which are claimed to be effective in suppressing methanogens or overall bacterial activities (Chalupa, 1984). Approved ionophores such as monensin, salinomycin are known as propionate enhancers and have been put to practical use. The suppressing effect of these antibiotics on rumen methanogenesis has been well recognized (Chalupa, 1980). However, there is some anxiety due to appearance of resistant bacteria from long-term use, residues in the food products, which has resulted in human safety concerns (Wallace., 1992; Fuller, 1992).

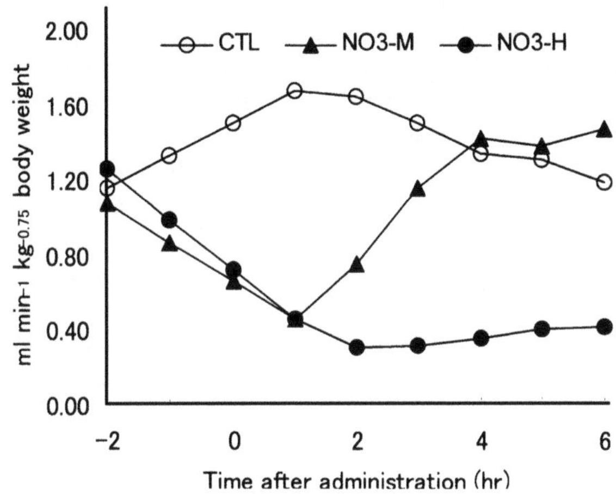

Fig. 3. The suppressing effect nitrate on rumen methanogenesis in sheep
CTL:control, NO3-M :0.65g $NaNO_3$ $kg^{-0.75}$, NO3-H :1.3g $NaNO_3$ $kg^{-0.75}$

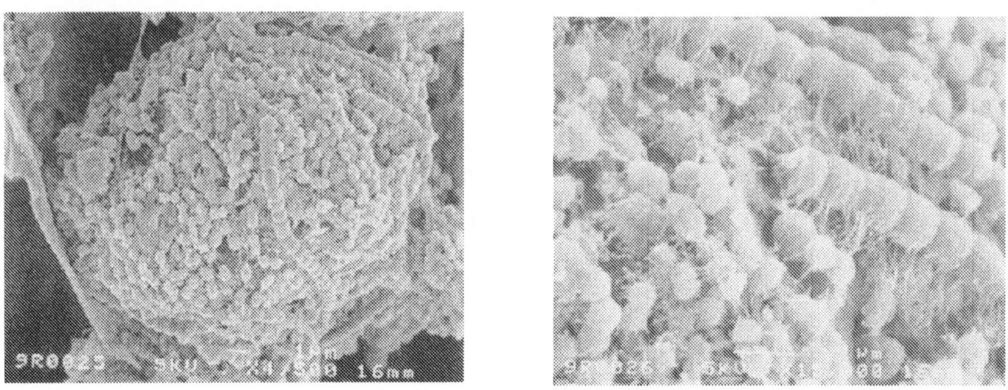

Photo.1. Electric scanning microscopy of symbioses of methanogens on the surface of Ciliate Protozoa.

Additionally, although suppression of rumen methanogenesis has been reported from halogen compounds (Davies, et al., 1982) or saturated and unsaturated fatty acids (Czerkawski et al., 1966a; Demeyer & Henderickx,1967), there are some adverse effects on fiber digestion (Czerkawski et al., 1966b). Therefore, practical use of these compounds is not indicated. In a word, to control only the rumen CH_4 production without adverse effect on the useful bacteria in the rumen and the manipulation method must be safe for animal and human bodies, and the environment. The effects of various probiotics such as *Saccharomyces cerevisiae*, the plant extracts and a sulfur containing amino acid are being examined as a feed additive of a natural material which suits these conditions (Chalupa, 1984; Van navel & Demeyer, 1988; Mackie et al., 1992; Fonty et al., 1993; Takahashi, et al., 1997, Kmet et al., 1993). Microbial preparations are commercially available as ruminant feed additives but their potential to reduce methane emission in ruminants needs *in vivo* assessment.

Fig. 4 shows the effects of L-cysteine and microbial probiotic preparation on methane emission from sheep using a ventilated hood system (Takahashi et al., 1999). In this trial, a minimum amount of L-cysteine and commercially purchased microbial probiotic preparation (Amnite S100, Cleveland Biotech Lted., UK, currently used in pollution water treatment) were tested according to 4×4 Latin square design. L-cysteine tended to decrease methane emission, but the probiotics preparation significantly ($P<0.05$) increased methane emission (Takahashi, et al.,1997). In addition, the effect of *Yucca Schidigera* (DK Sarsaponin 35, Desert King International, USA, a natural rumen manipulator which has a deodorant effect due to ammonia absorbing ability) on rumen methanogenesis has been investigated in vitro (Takahashi, et al., 2000). The quite small amount of *Yucca Schidigera* (YC125U: Yucca extract 125mg/500ml + Urea 5mM) may be effective on the rumen methanogenesis as a feed additive from the results of *in vivo* evaluation using sheep (unpublished data).

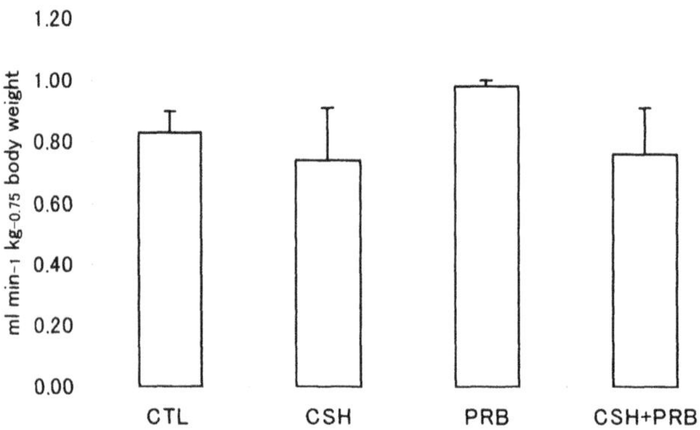

Fig. 4. Effects of L-cysteine (CSH) and/or microbial probiotics (PRB) on methane emission from sheep

3. OVERALL EVALUATION OF RECYCLED ENERGY FROM ANIMAL EFFLUENT

Animal effluent is another source of methane from animals, corresponding to 22.7% of total livestock methane emission. This CH_4 can be recycled as an energy resource by utilizing anaerobic fermentation processing. The CH_4 energy from the domestic animal excreta has already been made the capital at the stage of practical use. In the comparison of CH_4 productivity of major animal effluents from fermentation plants, poultry waste produced the highest CH_4 yields on a live weight basis while dairy waste was the least productive on a live weight and total solid loaded basis (Hill, 1984). These species differences, however, are primarily a reflection of the differences in their feeds consumed and digestive systems. Thus, the quantitative analyses of indigestible energy of feeds should be done in the evaluation of potential energy of animal effluents as a biomas energy. Comprehensive energy balance sheets including energy losses of rumen CH_4, feces, urine and heat production must be completed to evaluate the feed energy utilized and recycled through animal body. All energy converted from biogas must be evaluated as a part of gross energy of feed to extract the maximum recycled energy.

Animal effluents cause not only odor problems but also eutrophication pollution problems in lakes, ponds, rivers and sea water by nitrate nitrogen which derives from ammonia nitrogen. In addition, because nitrous oxide (N_2O), which derives from ammonia, is another greenhouse gas and contributes to acid rain, as well, animal effluent processing is a global issue. Some EU countries have legislation to maintain a severe restriction method for animal effluent processing. That is, the control of the CH_4 emission from animal is a big problem, which includes the control of the CH_4 fermentation in the rumen, the promotion of anaerobic fermentation of effluents to capture methane and use of the residue after fermentation as a fertilizer. Furthermore, an efficient conversion of methane energy remains to be elucidated for recycling. Not only the animal husbandry of one country but all over the world, correspondence is urged to contribute to the prevention of global warming by the solution of these problems.

In Japan, excess amounts of the untreated effluent have resulted in excess nitrate in underground water and pathogens such as criptosporidium in surface water. The complaints of pollution problems from animal agriculture industry have increased along with the drastic expansion of animal population in limited areas. The issues that should be resolved are the developments of various technologies for animal effluent management to control offensive odors and wastewater, exploitation of energy through anaerobic methane fermentation and feeding management to reduce the excretion of nitrogen and phosphorus. Promoting environmentally conscious agriculture that does not rely on chemical fertilizer can be realized by effective use of liquid manure and compost. Many research projects have been carried out to extract energy as methane through processes related to animal and organic effluent management. The fuel cell that uses animal effluents and organic wastes deserves attention as an epoch-making technology.

REFERENCES

Czerkawski, J. W., K. L. Blaxter & F. W. Wainman. 1966a. The metabolism of oleic, linoleic, and linolenic acids by sheep with reference to their effects on methane production. Brit. J. Nutr. 20:349-362.

Czerkawski, J. W., K. L. Blaxter & F. W. Wainman. 1966b.The effect of linseed oil and of linseed oil fatty acids incorporated in the diet on the metabolism of sheep. Brit. J. Nutr.20: 485-494.

Chalupa, W. 1980. Chemical control of rumen microial metabolism. In: Digestive Physiology and Metabolism of Ruminants.Y . Ruchenbusch & P. Thivend (editors). MTP Press Ltd, Lancaster, pp325-347.

Chalupa, W. 1984. Manipulation of rumen fermentation. In: Recent Advances in Animal Nutrition. W. Haresign & D.J.A. Cole (editors). Recent Advances in Animal Nutrition. Butterworths, London, pp. 143-160.

Davies, A., H.N. Nwaonu, G. Stanier & F. T. Boyle. 1982. Properties of a nivel series of inhiitors of rumen methanogenesis; *in vitro* and *in vivo* experiments including growth trials on 2,4-bis (trichloromethyl)-benzo [1,3] dioxin-6-carboxylic acid. Brit. J. Nutr. 47: 565-576.

Demeyer, D. I. & H.K. Henderickx. 1967. The effect of C_{18} unsaturated fatty acids on methane production and its control. In: Digestion and Metabolism in the Ruminant. I. W. MacDonald & A.C. Warner (editors). The University of New England Publishing Unit, Armidale, pp. 366-382.

Finlay, B. J. 1994. Some rumen ciliate have endosymbiotic methanogens. FEMS. Microbiol. Lett. 117: 157-161.

Fonty, G., P. Raibaud & Ph. Gouet. 1993. Manipulation the gut microflora: experimental approach in animals. Proc. Nutr. Soc. 52: 345-356.

Fuller, R. 1992. History and development of probiotics. In: Probiotics The scientific basis. R. Fuller (editor). Chapman & Hall, London, pp.1-8.

Hill, D.T., 1984. Methane productivity of the major animal waste types. TRANSACTION of the ASAE, 27:530-534.

Hill, D.T., S.A. Cobb & J. P. Bolte, 1987. Using volatile fatty acid relationship to predict anaerobic digester failure. TRANSACTIONS of the ASAE, 30:496-501.

Houghton, J. 1994. Global warming. pp29-45. Lion Publishing plc. Oxford. Intergovernmental panel on climate change (IPCC), 1994. Climate change 1994. (Ed. J.H. Houghton, L.G.Meria Filho, J.Bruce, Hoesung Lee, B.A. Callander, H.Haites, N.Harris & K.Maskell, Cambridge University Press, NewYork, pp25-27.

Kemt, V., H. J. Flint and R.J. Wallace. 1993. Review Probiotics and manipulation on rumen development and function. Arch. Anim. Nutr., 44:1-10.

Mackie R.I., B. A. White & M. P. Bryant. 1992. Methanogenesis, biochemistry. In: Encyclopedia of Microbiology, vol. 3. J. Lederberg (editor). Academic Press, San Diego, 97-109.

Newbold, C. J., B. Lassalas & J. P. Jouany. 1995. The importance of methanogens associated with ciliate protozoa in ruminal methane production in vitro. Letters in Applied Microbiology, 21: 230-234.

Stumm, C .K., H. J. Gilzen & G.D. Vogels. 1982. Association of methanogenetic bacteria with ovine rumen ciliates. Brit. J. Nutr. 47: 95-99.

Takahashi, J., A.S. Chaudhry, R.G. Beneke & B. A. Young. 1999. An open-circuit hood system for gaseous exchange measurements in small ruminants. Small Ruminant Res.32: 31-36.

Takahashi, J., A.S. Chaudhry, R.G. Beneke, Suhubdy & B.A. Young. 1997. Modification of methane emission in sheep by cysteine and a microbial preparation. Modification of methane emission in sheep by cysteine and a microbial preparation. Sci. Total Environ. 204:117-123.

Takahashi, J., T. Chiba, T. Encho, J. Ikariga & H. Fujita. 1983. The acute and chronic effect on heat production and methanogenesis in ruminants. Proc. 5th World Conf. on Animal Production vol. 2. 14-19 august 1983, Tokyo, Japan, Japanese Society of Zootechnical Science, Tokyo. pp 351-352.

Takahashi.J., M. Ikeda, S. Matsuoka & H. Fujita. 1998. Prophylactic effect of L-cysteine to acute and subclinical nitrte toxicity in sheep. Anim. Feed Sci. Technol. 74:273-280.

Takahashi, J., N. Johchi & H. Fujita. 1988. Inhibitory effect of sulphur compounds, copper and tungsten on nitrate reduction by mixed rumen micro-organisms. Brit. J. Nutr. 61: 741-748.

Takahashi, J., T.Miyagawa, Y. Kojima & K. Umetsu. 2000. Effects of *Yucca schidigera* extract, probiotics, monensin and L-cysteine on rumen methanogenesis. Asian-Aust. *J. Anim. Sci. 13(Suppl.): 499-501.*

Takahashi, J. & B.A. Young. 1991. Prophylactic effect of L-cysteine on nitrate-induced alterations in respiratory exchange and metabolic rate in sheep. Anim. Feed Sci. Technol. 35: 105-113.

Takahashi, J. & B.A. Young. 1992. The modulation of nitrate-enhanced hypothermia by sulphur compounds in cold-exposed sheep. Anim. Feed Sci. Technol.39: 347-355.

Takahashi, J. & B.A. Young. 1994. The regulation of energy metabolism in sheep by nitrate and L-cysteine. In: Energy Metabolism of Farm Animals. J.F. Aguilera (editor). EAAP publication No. 76, Granada, pp. 387-390.

Van Nevel C. S. & D. I. Demeyer. 1988. Manipulation of rumen fermentation. In: the Rumen Microbial Ecosystem. P. N. Hobson (editor). Elsevier Applied Science, London, pp.387-443.

Wallace, R.J. 1992. Rumen microbiology, biotechnology and ruminant nutrition: The application of research findings to a complex microbial ecosystem. FEMS Microbial. Lett. 100: 529-534.

Greenhouse Gases and Animal Agriculture
J. Takahashi & B. A. Young (Editors)

The use of sulfur hexafluoride for measuring methane emissions from farm animals

K. A. Johnson[1], H. H. Westberg[2], B. K. Lamb[2], & R. L. Kincaid[1]

[1]Dept. of Animal Sciences and [2]Laboratory for Atmospheric Research, Dept. of Civil and Environmental Engineering, Washington State University, Pullman, WA 99164, USA

Measurement of methane emission by ruminants in production situations is possible through the use of atmospheric tracers such as sulfur hexafluoride. Individual animal measurements are accomplished by use of a ruminal permeation tube that releases SF_6 at a known rate. Emissions from groups of animals in a room or groups on pasture are possible through the release of a tracer into the room or pasture area. Requirements of the tracer technique include: known constant tracer release from a permeation tube or gas cylinder; complete mixing of the tracer and methane prior to sampling; collection of exhaled ruminant methane; and sensitive detection of the tracer and methane. The use of tracer technology provides options for scientists to measure methane emissions for inventory purposes, to develop methane mitigation strategies and to examine whole animal energy nutrition.

Keywords: Methane, ruminants, tracer gas, sulfur hexafluoride

1. INTRODUCTION

Methane emissions from livestock have been measured as a part of studies on ruminal fermentation, energy balance, evaluation of feed additives and, most recently, to characterize and reduce the contribution of ruminants to the global methane burden. Historically, livestock methane emissions have been measured using respiration calorimetry systems such as head boxes, whole body chambers, ventilated hoods and face masks (Johnson and Johnson, 1995). Applicability of the data are limited by restriction of animal movement, behavior and feed intake associated with confinement. However, data obtained from these techniques have been the foundation of the prediction equations used to generate mathematical models and countrywide and global inventories (Blaxter and Clapperton, 1965; Baldwin et al., 1987; IPCC, 1990; Benchaar et al., 1998; Mills et al., 2001). Although these equations and models have usefulness in the prediction of methane production from animals under the conditions from which the equation or model was developed, they are of limited usefulness in the prediction of methane production when intake is unknown or when the rumen is perturbed. Johnson et al. (2000) compared calorimetry results to those predicted from the Blaxter and Clapperton equation and found a poor predictive relationship ($r^2 = .23$).

Measurement of methane from animals in production settings, e.g. grazing or feedlot, can be made using techniques developed by atmospheric scientists (Lamb et al., 1986; Lamb et al., 1995, Czepiel et al., 1996; Marik & Levin, 1996). These techniques use an inert gas tracer

(Johnson et al., 1994) or mass balance approach (Leuning et al., 1999). Of these two, the tracer approach is less expensive and offers more flexibility in use.

A tracer such as sulfur hexafluoride, SF_6, can be used to make measurements of individual animals (Johnson et al., 1994; Lassey et al., 1997; Ovenell-Roy et al., 1998; Pavao-Zuckerman et al., 1999; Nelson et al., 2001;), groups of animals in a room or groups of animals in a pasture setting (Johnson et al., 1998). The principle behind tracer use is the same regardless of application and is based on the assumption that the tracer gas and CH_4 are mixed and the dilution is the same for both gases prior to sample collection. Knowledge of the tracer release rate and the concentrations of tracer and methane in sampled gas are required to calculate methane production. When individual animals are measured, mixing of SF_6 and CH_4 occurs inside the rumen. When SF_6 is used as a room tracer, fans are placed around the room to ensure mixing; and when SF_6 is used in an outside environment, air turbulence overcomes differences in gas density and ensures proper mixing of SF_6 released during pasture measurements.

2. MATERIALS AND METHODS

2.1. Measurements

2.1.1. Individual animals.

To make animal measurements, a small permeation tube (Figure 1) containing pure SF_6 is constructed and its release rate of SF_6 (at 39°C) calibrated gravimetrically in the laboratory. After six to eight weeks of calibration, the tube is placed in the rumen and allowed to equilibrate for several days. A horse halter (Figure 2) modified with a 50-micron filter and length of capillary tubing is placed on the animal's head and connected to an evacuated sampling canister (Figure 3). As the vacuum in the canister dissipates, a steady stream of air from around the mouth and nose of the animal is collected. The length of the capillary tubing controls the sampling duration. After the sampling canister is filled to half an atmosphere (ATM), the canister is removed and pressurized with nitrogen to 1.2 ATM. An ambient air sample (24-h) is also collected at the same time as the animal samples which allows background sources of CH_4 to be measured.

The concentration of CH_4 and SF_6 is determined with gas chromatographic procedures. Methane concentration is measured using a gas chromatograph fitted with a 1 cc sampling loop and flame ionization detector. A 0.25 cm x 1.2 m Poropak-N column is maintained at 50° C. Sulfur hexafluoride is measured using a 1 cc sampling loop and an electron capture detector. A molecular sieve 5A column (0.215 cm x 1.83 m) for this separation is maintained at 50° C. Methane emission rate is calculated as the product of the permeation tube release rate and the ratio of CH_4 to SF_6 in the sampling canister. The equation used is: $Q_{CH4} = Q_{SF6} \times [CH_4] / [SF_6]$; where Q_{CH4} is the methane emission rate in g/hr, Q_{SF6} is the known release rate of the permeation tube, and $[CH_4]$ and $[SF_6]$ are the measured concentrations of CH_4 and SF_6 in the sampling canister. The following web site has a slide show illustrating the technique: www.epa.gov/outreach/rlep/interactive/methane/index.htm.

Several groups have modified the sampling canisters for their conditions. McCaughey et al. (1997; 1999) used a sampling canister suspended similar to a cowbell from a neck rope. Martin et al. (2000) used a PVC cylinder supported by netting placed in the center of the sheep's back. Ovenell-Roy et al. (1998), Nelson et al. (2001) used cylinders placed in pockets of a sheep backpack.

It must be noted the use of the SF_6 tracer technique in individual animals is limited in the U.S. by regulations of the Food and Drug Administration. These limitations include a long withdrawal time after the permeation tube is removed or has completed its release of SF_6. In addition, milk from treated dairy cows must be discarded.

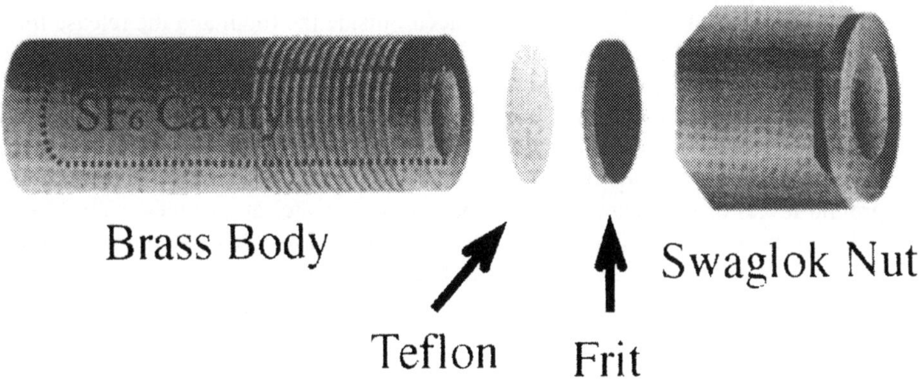

Figure 1. Permeation tube

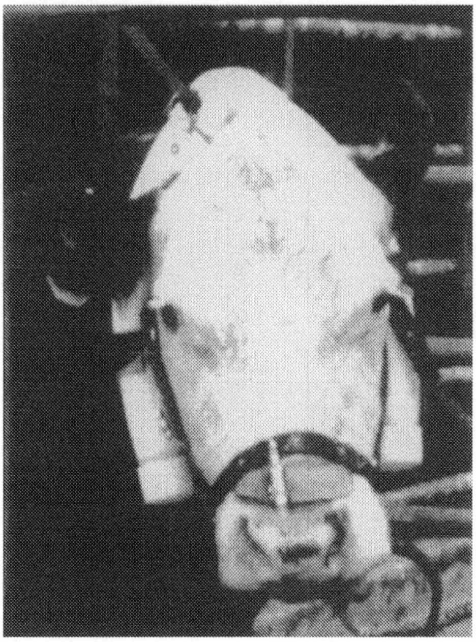

Figure 2. Cow wearing a sampling halter and canister

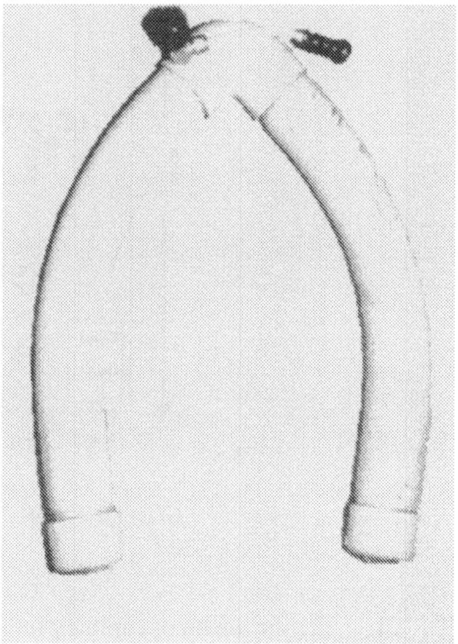

Figure 3. Sampling canister

2.1.2. Room measurements

Emission measurements can be made from an animal or groups of animals in a room using a tracer to account for the room ventilation rate (Marik & Levin, 1996; Kincaid et al., 2001). A preliminary study should be conducted to determine the tracer release rate needed to reach an appropriate steady state tracer concentration. This release rate can be used for subsequent studies in the same location. We used this technique to measure methane emissions from groups of dairy cattle fed different diets (Kincaid et al., 2001). The room was, 6.1 m x 8.8 m, and had three windows on one side that remained open. Two box fans were suspended from the ceiling in opposite corners of the room. A tank containing 1024 ppm SF_6 fitted with a mass flow controller (Tylan FC-260) was placed outside the room and the release line passed through the window and dispersed tracer in front of one of the fans (Figure 4). A release rate of 50.2 μg SF_6/min was determined to be adequate for this room. A release time of four hours prior to making measurements was used to ensure steady-state concentrations were reached and room air turned over completely three times. Evacuated sampling canisters were suspended around and outside the room to determine room SF_6 and CH_4 concentrations and background levels of both gases. Cattle had access to water at all times, were free to move about the room and the feeding regimen was maintained as when the animals were housed normally. The room was completely cleaned between measurements.

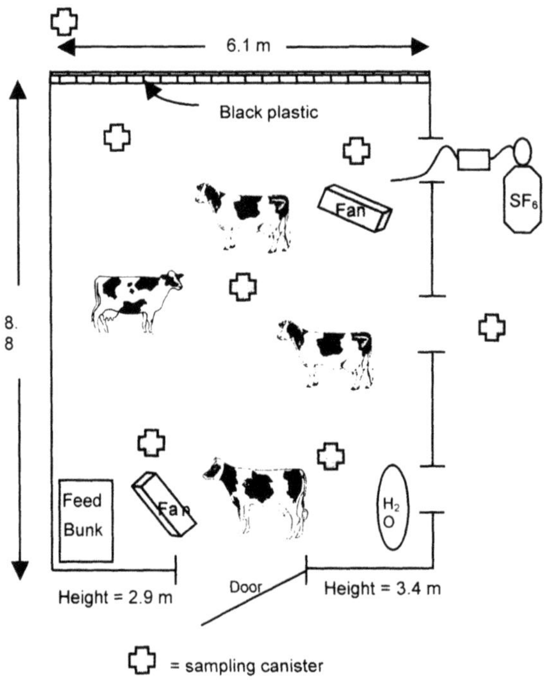

Figure 4. Room tracer diagram

Changing the length of capillary tubing on the sampling canister may alter the duration of a sample collection. When the sampling duration is completed the canisters are removed, pressurized with nitrogen (1.2 ATM) and concentrations of CH_4 and SF_6 determined. Methane emission rate for the group of animals is calculated as is done with individual animals. The equation used is: $Q_{CH4} = Q_{SF6} \times [CH_4] / [SF_6]$; where Q_{SF6} is the known release rate of the tracer gas source. Correction by subtraction of background methane or SF_6 concentrations should be considered.

2.1.3. Pasture measurements

Measurements can be made from groups of animals in a in a pen or pasture using SF_6 as the tracer gas. An enclosure is constructed with electric fencing or panels that provides sufficient space for the number of animals to be measured. If grazing measurements are to be made, the feed supply should reflect the conditions in the pasture and provide adequate feed and water for the number of animals and duration of the measurements. Animals should be relatively docile and accustomed to the sampling area and the presence of people. Airflow patterns and conditions in the area should be investigated prior to sampling to ensure adequate wind speed and no significant upwind background sources of SF_6 and CH_4. Background sources can include oil wells, landfills, lagoons, and other ruminants.

Canisters containing tracer gas with calibrated flow meters or a tank of tracer outfitted with a mass flow meter and tracer release line are placed on the upwind side of the enclosure. Evacuated sampling canisters fitted with restriction tubes appropriate for the sampling duration are placed in two rows on the downwind side of the enclosure. The canisters are placed far enough downwind to ensure adequate mixing of tracer and CH_4 but close enough to ensure concentrations within analytical range. A canister to measure upwind SF_6 or methane sources is placed well upwind of the release system (Figure 5). One person deals exclusively with the release system and another with the sample collection system to prevent contamination of the air samples. Ambient weather condition such as temperature, barometric pressure, and wind speed and direction should be recorded during the sampling period.

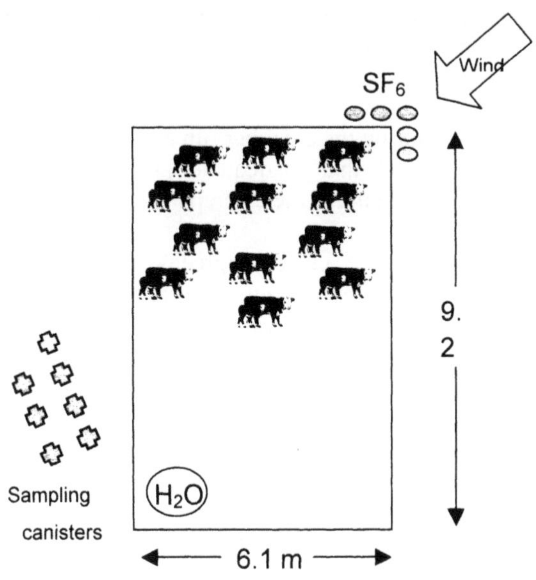

Figure 5. Pasture tracer diagram

We found a release rate of six g/min SF_6 is adequate to measure methane emissions from twelve cow-calf pairs in a 56 m^2 area of pasture. Tracer release is initiated after the cows are moved to the pasture area. After a ten-minute equilibration period, the collection of air begins. The sample canisters are pressurized to 1.2 ATM and SF_6 and CH_4 concentrations determined by gas chromatography. Methane emissions by the cattle are calculated as described above for the room tracer method.

3. RESULTS AND DISCUSSION

3.1. Individual animal measurements

Validation. Methane emissions have been successfully measured from sheep (Lassey et al., 1997, 1999; Ulyatt et al., 1999; Martin et al., 2000; Nelson et al., 2001), beef cattle (Johnson et al., 1994; Westberg et al., 2001; McCaughey 1997 & 1999; Pavao-Zuckerman et al., 1999), dairy cattle (Johnson et al., 1998), and water buffalo (Turnbull et al., 2000) using the SF_6 technique. The method and validation are described in Johnson et al. (1994). Validation consisted of measurement of animals using open circuit indirect calorimetry chambers and the tracer technique. Measurements were made by the tracer technique outside in the animal's outside pen for 4 days prior to respiration chamber measurements. The cattle were held in the chambers for 2 days of measurements, then returned to their outside pens for an additional 3 days of tracer measurements. Statistical comparisons indicated no difference between the chamber and outside measurements. The cattle measured in this manner represented cattle fed diets ranging from high forage (100%) to 90% grain: 10% forage. Chamber estimates of methane production represented a pooled 22 h air sample from which the methane concentration was analyzed using an infrared detector and the concentration multiplied by measured airflow from the chamber. Additionally, syringe samples of the collected gas samples were analyzed for CH_4 and SF_6 concentrations using flame ionization (FID) and electron capture detectors, respectively and then CH_4 production was calculated using the tracer approach. These two methods resulted in identical estimates of CH_4 production. However, similar to the findings of Ulyatt et al. (1999), when air samples were collected in the chamber with a syringe sampler during the chamber run and analyzed for SF_6 and CH_4, we did not get good agreement and feel that inadequate air mixing in the chambers or recirculation through the cattle may account for the discrepancy.

Results from the SF_6 tracer technique closely agree with those from mass balance techniques for grazing cattle (Judd et al., 1999; Leuning et al., 1999). Leuning et al. (1999) found mean CH_4 emissions from sheep grazing a grass-legume pasture to be 11.7 ± 0.4 g CH_4 sheep $\cdot$ d^{-1} and 11.9 ± 1.5 g CH_4 sheep $\cdot$ day^{-1} from the tracer and mass balance techniques, respectively. No significant differences were found in CH_4 emissions estimates using the tracer technique between sheep or between days of measurement. The inter-sheep variability ranged from 9.5 ± 2.1 to $14.8 \pm$ g CH_4/d and inter-day emission measurements ranged from 9.6 ± 1.3 to 14.4 ± 1.3 g CH_4/d (Luening et al., 1999). Judd et al. (1999) also examined the tracer technique and mass balance techniques and found good agreement. In this study, emissions estimates gathered using the tracer method varied less from day to day (SD = 2.7 g/d) than from animal to animal (SD = 4.2 g/d).

Westberg et al. (2001) conducted an error analysis of the individual animal tracer technique. They found the average coefficient of variation in methane emissions measurements between groups of cattle was approximately 18%. This estimate included analytical errors and inter- and intra-variability in methane emissions from the animals in a measurement group over a five day measurement period. The animals and diets in this estimate included lactating dairy cattle fed corn silage or alfalfa haylage diets, beef cows on good quality pasture, poor quality pasture and good quality hay, growing beef heifers grazing good quality pasture or good quality hay and growing steers fed 40% grain, 60% forage or 85% grain, 15% forage. The analytical error rate of SF_6 and CH_4 measurement was estimated to be < 2%, and error associated with the measurement of permeation tube release rate to be < 5%. Combining these factors yields an uncertainty rate of approximately 20%. Lassey et al. (2001) recently reviewed permeation tube performance and suggested methods to reduce the error associated with permeation tube release rate measurements.

3.2. Room measurements

Emissions of methane from groups of cattle in a room have been measured from lactating dairy cows, growing heifers, bulls and suckling calves. In all cases, repeat samplings over several days were used to examine day-to-day variability in the technique. In the dairy cow study (Kincaid et al., 2001), measurements from the same cows over three days were 12.6 g CH_4 cow· h^{-1}, 13.6 g CH_4 cow · h^{-1} and 15.8 g CH_4 cow · h^{-1}. These emissions measurements agree with Westberg et al. (2001) who used the individual cow tracer methodology, Marik and Levin (1996) who used an SF_6 tracer approach, whole barn, to measure CH_4 emissions from lactating cows, and Kinsman et al. (1995) who measured CH_4 emissions from lactating dairy cattle housed in a barn fitted with flow meters to measure air input and output. Growing heifers fed diets differing in the content of apple pomace were measured three times (there were no differences in CH_4 emissions due to diet) and mean CH_4 emissions found to be 11.1 ± .4 g heifer·h^{-1} (Kincaid et al., 1998). These values agree with other measurements for animals fed similar diets.

We generally have used four animals per treatment in room tracer studies. This number was determined by the size of the pens available. The emissions measured are the result of all of the animals and measurement of individual emissions is not possible. It is possible to make room measurements for individual animals, however, behavioral considerations must be incorporated into the sampling plan. Other considerations are the length of time the sampling period encompasses and feed intake. We have always allowed ad libitum intake prior to and during the sampling periods.

3.3. Pasture measurements

The measurements we have made from groups of cattle on pasture have resulted in estimates of methane production that are similar to those observed when individual animals are grazing the same pasture and measured with the individual tracer technique (Westberg et al., 2001). Emissions estimates of 9.83 ± 1.7 g CH_4 cow·h^{-1} were obtained from a group of twelve cows.

One of the challenges to the outside tracer method is illustrated by the data obtained when the wind died during a pasture measurement. Apparent methane production was 4.1 g CH_4 cow·h^{-1}, which is 58 % lower than the value observed when conditions were favorable for measurement and there was adequate wind for mixing of the tracer and methane.

REFERENCES

Baldwin, R.L., J.H.M. Thornley & D.E. Beever. 1987. Metabolism of the lactating cow. II. Digestive elements of a mechanistic model. J. Dairy Res. 54:107-131.

Benchaar, C., J. Rivest, C. Pomar & J. Chiquette. 1998. Prediction of methane production from dairy cows using existing mechanistic models and regression equations. J. Anim. Sci. 76:671-627.

Blaxter, K.L. & J.L. Clapperton. 1965. Prediction of the amount of methane produced by ruminants. Br. J. Nutr. 19:511.

Czepiel, P., B. Mosher, J. Shorter, B. McManus, C. Kolb, E. Allwine, B. Lamb & R. Hariss. 1996. Landfill methane emissions measured by static enclosure and atmospheric tracer techniques. J. Geophys. Res. 101:16711.

IPPC, 1990. Intergovernmental Panel on Climate Change. Climate Change-The IPCC Scientific Assessment.

Johnson, D.E., K.A. Johnson, G.M. Ward & M.E. Branine. 2000. Ruminants and other animals. In: Atmospheric Methane: Its Role in the Global Environment, M.A.K. Khalil (editor), Springer-Verlag, Berlin Heidelberg, pp. 112-133.

Johnson, K. A., M. T. Huyler, H. H. Westberg, B. K. Lamb & P. Zimmerman. 1994. Measurement of methane emissions from ruminant livestock using a SF_6 tracer technique. Environ. Sci. and Technol. 28:359.

Johnson, K.A. & D.E. Johnson. 1995. Methane emissions from cattle. J. Anim. Sci. 73:2483-2492.

Johnson, K.A., H.H. Westberg, B.K. Lamb, & R.L. Kincaid. 1998. The use of sulfur hexafluoride for measuring methane production by cattle. In. Energy Metabolism of Farm Animals, K. McCracken, E.F. Unsworth and A.R.G. Wylie (eds). CAB International, Cambridge U.K. pp 198-192.

Judd, M.J., F.M. Kelliher, M.J. Ulyatt, K.R. Lassey, K.R. Tate, I.D. Shelton, M.J. Harvey & C.F. Walker. 1999. Net methane emissions from grazing sheep. Global Change Biology. 5:647-657.

Kincaid, R. L. & K. A. Johnson. 1997. Feeding full fat oilseeds improves economic net returns and reduces methane losses. Prog. Dairyman. 11(2):2-4.

Kincaid, R., K. Johnson, S. Landblom, J. Cronrath, A. Guuma. 1998. Apple pomace for growing beef calves. Farmer-Stockman, September, 1998. p. 19

Kincaid, R.L., K.A. Johnson, H.H. Westberg, B.K. Lamb. 2001. The effect of oilseeds in diets of lactating cows on milk yield, composition and methane emissions. J. Dairy Sci. (submitted).

Kinsman, R., F. D. Sauer, H. A. Jackson, and S. Wolynetz. 1995. Methane and carbon dioxide emissions from dairy cows in full lactation monitored over a six-month period. J. Dairy Sci. 78:2760-2766.

Lamb, B., H. Westberg, & G. Allwine. 1986. Isoprene emission fluxes determined by atmospheric tracer technique, Atmos. Environ. 20:1-8.

Lamb, B., J. McManus, J. Shorter, C. Kolb, B. Mosher, R. Harriss, E. Allwine, D. Blaha, T. Howard, A. Guenther, R. Lott, R. Siverson, H. Westberg, & P. Zimmerman. 1995. Development of atmospheric tracer methods to measure methane emissions from natural gas facilities and urban areas. Environ. Sci. Tech. 29:1468.

Lassey, K. R., M. J. Ulyatt, R. J. Martin, C. F. Walker & I. D. Shelton. 1997. Methane emissions measured directly from grazing in New Zealand. Atmospheric Environment. 31:2905-2914.

Lassey, K. R. & M. J. Ulyatt. 1999. Methane emission by grazing livestock: A synopsis of 1000 direct measurements. Non-CO_2 Greenhouse Gases: Scientific Understanding, Control and Implementation, J. van Ham, A.P.M. Baede, L.A. Meyer, R. Ybema (editors), Kluwer, Dordrecht, Netherlands, pp. 101-166.

Lassey, K.R., C.F. Walker, A.M.S. McMillan & M.J. Ulyatt. 2001. On the performance of SF_6 permeation tubes used in determining methane emission from grazing livestock. Chemosphere (in press).

Leuning, R., S.K. Baker, I.M. Jamie, C.H. Hsu, L. Klein, O.T. Denmead, D.W.T. Griffith. 1999. Methane emission from free-ranging sheep: a comparison of two measurement methods. Atmos. Environ. 33:1357-1365.

Marik, T. & I. Levin. 1996. A new tracer experiment to estimate methane emissions from a dairy cow shed using sulfur hexafluoride (SF_6). Global Biogeochemical Cycles. 10:413.

Martin, C., Y. Rochette, M. Fabre, & J.P. Jouany. 2000. Influence of cereal supplementation on methane production by sheep measured by SF_6 tracer method. In: Challenges for microbial digestive ecology at the beginning of the third millenium. Repro. Nutr. Dev. 40 (2), 211.

McCaughey, W.P., K. Wittenberg, & D. Corrigan. 1997. Methane production by steers on pasture. Can. J. Anim. Sci. 77:519-524.

McCaughey, W.P., K. Wittenberg, & D. Corrigan. 1999. Impact of pasture type on methane production by lactating beef cows. Can. J. Anim. Sci. 79:221-226.

Mills, J.A.N., J. Dijkstra, A. Bannick, S.B. Cammell, E. Krebreab, & J. France. 2001. A mechanistic model of whole-tract digestion and methanogenesis in the lactating dairy cow: model development, evaluation and application. J. Anim. Sci. 79:1584-1597.

Nelson M.L., H.H. Westberg, & S. M. Parish. 2001. Effects of tallow on the energy metabolism of wethers fed barley finishing diets. J. Anim. Sci. 79:1892-1904.

Ovenell-Roy, M.L. Nelson, H.H. Westberg & J.A. Froseth. 1998. Effects of barley cultivar on energy and nitrogen metabolism of lambs. Can. J. Anim. Sci. 78:389-397.

Pavao-Zuckerman, M. A., J.C. Waller, T. Ingle, & H. A. Fribourg. 1999. Methane emissions of beef cattle grazing tall fescue pastures. J. Environ. Qual. 28:1963-1969.

Turnbull, G.W., B. Ducharme, R. Livingston & R. Bowman. 2000. The effect of molasses-urea supplementation on high fiber buffalo diets in Gujarat State, India on production parameters and methane losses. J. Anim. Sci. 78(Suppl. 1):153.

Ulyatt, M.J., S.K. Baker, G.J. McCrabb & K.R. Lassey. 1999. Accuracy of SF_6 tracer technology and alternatives for field measurements. Aust. J. Agric. Res. 50:1329-1334.

Westberg, H., B. Lamb, K.A. Johnson, & M. Huyler. 2001. Inventory of methane emissions from U.S. cattle. J. Geophys. Res. 106:633-642.

Greenhouse Gases and Animal Agriculture
J. Takahashi & B. A. Young (Editors)

Nutritional options for abatement of methane emissions from beef and dairy systems in Australia

G. J. McCrabb

CSIRO Livestock Industries, Rockhampton, Queensland 4702, Australia (Present Address: International Livestock Research Institute, PO Box 5689, Addis Ababa, Ethiopia)

Greenhouse gas emissions associated with ruminant livestock production, particularly methane and nitrous oxide, are a major contributor to Australia's national greenhouse gas emissions. Even though there are presently no financial signals to the livestock industries in Australia to reduce emissions, the beef and dairy industries are beginning to search for solutions. Through the abatement of greenhouse gases there is also an opportunity for the livestock industries to produce meat and milk more efficiently, and to demonstrate stewardship of the production environment. The development of viable livestock production systems with minimal undesirable environmental impacts will ensure that food production, quality of product and enterprise profitability are maintained. A range of nutritional and management options for abatement of enteric methane emissions are cited in the literature, however, the effectiveness of these options is yet to be critically evaluated. This paper considers approaches required to evaluate the effectiveness of options for the abatement of enteric methane emissions, and questions whether abatement of methane emissions from Australian beef and dairy industries is achievable with existing nutritional options.

Keywords: greenhouse gases, methane emission, beef, dairy, Australia

1. AUSTRALIA'S GREENHOUSE GAS EMISSIONS

Australia's national greenhouse gas emissions have increased by 17 % since 1990, increasing from 390 million tons carbon dioxide equivalents (Mt CO_2-e) in 1990 to 458 Mt CO_2-e in 1999 (Australian Greenhouse Office, 2001). The Energy sector was responsible for 80 % of national emissions in 1999, and has increased by 22 % from 300 Mt CO_2-e in 1990 to 365 Mt CO_2-e in 1999. The Agriculture sector is the major source of greenhouse gas emissions in Australia. Emissions from the Agriculture sector were responsible for 20 % of national greenhouse gas emissions in 1999, and has increased from 91 Mt CO_2-e in 1990 to 94 Mt CO_2-e in 1999.

In 1999, livestock were responsible for 63 Mt CO_2-e or 67 % of greenhouse gas emissions from the Agriculture sector in Australia, which represented 14 % of total greenhouse gas emissions (Australian Greenhouse Office, 2001). Emissions from livestock in 1999 were 6 % lower than emissions in 1990 (66 Mt CO_2-e), which was due in part to a 31% reduction in the Australian sheep population. Figure 1 presents methane emissions from the Agriculture sector of which enteric emissions from the dairy, beef and sheep populations were the greatest sources. Enteric methane emissions from beef cattle have increased by 4 % since 1990, due to a minor increase in beef cattle population, whereas increases in the dairy cattle population has resulted in a 28% increase in methane emissions from dairy cattle. In contrast, methane emissions from the sheep population have declined by 33% since 1990, due to changes in market conditions and drought.

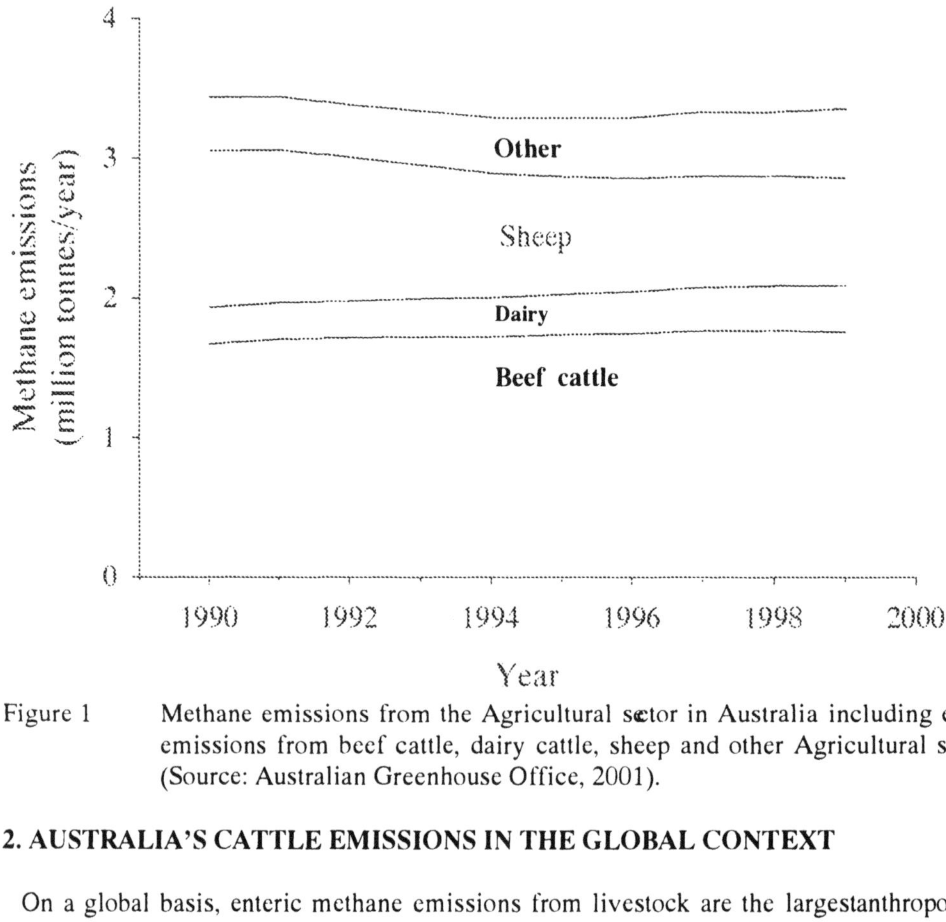

Figure 1 Methane emissions from the Agricultural sector in Australia including enteric emissions from beef cattle, dairy cattle, sheep and other Agricultural sources (Source: Australian Greenhouse Office, 2001).

2. AUSTRALIA'S CATTLE EMISSIONS IN THE GLOBAL CONTEXT

On a global basis, enteric methane emissions from livestock are the largestanthropogenic source of methane gas. In 1990, global enteric methane emissions from livestock were 495 million Mt CO_2-e or 23 % of all anthropogenic sources (IPCC 1995, 1996). Rice Paddies and Natural Gas & Oil were the next largest sources, and accounted for 16 and 15 % of global anthropogenic sources, respectively. In 1990, enteric methane emissions from cattle alone accounted for 73 % of enteric emissions from all livestock species, and 82 % of enteric emissions when the cattle and buffalo populations were combined. Figure 2 presents cattle populations in different countries in 2000 (FAO 2001), and indicates that Australia's cattle population of 27 million is relatively small on a global scale. The world's cattle population has increased by 6 %, from 1.38 billion in 1990 to 1.46 billion in 2000. Since 1990 there has been an 18 % decline in the cattle population in developed countries, in contrast to a 14 % increase in developing countries (FAO 2001). The cattle population for China alone has increased by 32% since 1990, and is presently 105 million head. Projections to the year 2020 suggest that meat and milk consumption in the developing world will increase at a rate of 2.7 and 3.2 % per year, respectively, whereas the projected increases in developed countries are 0.7 and 0.4 % per year respectively (Delgado *et al.* 1999). These changes may lead to increased greenhouse gas emissions from the global cattle and buffalo population, even under conditions where production efficiency is improved.

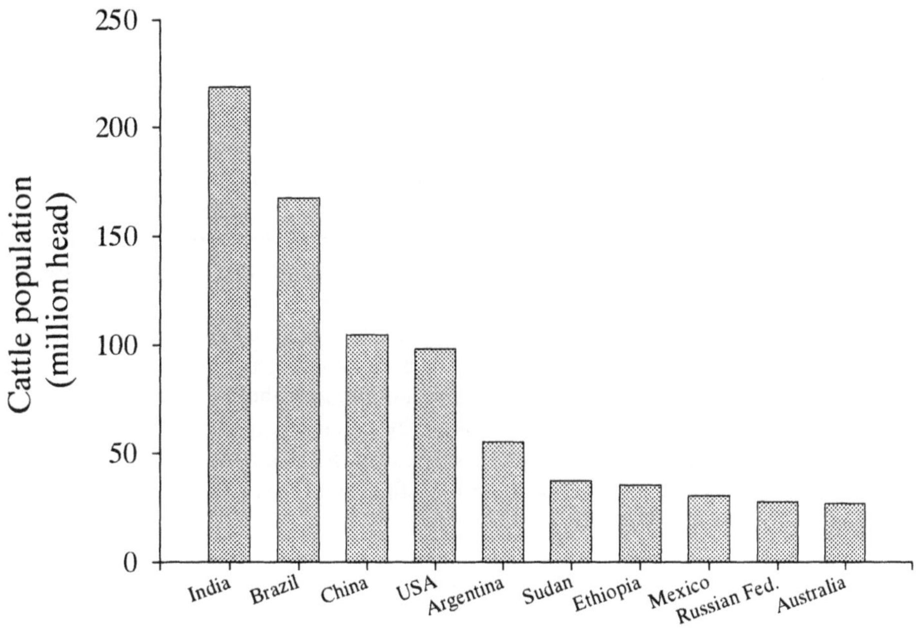

Figure 2 The ten largest cattle populations in the world in 2000, on a country basis (Source: FAO, 2001).

3. THE AUSTRALIAN GOVERNMENT'S SUPPORT FOR GREENHOUSE ABATEMENT

The Australian Government is encouraging its livestock industries to explore approaches for the abatement of greenhouse gas emissions, particularly from the Energy sector. Programs established by the Australian Greenhouse Office to facilitate greenhouse gas abatement include the Greenhouse Gas Abatement Program (http://www.greenhouse.gov.au/ggap) and the Greenhouse Challenge Program (http://www.greenhouse.gov.au/challenge).

The Greenhouse Gas Abatement Program has allocated funding of $400 million Australian dollars between 2000 and 2004, and aims to reduce Australia's greenhouse gas emissions by funding activities that will result in substantial reduction in emissions or sink enhancement. This program targets opportunities for large scale, cost-effective and sustained greenhouse gas abatement, and projects are only being supported if the level of abatement is above the threshold of 250,000 tons CO_2-e per annum. CSIRO (2001) is developing a vaccine targeted at reducing methane enteric emissions from ruminant livestock, and has set up a program aiming to abate 300,000 tons CO_2-e per annum by including over 1 million cattle and 2 million sheep in the program.

The Greenhouse Challenge Program is a voluntary initiative between the Australian Government and industry, and was established in 1995. In March 2001, participants in the program covered 100 % of aluminum and cement production, 98 % of oil and gas production and 91 % of coal mining in Australia. In contrast, there has been little active participation by the livestock sector in Australia. The Queensland Dairyfarmers Organization is the only livestock related organization of 547 participating organizations in the program.

4. SOURCES OF GREENHOUSE GASES FROM LIVESTOCK SYSTEMS

Although methane is the major source of greenhouse gas from ruminant livestock systems, other greenhouse gasses such as nitrous oxide and carbon dioxide are also produced. The Intergovernmental Panel on Climate Change (IPCC, 1997) has developed guidelines for estimating greenhouse gas emissions from various sectors responsible for greenhouse gas emissions including Energy, Industrial Processes, Solvent & Other Product Use, Agriculture, Land-Use Change and Forestry, and Waste. To effectively consider approaches for greenhouse gas abatement from livestock systems, all greenhouse gases must be considered, and not only enteric methane emissions. When conducting greenhouse gas inventories of livestock production systems or industries, care should be taken when attributing greenhouse gas emissions to the individual production systems, because the risk of double-accounting of greenhouse gas emissions that should be attributed to other enterprises (IPCC, 1999). Major sources of greenhouse gas from livestock systems include:

i. Carbon dioxide from the combustion of fuels used to generate electricity that is consumed on farm.
ii. Carbon dioxide generated from the combustion of fuel in farm vehicles, tractors and pumps.
iii. Methane from the digestive tract of cattle, and from livestock effluent management.
iv. Nitrous oxide production from agricultural soils, including nitrogen fertilizer application and from nitrogen excreted in livestockfaeces and urine.

5. GREENHOUSE GAS EMISSION PROFILE OF A DAIRY FARM

Johnson *et al.* (2000) determined the output of methane, nitrous oxide and carbon dioxide from dairy production systems in the US and New Zealand. In the US systems analysed, annual milk production was 7000-9000 kg milk per cow compared to 3400 kg milk per cow in the New Zealand system. Figure 3 reports the greenhouse gas emission profile for a Californian dairy farm with 98 milking cows, each producing 9000 kg per lactation. Greenhouse gas emissions from this farm were 1.2 kg CO_2-e per L milk (1037 tons CO_2-e) compared to 0.9 kg/L for the Wisconsin farm and 1.3 kg/L for the New Zealand farm. These data also illustrate that nitrous oxide was a significant source of greenhouse gas from dairy farms, and accounted for 38% of farm emissions. Significant abatement may therefore be achieved by introduction of technologies and management practices that reduce nitrous oxide production. In contrast, reducing carbon dioxide emissions would only have minor impact on abatement of total farm emissions. Most of the greenhouse gas emissions from farms are related to nutritional management of livestock, indicating that nutritional management is likely to play an important role in greenhouse gas abatement.

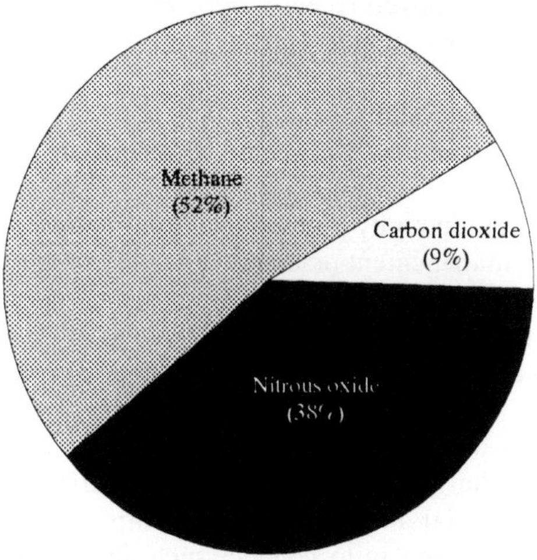

Figure 3 Emissions of methane, nitrous oxide and carbon dioxide for a Californian dairy farm, weighted by global warming potentials (Johnson *et al.* 2000).

The greenhouse emission profile of a livestock system is related to the degree of intensification in the system. Enteric methane emissions account for the majority of emissions from extensive beef production systems in northern tropical Australia, because there are relatively few other greenhouse emitting sources (e.g. energy generation) in the system. In contrast, emissions contributed from non-livestock sources in more intensive beeffeed-lot and dairy production systems are greater. Clearly the most reliable approach to greenhouse abatement from livestock systems is to develop a range of options that can be applied to different systems when appropriate.

6. EXISTING OPTIONS FOR GREENHOUSE ABATEMENT FROM LIVESTOCK SYSTEMS

Livestock systems have a range of unique characteristics that make abatement difficult. Ruminant livestock are typically managed by many independent farm managers, each responsible for a small number of livestock, and hence a minor proportion of total industry emissions. A major reduction in greenhouse gas emissions from a national livestock population requires the adoption of abatement strategies by the majority of individual farm managers. Greenhouse gas abatement from livestock may be associated with increased feed conversion efficiency and reduced production costs. Such benefits will encourage the adoption of abatement strategies. Projects that demonstrate abatement should be accompanied by economic analyses to demonstrate the joint biological and economic benefits of abatement strategies to the livestock industries (EPA, 2000).

There are a number of researchable approaches presently available that may result in the development of technologies to reduce enteric methane emissions from livestock Reyenga & Howden 1999). These approaches do not yet deliver practical solutions to the livestock industries. This review will therefore only consider existing options for greenhouse gas abatement, which have been reviewed recently byMeeks and Bates (1999), EPA (1999) and Bates (2001). The existing options include:

i. Reduce livestock population size.
ii. Improve feed conversion efficiency by increasing diet quality and correctly balancing diets.
iii. Increase animal productivity with no increase in population size.
iv. Modify grazing management practices.
v. Treat livestock with production enhancing agents such as bovine somatotrophin, anabolic steroids and ionophores.
vi. Introduce market-based strategies including trade and price.

To facilitate the adoption of existing abatement options it is necessary for industry and government to demonstrate the impacts on the biological efficiency and profitability of the livestock system, and other social and environmental implications of the various approaches (EPA, 2000). A suggested strategy is to conduct such simulation analyses to describe the greenhouse gas abatement possible, and the associated effects on livestock production and profitability. Once effective abatement options have been identified, pilot projects should be established to demonstrate these benefits.

7. EXAMPLES OF ABATEMENT OPTIONS

7.1. Improve the quality of diets for beef cattle

Beef production enterprises in tropical Australia are typically operated under extensive grazing conditions on pastures dominated by native and improved tropical forage species. A proposed abatement option is to improve the quality of the diet (EPA 1999, Bates 2001). Diet improvement could be achieved, for example, by moving pasture fed animals to intensive feed-lot systems to be fed on diets containing high levels of cereal grain. Figure 4 presents the results of a simulation analysis of lifetime methane production from beef cattle managed under extensive grazing conditions to slaughter weights of 400, 520 or 640 kg (McCrabb *et al.* 1998). For each of these slaughter weights, the impact of feeding cattle on a high cereal grain diet during the final 70-150 days prior to slaughter was evaluated. During grain feeding, daily liveweight gain is increased and methane production per kg liveweight gain is reduced (Kurihara *et al.* 1999). Changing the diet of cattle to a cereal grain diet for 70 – 150 days was associated with a 34 – 54 % reduction in lifetime methane output per kg beef production. This simulation demonstrates that there is potential to reduce methane emissions in northern Australia by dietary changes, however a recommendation to change the diet of an entire slaughter generation of pasture fed cattle to a cereal grain diet is a very simplistic conclusion. More broader analyses of emissions from beef systems are still required if effective abatement options are to be identified. These analyses need to quantify the production of other greenhouse gases including nitrous oxide and methane from the management of excreta (faeces and urine) in intensive production systems, carbon dioxide from the intensive feed-lot enterprises, and methane emissions from livestock not directly producing beef in the system (e.g. reproductive herd).

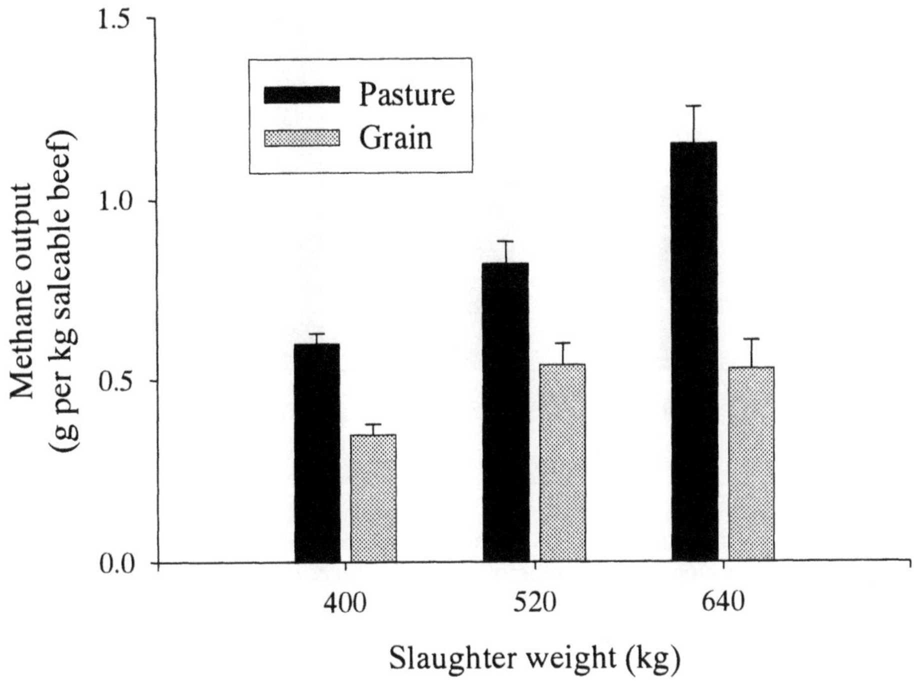

Figure 4 The effect of dietary finishing system (pasture *vs.* cereal grain) on lifetime methane output from beef cattle in northern Australia (McCrabb *et al.* 1998).

7.2. Treatment of beef cattle with hormonal growth promotants

The US EPA (1999) has identified the use of hormonal growth promotants (e.g. oestradiol) as an abatement option for beef cattle, because they reduce the time required for cattle to reach market weight by enhancing liveweight gain, feed efficiency and lean tissue accretion. We have conducted simulation analyses of lifetime methane output from pasture finished beef cattle repeatedly treated with oestradiol to 400, 520 and 640 kg liveweight at slaughter under extensive grazing conditions in northern Australia (McCrabb & Hunter, unpublished). Lifetime methane output per kg beef production for these finishing systems is presented in Figure 5. Repeated treatment of cattle from one year of age to the time of slaughter was associated with a 9 to 16 % reduction in lifetime methane production per kg beef production. Even though this level of abatement per animal appears large, the impact on greenhouse emissions from the whole livestock system will be much lower when greenhouse gases emissions from all sources are included. The effectiveness of hormonal growth promotants as a tool for greenhouse gas abatement from commercial beef production systems, not individual beef cattle, has not yet been evaluated.

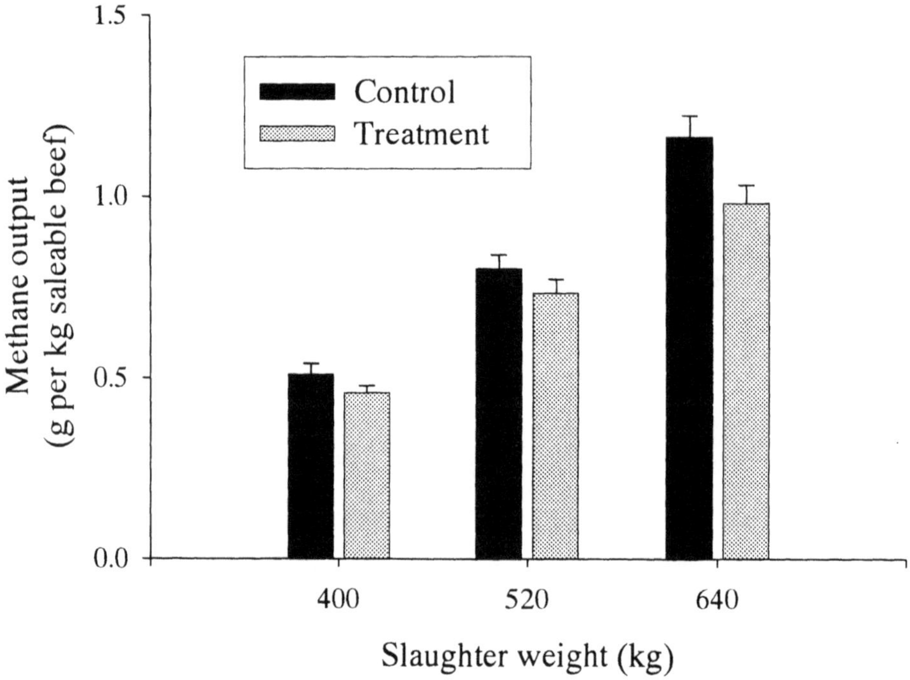

Figure 5 Lifetime methane output is reduced when beef cattle are repeatedly treated with an hormonal growth promotant (McCrabb & Hunter, unpublished data).

7.3. Increase milk production per dairy cow

Milk production per lactation for dairy cows in Queensland has increased from 2924 L/head in 1988 to 4046 L/head in 1996 through improvements in feed quality and management, and was associated with a 26 % reduction in methane emissions per L milk (Howden & Reyenga, 1999). However, during this eight year period average feed intake and methane emissions (kg/head/year) have increased by 4 and 6 % respectively. This data demonstrates that nutritional approaches to reduce methane emissions per unit of milk production are not associated with abatement from the livestock population.

8. CONCLUSION

Effective nutritional options for greenhouse gas abatement from beef and dairy systems have not yet been demonstrated, even though a number of possibilities have been proposed in the literature. There exist a number of researchable options that may lead to the development of effective abatement technologies. The use of simulation analyses of greenhouse emissions from livestock systems at both the enterprise and industry level will assist in identifying effective options for greenhouse gas abatement. Demonstration of proposed abatement options associated with more efficient feed management and increased efficiency of meat and milk production should first be demonstrated in pilot projects before the livestock industries can be expected to adopt such options.

ACKNOWLEDGEMENT

I gratefully acknowledge the assistance of my CSIRO colleagues during the preparation of this paper, particularly Bob Hunter and ChrisMcSweeney.

REFERENCES

Australian Greenhouse Office, 2001. *National Greenhouse Gas Inventory 1999.* Commonwealth of Australia, Canberra, Australia. (Available at: http://www.greenhouse.gov.au/inventory/inventory/latest_inventory.html)

Bates, J., 2001. *Economic evaluation of emission reductions of nitrous oxides and methane in agriculture in the European Union. Bottom-up analysis.* AEA Technology Environment, Oxfordshire, United Kingdom.

CSIRO, 2001. *CSIRO Media Release (June 6, 2001). Graziers flock to block burps.* (Available at: http://www.csiro.au/page.asp?type=mediaRelease&id=prmethanevax)

Delgado, C., M. Rosegrant, H. Steinfeld, S. Ehui, & C. Courbois, 1999. *Livestock to 2020. The next food revolution.* Food, Agriculture and the Environment Discussion Paper 28, IFPRI, FAO and ILRI. (Available at: http://www.ifpri.cgiar.org/gifs/pdf.gifs)

EPA, 1999. *U.S Methane Emissions 1990 – 2020: Inventories, Projections, and Opportunities for Reductions*. EPA Publication 430-R-99-013. US EPA Office for Air and Radiation, Washington DC. (Available at: http://www.epa.gov globalwarming/publications/emissions/methane_emissions.pdf)

EPA, 2000. *Proceedings of the Second International Methane Mitigation Conference. Conference Outcomes.* US EPA, Washington DC. (Available at: http://www.ergweb.com/methane/outcomes.html)

FAO, 2001. *FAOSTAT Database*. FAO, Rome, Italy. (Available at:http://apps.fao.org)

Howden, S.M. & P.J. Reyenga, 1999. *Methane Gas Emissions and Australia's Livestock Industries.* In Reyenga, P.J. & S.M. Howden (eds), 1998. *Meeting the Kyoto Target – Implications for the Australian Livestock Industries.* Bureau of Resource Sciences, Canberra (Available at: http://www.affa.gov.au/docs/rural_science/greenhouse_climate /livestock _methane.html)

IPCC, 1995. *Climate Change 1994: Radiative forcing of Climate Change.* Intergovernmental Panel on Climate Change. Cambridge University Press, Cambridge, United Kingdom.

IPCC, 1996. *Climate Change 1995: The Science of Climate Change.* Intergovernmental Panel on Climate Change. Cambridge University Press, Cambridge, United Kingdom.

IPCC, 1997. *Revised 1996 Guidelines for National Greenhouse Gas Inventories. Workbook (Volume 2).* Intergovernmental Panel on Climate Change, Geneva, Switzerland. (Available at: http://www.ipcc-nggip.iges.or.jp/public /gl/invs5.htm)

IPCC, 1999. *Good practice guidance and uncertainty management in national greenhouse gas inventories* Intergovernmental Panel on Climate Change, Geneva, Switzerland. (Available at: http://www.ipcc-nggip.iges.or.jp/public/gp /gpgaum.htm)

Johnson, D., H. Phetteplace, & M. Ulyatt, 2000. Variations in the proportion of methane of total greenhouse gas emissions from US and NZ dairy production systems. *Proceeding of the Second International Methane Mitigation Conference.* 249-254. US EPA, Washington DC. (Available at: http://www.ergweb.com /methane/pdf/johnson_var.pdf)

Kurihara, M., R.A. Hunter, T. Magner, & G.J. McCrabb, 1999. Methane production and energy partition of cattle in the tropics. *British Journal of Nutrition* 81, 227-234.

McCrabb, G.J., M. Kurihara & R.A. Hunter, 1998. The effect of finishing strategy on lifetime methane production for beef cattle in northern Australia. *Proceedings of the Nutrition Society of Australia* 22, 55.

Meeks, G. & J. Bates, 1999. *Cost effectiveness options for reducing UK methane emissions.* AEA Technology Environment, Oxfordshire, United Kingdom.

Reyenga, P.J. & S.M. Howden, 1999. *Meeting the Kyoto Target – Implications for the Australian Livestock Industries.* Bureau of Resource Sciences, Canberra. (Available at: http://www.affa.gov.au/docs/rural_science/greenhouse_climate /livestock_methane.html)

Greenhouse Gases and Animal Agriculture
J. Takahashi & B. A. Young (Editors)

Detection of methanogens from the rumen by PCR-based techniques

M. Mitsumori[1], K. Tajima[2] & H. Itabashi[3]

[1]National Institute of Livestock and Grassland Science, Ikenodai 2, Kukizaki, Ibaraki, 305-0901 Japan, [2]STAFF Institute, 446-1, Kamiyokoba, Tsukuba, Ibaraki, 305-0854 Japan, [3]Tokyo University of Agriculture and Technology, 183-8509 Japan

Traditionally, methanogens have been identified in the rumen using *in vitro* cultivation methods, which require a great deal of time and labor. Recently, PCR-based techniques, which are convenient and do not require bacterial cultivation, have been used to identify methanogens inhabiting the rumen. Depending on the PCR-based technique used, *Methanomicrobium mobile*, *Methanobrevibacter smithii*, *Methanobrevibacter ruminantium*, *Methanosphaera stadtmanae*, and unidentified methanogens can be detected. These results are not completely in agreement with the results obtained by *in vitro* cultivation methods. The cultivation methods were unable to detect *Methanosphaera stadtmanae* and a number of unidentified methanogens, which were classified as "not culturable". Therefore, PCR-based techniques are expected to be simple and useful tools for detecting methanogens in the rumen.

Keywords: methanogens, PCR, rumen

1. INTRODUCTION

Identification and quantification of bacteria, which constitute the microbial ecosystem in any environment, provide the key parameters in diversity studies. Culture-dependent analyses are routinely performed but are still difficult tasks in so far as they require microbial cultivation. Over the last 10 years, several molecular techniques have been developed to analyze natural microorganisms (Wintzingerode *et al.*, 1997). These techniques aid in identifying microorganisms without the need to cultivate and isolate them. Furthermore, in many cases, the techniques have shown a diversity among microorganisms that had not been seen through cultivation methods. Indeed, it has become apparent that most natural bacteria are not accessible by the cultivation methods being used even today (Colwell & Grimes, 2000). However, it is very probable that molecular techniques also provide a biased view of bacterial diversity.

2. DETECTION OF METHANOGENS FROM THE RUMEN BY CULTIVATION METHODS

Methanogens, which inhabit the rumen, are detectable by cultivation methods. Since methanogens are very strictly anaerobes, media for them have to maintain Eh values below -300 mV. Stewart (1997) reviewed cultivation methods for detecting methanogens from the rumen. *Methanobrevibacter ruminantium*, *Methanomicrobium mobile*, *Methanosarcina mazei*, *Methanosarcina barkeri* and *Methanobacterium formicicum* were isolated from the rumen by cultivation.

In the rumen, methane is mainly synthesized from H_2 and CO_2. Other substrates, such as formate, acetate, methylamine and methanol also participate in methane production (Stewart *et al*., 1997). *M. ruminantium*, *M. formicicum* and *M. mobile* utilize H_2/CO_2 and formate to produce methane. On the other hand, *M. mazei* synthesizes methane from acetate, methanol and methylamines. *M. barkeri* utilizes H_2/CO_2, acetate, methanol and methylamines for methane synthesis (Jarvis *et al*., 2000).

3. DETECTION OF METHANOGENS FROM THE RUMEN BY PCR-BASED TECHNIQUES

The analysis of microbial ecosystems has changed more and more toward molecular biological methods. In these studies, ribosomal RNA and their genes (rDNAs) are used as reliable phylogenetic markers for the relatedness between isolated and uncultured bacteria (Wintzingerode *et al*., 1997). The small-subunit (SSU) rRNA gene (16S rDNA) types are most useful for this purpose. The 16S rDNAs, amplified by PCR, serve for the following analysis. Molecular genetic analysis of rDNA amplified from DNA extracted from natural habitats has been commonly used in many laboratories to investigate microbial ecosystems such as those in soil and marine environments (Colwell & Grimes, 2000). Although there are few reports concerning rumen microbial ecosystems, these reports show new aspects of methanogens in the rumen.

4. PCR AMPLIFICATION

PCR amplification has been used to obtain rDNA sequence data from microbial ecosystems (Wintzingerode *et al*., 1997). The 16S rDNAs can be amplified directly from extracted DNAs with a set of primers binding to conserved regions of the 16S rDNA.

The first step in PCR amplification is to prepare DNA templates from microbial ecosystems or pure cultures. In comparing between microbial ecosystems and pure cultures, several problems depend upon the method chosen for the microbial ecosystems. (i) Extracting DNA from the microbial ecosystems is not easy, because natural microbial resources contain various materials that are difficult to remove and that somewhat inhibit PCR amplification. Moreover, the efficiency of cell lysis in different bacterial species during the DNA extraction would reflect on the quantitative composition of the obtained DNA mixture. (ii) Differential PCR amplification would occur. Because extracted DNA, serving as a template, is a mixture of DNA from different bacterial species, the efficiency of PCR amplification would differ in each DNA from each different species. The underlying reasons for this are efficiency of primer-template formation, the number of rRNA gene regions (rrn operons) in the genome, etc. (Wintzingerode *et al*., 1997).

5. PCR PRIMERS FOR DETECTION OF METHANOGENS FROM THE RUMEN

Sets of primers for detecting methanogens from the rumen are listed in Table 1. Although set B targets methanogens, other sets aim at Archaea.

Table 1. Primers for the detection of methanogens from the rumen.

sets	primers	sequences	References
A	Ar1000	5'-AGTCAGGCAACGAGCGATTACGACTT	Yanagita *et al.*, 2000
A	Ar1500	5'-GGTTACCTTGTTACGACTT	Yanagita *et al.*, 2000
B	M301F	5'-TACGGGTTGTGAGAGCAAGA	Tokura *et al.*, 1999*a*
B	M915R	5'-TGCTCCCCCGCCAATTCCT	Tokura *et al.*, 1999*a*
C	1Af	5'-TCYGKTTGATCCYGSCRGAG	Whitford *et al.*, 2001
C	1100AR	5'-TGGGTCTCGCTCGTTG	Whitford *et al.*, 2001
D	D30	5'-ATTCCGGTTGATCCTGC	Tajima *et al.*, 2001
D	D33	5'-TCGCGCCTGCGCCCCGT	Tajima *et al.*, 2001
E	0025e	5'-CTGGTTGATCCTGCCAG	Tajima *et al.*, 2001
E	1429	5'-GGTTACCTTGTTACGACTT	Tajima *et al.*, 2001

The methanogens and archaea detected using the above primers are listed in Table 2. These microorganisms were detected from the molecular libraries constructed from DNA fragments, amplified with specific primers. It should be noted that different primer sets were able to amplify DNA fragments that exhibited different molecular diversities. Because the unidentified archaea in Table 2 would be methanogens, almost all archaea inhabiting the rumen would be methanogens. Since these unidentified methanogens were not detected by cultivation methods, they would be "viable but not culturable (VBNC)".

Although *M. ruminantium* and *M. mobile* were detected by both cultivation methods and PCR-based methods, *M. mazei*, *M. barkeri* and *M. formicicum* were detected by only cultivation. On the other hand, PCR-based methods were able to detect *M. smithii*, *M. stadtmanae*, and unidentified archaea. These results show that both methods have biases, i.e., tendencies to detect specific microorganisms.

Table 2. Methanogens detected by PCR-based methods.

primer set	detected microorganisms	sources
A	*Methnomicrobium mobile*, *Methanobrevibacter smithii*, *M. ruminantium*	sheep
B	*M. smithii*, *M. ruminantium*, *Methanosphaera stadtmanae*	sheep
C	*M. ruminantium*, *M. stadtmanae*, unknown methanogens	cow
D	*M. mobile*, unidentified archaea	cow
E	*M. mobile*, *Methanobrevibacter* sp., unidentified archaea	cow

It is observed that some methanogens closely associate with rumen ciliates. Defaunation or elimination of protozoa from the rumen could eliminate 30-45% of methane production from the rumen (Tokura *et al.*, 1999*a*). It is presumed that the rumen ciliates supply H_2 to methanogens closely associated with them. Tokura (1999*a*) demonstrated that the methanogen associated with rumen ciliates was *M. smithii* by a PCR-based method. However, those authors isolated strains of *M. ruminantium* from a ciliate fraction by a cultivation method (Tokura *et al.*, 1999*b*). Therefore, the methanogens associated with rumen ciliates would be *M. smithii* and *M. ruminantium*. According to an analysis of DNA extracted from rumen fluid, many sequences affiliated with *M. smithii* were detected from the sheep rumen, and many sequences affiliated with *M. ruminantium* were detected from the bovine rumen. It is unclear as to whether the difference in these results is caused by different samples, different primers, and /or other differences.

6. FUTURE OF THE PCR-BASED TECHNIQUES

Using PCR-based techniques, it has been clear what kinds of methanogens would be VBNC in the rumen, and rough compositions of these have been made. In most studies of PCR-based techniques, the analysis of 16S rDNA was accomplished after clone libraries were constructed. This analysis therefore consumed much time and cost, and was unable to analyze many samples at once. In order to analyze methanogens in the rumen in detail, it is necessary to solve these problems. For this purpose, several kinds of electrophoresis, such as denaturing gradient gel electrophoresis (DGGE), temperature gradient gel electrophoresis (TGGE), and single strand conformation polymorphism (SSCP), will apply to the analysis of the DNA fragments (Westermeier, 1997). Moreover, new sets of primers should be designed to detect wide or narrow ranges of methanogens.

REFERENCES

Colwell, R. R. & D. J. Grimes, 2000. Semantics and strategies. In: Nonculturable Microorganisms in the Environment, Colwell, R. R. & D. J. Grimes (editors), ASM Press, Washington, DC, USA. 1-6.

Jarvis, G. N., C. Strompl, D. M. Burgess, L. C. Skillman, E. R. B. Moore & K. N. Joblin, 2000. Isolation and identification of ruminal methanogens from grazing cattle. Curr. Microbiol. 40: 327-332.

Stewart, C. S., H. J. Flint & M. P. Bryant, 1997. The rumen bacteria. In: The Rumen Microbial Ecosystem, 2nd edition, P. N. Hobson & C. S. Stewart (editors), Blackie Academic & Professional, London, UK. 10-72.

Tajima K., T. Nagamine, H. Matsui, M. Nakamura & R. I. Aminov, 2001. Phylogenetic analysis of archaeal 16S rRNA libraries from the rumen suggests the existence of a novel group of archaea not associated with known methanogens. FEMS Microbiol. Lett. (in press).

Tokura, M., I. Chagan, K. Ushida & Y. Kojima, 1999*a*. Phylogenetic study of methanogens associated with rumen ciliates. Current Microbiology. 39: 123-128.

Tokura, M., K. Tajima & K. Ushida, 1999*b*. Isolation of Methanobrevibacter sp. as a ciliate-associated ruminal methanogen. J. Gen. Appl. Microbiol. 45: 43-47.

Westermeier, R., 1997. Electrophoresis in practice, 2nd edition. VCH Verlagsgesellshaft mbH, Weinheim, Germany. pp25-26.

Whitford, M. F., R. M. Teather & R. J. Forster, 2001. Phylogenetic analysis of methanogens from the bovine rumen. BMC Microbiology. 1: 5.

Wintzingerode, F. v., U. B. Gobel & E. Stackebrandt, 1997. Determination of microbial diversity in environmental samples: pitfalls of PCR-based rRNA analysis. FEMS Microbiol. Reviews. 21: 213-229.

Yanagita, K., Y. Kamagata, M. Kawaharasaki, T. Suzuki, Y. Nakamura & H. Minato, 2000. Phylogenetic analysis of methanogens in sheep rumen ecosystem and detection of *Methanomicrobium mobile* by fluorescence *In Situ* Hybridizaion. Biosci. Biotechnol. Biochem. 64: 1737-1742.

Greenhouse Gases and Animal Agriculture
J. Takahashi & B. A. Young (Editors)

Inhibition of rumen methanogenesis and its effects on feed intake, digestion, and animal production

C. S. McSweeney[1] & G. J. McCrabb[2]

[1]CSIRO Livestock Industries, Long Pocket Laboratories, Brisbane 4068; [2]Rendel Laboratory, Rockhampton, Queensland 4702, Australia. [2]Present address: International Livestock Research Institute, PO Box 5689, Addis Ababa, Ethiopia

Domesticated large ruminants are major contributors to global anthropogenic methane emissions. Apart from environmental benefits, a reduction in rumen methanogenesis is an attractive proposition in terms of productivity since 8 to 10% of the energy of the diet can be lost as methane. Animal trials involving chemical agents that specifically inhibit microbial enzymes associated with methane production probably provide the most reliable data for interpretation of the effects of inhibition of methanogenesis on digestive and animal performance parameters. Published studies indicate that a reduction in methanogenesis in the rumen may not affect feed intake but, depending on the level of methane reduction, the type of diet and physiological state of the animal, intake can be reduced by 0 to 13%. However at this stage it is not possible to accurately predict the effect of methane inhibitors on feed intake. Feed conversion efficiency increased by 0 to 11% in the majority of *in vivo* studies while only one study reported a significant reduction. Although feed:gain ratio declined in many cases, live weight gain tended to be 0 to 5% lower because of effects on intake even though improved weight gains of 10% have been reported. If increased growth rates and further improvements in feed efficiency are to be achieved, then the mechanism causing reduced feed intake in animals treated with chemical methane inhibitors needs to be determined. Complete inhibition of methanogenesis is likely to result in the accumulation of H_2 in the rumen with detrimental effects on microbial metabolism. However the use of technologies such as vaccines or chemical compounds which partially inhibit methanogenesis could have beneficial effects on animal production. The potential to utilise H_2 via an alternative pathway to methanogenesis in the rumen, such as reductive acetogenesis which converts H_2 and CO_2 into acetate, is also discussed. However, any attempt to reduce methane emissions from livestock is unlikely to be adopted unless production efficiency is at least maintained, if not enhanced.

Keywords: rumen, methanogenesis, feed intake, acetogenic bacteria, methane inhibitors

1. INTRODUCTION

Methane concentration in the atmosphere was rising at greater than 1% per year but recent data suggest that the rate of increase is levelling off at <1% per year (Augenbraun *et al.*, 1999). It has also been suggested that the contribution of methane to the "greenhouse" effect, is a threat to the sustainability of the present life systems on the planet (Hogan *et al.*, 1991). Livestock populations worldwide (excluding wild herbivores) are the major contributors to anthropogenic methane emissions with cattle and buffalo populations being responsible for 82% of the total emissions from all domesticated livestock species (USEPA, 1994). Methane emissions from wild ruminants and large herbivores have been estimated at less than 10% of the emissions from domesticated animals (Leng, 1993). Recent studies in respiration chambers have confirmed that methane emissions from ruminants fed on fibrous tropical diets are higher than outputs from better quality temperate forages (Kurihara *et al.*, 1999). Even greater reductions would probably accompany the feeding of mixed or high-grain diet.

Methanogenesis is a critical component of the energetically efficient process of anaerobic microbial degradation of organic material in diverse environments (e.g., the rumen, sludge, ponds, and rice fields). In ruminant animals, end products of rumen digestion such as short chain volatile fatty acids (VFA) are absorbed and used as sources of energy while the methane produced escapes to the atmosphere. The excretion of methane from the rumen can represent a loss of 8-10% of the energy of the diet (Blaxter and Clapperton, 1965) depending on the type of diet. Therefore reducing methane production could benefit the ruminant energetically provided the efficiency of ruminal metabolism is not compromised. CSIRO Livestock Industries is investigating the use of a vaccine against rumen methanogens as a strategy for reducing methane emissions (Sue Baker, pers. comm.). However any attempt to reduce methane emissions from livestock is unlikely to be adopted unless production efficiency is at least maintained if not enhanced. The challenge therefore is to devise strategies that reduce methane emissions to a level which also results in improved production efficiency from ruminants.

2. DOES IN VIVO INHIBITION OF METHANOGENESIS AFFECT ANIMAL PRODUCTION?

2.1. Inhibitors of methane in the rumen

Compounds that inhibit methanogenesis in the rumen have been extensively reviewed (see Nagaraja et al., 1997). These agents have three different modes of action: (1) selective inhibition of specific microorganisms, (2) alternative electron acceptors which reduce the extent of CO_2 reduction to methane, and (3) inhibition of the enzymes involved in the methyl transfer reaction. Ionophores and defaunating agents are specific inhibitors of Gram positive bacteria and protozoa respectively and one of the indirect responses is a reduction in methane production. These agents also have effects on nitrogen metabolism and the pathways of VFA production in the rumen. It is therefore difficult to apportion animal production responses to changes in methanogenesis in animal trials which involve these agents. Similarly, agents such as unsaturated long chain fatty acids (LCFAs) and sulfate, which act as alternative electron acceptors to CO_2, also have nutritional and metabolic effects on animal metabolism. Therefore animal trials involving agents which specifically inhibit enzymes involved in methane production are probably the most reliable data for interpretation of the effects of inhibition of methanogenesis on digestive and animal performance parameters.

2.2. Animal studies involving methane inhibitors

The most potent and well researched inhibitors of ruminal methane production are the halogenated methane analogues (e.g. bromochloromethane, BCM) and related compounds such as amichloral (a hemiacetyl of chloral and starch), chloroform and chloral hydrate.

Table 1. A summary of publications reporting the effects of methane inhibition, using bromochloromethane and hemiacetyl of chloral and starch, on feed utilization and animal performance in cattle and sheep.

Reference	Animal/diet	Inhibitor	Duration of Treatment	Feed Intake	Feed Digestibility	Daily Liveweight Gain	Feed Conversion Efficiency	Methane inhibition (%)
Johnson *et al.* (1972)	Cattle/ Fibrous by-products	BCM	28 days	NS	NS	NS	Not reported	50
Cole & McCroskey (1975)	Cattle/ Concentrate	HCS	120 days	9% reduced	Reduced (d30) NS (d120)	4% reduced	6% increased	30-40
Leibholz (1975)	Calves/ Roughage	HCS	11 weeks	3% reduced	NS	10% increased	7% increased	Not measured
McCrabb *et al.* (1997)	Cattle/ Roughage	BCM	12 weeks	10% reduced	Not measured	5% increased	11% increased	Not measured
Sawyer *et al.* (1971)	Lambs/ Diet not described	BCM	105 days	Not cited	Increase	12% reduced	5% reduced	85%
Johnson (1972)	Sheep Roughage: concentrate	HCS	2 weeks	NS	Not measured	Not measured	Not measured	50-82
Trei *et al.* (1971)	Lambs Roughage: concentrate	TCA	90 days	NS	Not measured	1% increase	NS	66
Trei *et al.* (1972)	lambs/ Concentrate	HCS	90 days	0-4% reduced	Not measured	5% increased	7% increased	39-54
Trei *et al.* (1972)	lambs/ Concentrate	HCS	90 days	NS NS NS	Not measured	NS 2% increased 1% increased	NS 6% increased 7% increased	27 39 54
Johnson (1974)	lambs/ Concentrate	HCS	30 days	13% reduced	Increased	Not reported	Not reported	64

BCM, bromochloromethane; HCS, hemiacetyl of chloral and starch (Amichloral); TCA, trichloroacetamide
NS, Not significantly affected ($P > 0.05$).

Chloroform and chloral hydrate are not reliable experimental tools since they can also have acute toxic effects on the animal. Studies, which have employed these compounds to inhibit methanogenesis in animals, are summarised in Table 1. Collectively the data are conflicting and difficult to interpret in relation to the benefits or detrimental effects of reduced methane production on feed intake, digestibility, feed conversion efficiency and liveweight gain. Some experiments show a definite improvement in animal performance when methane production is reduced while other studies have observed the opposite effect. On closer examination, the studies cited in Table 1 have been undertaken in juvenile to adult animals from different species fed diets which range in quality from concentrates to roughage. One would expect varied responses due to the different microbial populations and pathways of metabolism which would predominate under these varying conditions. Also most studies have been conducted with diets that are relatively low in dietary fibre compared with tropical forages.

A more recent study by McCrabb and co-workers (1997) has provided insight into the potential effects of reduced methane production on animal performance in cattle fed tropical roughage diets of low and medium quality. Methane production was not measured in that study but subsequent work indicates the degree of inhibition would have been about 25%. In this study feed intake was reduced by 7 to 9% on both the low and medium quality diet when animals had been fed the methane inhibitor for 2 months (Table 2). Even though feed intake declined, liveweight gain was not significantly affected because feed conversion efficiency increased by 4 and 11% respectively on the low and medium quality diets. Animal performance on the lower quality diet may have been confounded by a nitrogen deficiency in the diet.

Table 2. Effect of the methanogen inhibitor bromochloromethane (1.2g/100kg of live weight /day) on feed intake, live weight gain, and feed conversion efficiency in steers fed a low and medium quality roughage (McCrabb et al., 1997).

	Low quality			Medium quality	
	Control	Treatment		Control	Treatment
Roughage intake (g of DM/kg of LW)					
Week 1-4	18.0	17.8	Week 1-7	19.1	17.2**
Week 5-8	18.4	17.9	Week 8-10	17.7	16.1*
Week 9-12	18.5	17.2**			
Liveweight gain (kg/day)	0.23	0.22		0.59	0.62
Feed: gain (kg/kg)	21.1	20.3		8.0	7.1**

**, P<0.01; *, P<0.05

It can be concluded from published studies that a reduction in methanogenesis in the rumen may not affect feed intake but, depending on the level of methane reduction, the type of diet and physiological state of the animal, intake can be reduced by 0 to 13%. However, at this stage, it is not possible to accurately predict the effect of methane inhibitors on feed intake. Feed conversion efficiency increased by 0 to 11% in seven of the *in vivo* studies (Table 1) where this measurement was made while only one study (Sawyer *et al.*, 1971) reported a significant reduction. Although feed:gain declined in many cases, live weight gain tended to be 0 to 5% lower because of effects on intake, although improved weight gains of 10% have been reported (Leibholtz, 1975).

If increased growth rates and further improvements in feed efficiency are to be achieved, the mechanism causing reduced feed intake in animals treated with methane inhibitors needs to be determined.

3. CONSEQUENCES OF INHIBITING METHANOGENEIS ON RUMINAL METABOLISM

Formation of CH_4 is the sole energy generating process for methanogenic microorganisms and the main substrates used by rumen methanogens are H_2 and CO_2 (Hungate, 1967). In some anaerobic environments, CH_4 is produced from the slow metabolism of VFA but the relatively rapid turnover time of the rumen does not favour this reaction.

All the animal studies described in the previous section are examples of the effect of directly inhibiting methanogenic microorganisms. A consequence of this strategy is the accumulation of H_2 in the rumen that is a major metabolic end product of forage digestion. The management of H_2 accumulation in the rumen under these circumstances is a critical determinant of the efficiency of digestion and animal performance. In a normally functioning bovine rumen, approximately 800 litres of H_2 are produced daily but only traces of H_2 can be detected in rumen fluid due to the rapid conversion of this H_2 to about 200 litres of CH_4 (Hungate, 1967).

The consequences of inhibiting methanogens on rumen fermentation will depend mainly on the impact on populations of microorganisms that have metabolic activities that can be described by equation 1.

$$\text{Hexose} \longrightarrow 2\ \text{Acetate} + 4\ H_2 + 2\ CO_2 \qquad (1)$$

This metabolic pathway only occurs when certain important functional groups of microorganisms such as the major cellulolytic organisms *Ruminococcus albus*, *R. flavefaciens* and ruminal fungi are grown in the presence of methanogens.

The adverse effects of H_2 accumulation include inhibition of reoxidation of NADH, the accumulation of reduced fermentation end products such as lactate, and reduced efficiency of microbial growth (Wolin *et al.*, 1997). For example, the partial pressure of H_2 influences the metabolism of two of the important species (*R. albus and R.flavefaciens*) of fibre degrading bacteria in the rumen. Accumulation of H_2 (made subscript) inhibits NADH oxidation in both species that results in the diversion of hydrogen from the formation of ethanol and succinate to acetate with a concomitant decrease in yield of ATP per unit of hexose fermented (Wolin *et al.*, 1997). Other rumen bacteria also shift their fermentation pattern to acetate from more reduced end products such as propionate due to interspecies H_2 transfer with methanogens. These effects have been observed in ruminal bacteria grown in continuous culture on cellulose (Pavlostathis *et al.*, 1990) and thus support the notion that similar responses would be pronounced on fibrous diets in which methanogens were inhibited. The adaptive changes in rumen microbial ecology to inhibition of methanogens are relatively unknown although enhanced propionate production is a consistent response and consequence of disruption to interspecies hydrogen transfer (Wolin, 1975). It is also often predicted that inhibition of methanogenesis will reduce the efficiency of growth and fibrolytic activity of cellulolytic bacteria due to the accumulation of H_2. Unfortunately, the concentration of H_2 in the rumen and ruminal digestion has been rarely measured when an antimethanogen was used. Although many have predicted that fibrolytic microorganisms would decrease due to inhibition of methanogenesis there have been no quantitative estimates of these populations in animals treated with antimethanogenic compounds. Molecular ecology methods are now available to accurately measure these populations to the species level and thus should be employed to determine the effect of elevated H_2 partial pressure in the rumen (McSweeney *et al.*, 2000).

4. OPTIONS FOR UTILISING H_2 WHICH ACCUMULATES FROM REDUCED METHANOGENESIS

Joblin (1999) states that management of H_2 in the rumen is the most important factor to be considered when developing strategies to control ruminant methane emissions. Unfortunately the focus tends to be primarily on the amounts of methane produced rather than also considering the disposal of H_2 that is toxic to the rumen if it accumulates. However, a reduction of methane production by 20 to 50% is known to be achievable without side effects (i.e. by using bromochloromethane; McCrabb, 2000) and this should increase the energetic efficiency of digestion by 2 to 5%, and possibly more on high roughage diets. Therefore, an understanding of the microorganisms and main metabolic pathways in the rumen is helpful in devising strategies to manage proton flow in the rumen for reduced methanogenesis.

Methanogenesis in complete fermentation is performed by microorganisms that utilise acetate, H_2 and CO_2 produced by two other classes of bacteria (Bryant, 1979). Fermentative bacteria hydrolyse polysaccharides such as cellulose and hemicellulose, and degrade the products of these to organic acids, alcohols, H_2 and CO_2. These bacteria also ferment proteins and lipids, producing similar products. Another group, collectively called the H_2-producing acetogenic bacteria, obtain energy for growth by producing acetate and H_2, and sometimes CO_2, from the organic acids and alcohols produced in the first stage. In the final stage, methanogenic bacteria utilise the products of the first two stages; i.e., mainly H_2, CO_2, and acetate to produce CH_4 and CO_2 (Figure 1). However, the short retention time of digesta in the rumen means that the H_2-producing acetogenic bacteria and the acetate-catabolising methanogenic bacteria, which are slow growing, are not maintained at high numbers in the rumen (Genthner *et al.*, 1981). Acetate and other VFA accumulate and are absorbed from the digestive tract and utilised as major energy sources in ruminant animals. The H_2 that is produced from primary fermentation is, in effect, oxidised using CO_2 to produce the CH_4, which is an energetically efficient strategy for the anaerobic bacterial flora.

An alternative pathway to methanogenesis (equation 2) in the rumen is reductive acetogenesis (equation 3) that converts H_2 and CO_2 into acetate as outlined in the following equations:

Methanogenesis:

$$4\,H_2 + CO_2 \longrightarrow CH_4 + 2\,H_2O \qquad \Delta G_o' = -135.6\ \text{kJ} \qquad (2)$$

Acetogenesis:

$$4\,H_2 + 2\,CO_2 \longrightarrow CH_3COOH + 2\,H_2O \qquad \Delta G_o' = -104.6\ \text{kJ} \qquad (3)$$

Until recently, little attention was paid to this process as it was regarded as relatively insignificant in ruminal metabolism (see Joblin, 1999). However, reasonably high populations of reductive acetogens have since been discovered in cattle (Leedle and Greening, 1988; Greening and Leedle, 1989). It has been suggested therefore that acetogens may be able to occupy the niche of methanogens as terminal sinks for ruminal H_2 (Mackie and Bryant, 1994;

see Joblin, 1999). While this proposition seems feasible there are some reasons why methanogenesis predominates as the favoured pathway for disposal of H_2. Firstly, the conversion of H_2 and CO_2 into methane yields more energy and is thermodynamically more attractive than their metabolism to acetate (Thauer *et al.*, 1977; Cord-Ruwisch *et al.*, 1988). Secondly methanogens tend to have lower utilisation thresholds for H_2 than acetogens although some ruminal acetogens grow at thresholds below 1uM H_2 (Joblin, 1999). Thirdly, ruminal acetogens do not appear to be obligate hydrogenotrophs and can also use other compounds such as simple sugars as energy yielding substrates (Genthner *et al.*, 1981, Grenning and Leedle, 1989; Rieu-Lesme *et al.*, 1998; Joblin, 1999). Although reductive acetogenesis is dominated by methanogenesis in the rumen, this does not appear to be the case in hind gut fermentations (Demeyer and de DeGraeve, 1991). The ecological reasons for this difference in metabolism between fore- and hind-gut compartments are unknown and require examination. In the termite hindgut, acetogens may dominate by occupying a microniche that is different from that of the methanogens (Leadbetter *et al.*, 1999).

Promoting acetogenesis in the rumen while employing strategies to inhibit methanogens is an attractive approach to reduce methane emissions but currently our understanding of the ecology of reductive acetogens is limited. If we are to employ this strategy successfully we need a more complete understanding of the physiology and ecology of ruminal acetogens *in vivo*. Recent studies indicate that selected acetogens can lower H_2 concentration when methanogens are inhibited in rumen contents that are incubated *invitro* (Nollett *et al.*, 1998; Le Van *et al.*, 1998). It has been suggested that acetogenesis will only become a significant metabolic pathway in the rumen if methanogenesis is inhibited which will allow the partial pressure of hydrogen to increase above the required threshold for reductive acetogenesis (Nollett *et al.*, 1977). The native acetogens may then increase in number or it may be necessary to introduce selected acetogens that are more active. Fortunately ruminal acetogens are members of a broad taxonomic group, suggesting that they could be adaptable to a variety of ruminant species and feeding conditions (Joblin, 1999).

In conclusion, any attempt to reduce methane emissions from livestock is unlikely to be adopted unless production efficiency is at least maintained, if not enhanced.

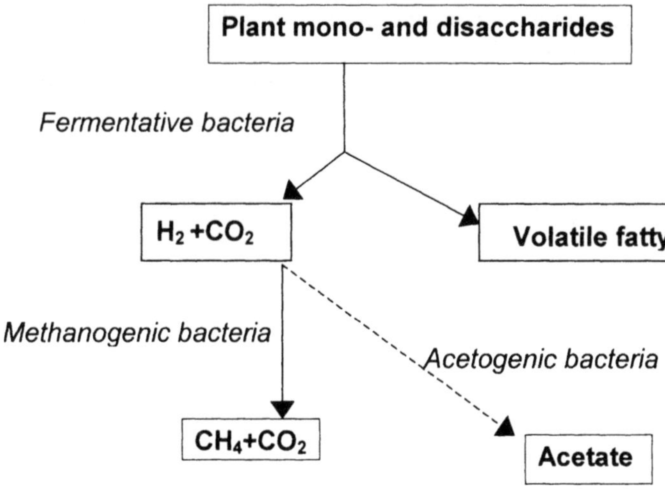

Figure1. Methanogenesis and acetogenesis in the rumen.

REFERENCES

Augenbraun, H., T. Smith, & E. Matthews, 1999. Assessing potential contributions to downward trend atmospheric methane concentrations. Retrieved 6th of July 2001 from the Institute on Climate and Planets Web site: http://www.icp.giss.nasa.gov/research/posters/ch4atmtrend.html

Blaxter, K. & J.L. Clapperton, 1965. Prediction of the amount of methane produced by ruminants. Br. J. Nutr. 19: 511-522.

Bryant, M.P., 1979. Microbial methane production – theoretical aspects. J. Anim. Sci. 48: 193-201.

Cole, N.A. & J.E. McCroskey, 1975. Effects of hemiacetyl of chloral and starch on the performance of beef steers. J. Anim. Sci., 41: 1735-1741.

Cord-Ruwisch, R., H-J. Seitz, & R. Conrad, 1988. The capacity of hydrogenotrophic anaerobic bacteria to compete for traces of hydrogen depends on the redox potential of the electron acceptor. Arch. Microbiol. 149: 350-357.

Demeyer, D.I. & K. De Graeve, 1991. Difference in stoichiometry between rumen and hind gut fermentation. J. Anim. Physiol. Anim. Nutr. 22: 50-61.

Genthner, .R., C.L. Davis, & M.P. Bryant, 1981. Features of rumen and sewage sludge strains of *Eubacterium limosum*, a methanol- and H_2-CO_2-utilizing species. Appl. Environ. Microbiol. 42: 12-19.

Greening, R.C. & J.A.Z. Leedle, 1989. Enrichment and isolation of *Acetitomaculum ruminis*, gen. Nov., acetogenic bacteria from the bovine rumen. Arch. Microbiol. 152: 399-406.

Hogan, K.B., Hoffman, J.S. and Thompson, A.M. (1991). Methane on the greenhouse agenda. Nature 354, 181-182.

Hungate, R.E., 1967. Hydrogen as an intermediate in the rumen fermentation. Arch. Microbiol. 59: 158-164.

Joblin, K.N., 1999. Ruminal acetogens and their potential to lower ruminant methane emissions. Aust. J. Agric. Res. 50: 1307-1313.

Johnson, D.E., 1972. Effects of a hemiacetyl of chloral and starch on methane production and energy balance of sheep fed a pelleted diet. J. Anim. Sci. 35: 1064-1068.

Johnson, D.E., 1974. Adaptional responses in nitrogen and energy balance of lambs fed a methane inhibitor. J. Anim. Sci. 38: 154-157.

Johnson, E.D., A.S. Wood, J.B. Stone, & E.T. Moran, 1972. Some effects of methane inhibition in ruminants (steers). Can. J. Anim. Sci. 52: 703-712.

Kurihara, M., T. Magner, R.A. Hunter, & G.J. McCrabb, 1999. Methane production and energy partition of cattle in the tropics. Br. J. Nutr. 81: 227-234.

Leadbetter, J.R., T.M. Schmidt, J.R. Graber & J.A. Breznak (1999). Acetogenesis from H_2 plus CO_2 by spirochetes from termite guts. Science 283: 686-689.

Leedle, J.A.Z. & R.C. Greening, 1988. Postprandial changes in methanogenic and acidogenic bacteria in the rumens of steers fed high- or low-forage diets once daily. Appl. Environ. Microbiol. 54: 502-506.

Leibholtz, J., 1975. Ground roughage in the diet of the early-weaned calf. Animal Production 20, 93-100.

Leng, R.A., 1993. Quantitative ruminant nutrition – a green science. Aust. J. Agric. Res. 44: 363-380.

Le Van, T. D., J.A. Robinson, J. Ralph, R.C. Greening, W.J. Smolenski, J.A.Z. Leedle, & D.M. Schaefer, 1998. Assessment of reductive acetogenesis with indigenous ruminal bacterium populations and *Acetitomaculum ruminis*. Appl. Environ. Microbiol. 64: 3429-3436.

Mackie, R.I., & M.P. Bryant, 1994. Acetogenesis and the rumen: sybtrophic relationships. In: Acetogenesis, H.L. Drake, Chapman and Hall, New York. 331-364.

McCrabb, G.J. 2000. The relationship between methane inhibition, feed digestibility and animal production in ruminants. Proceedings of the 2nd

International Methane Mitigation Conference, Novosibirsk, Russia, June 2000, (US Environmental Protection Agency, Washington DC and Russian Academy of Sciences, Novosibirsk, Siberia). 125 – 132.

McCrabb, G.J., K.T. Berger, T. Magner, C. May, & R.A. Hunter, 1997. Inhibiting methane production in Brahman cattle by dietary supplementation with a novel compound and the effects on growth. Aust. J. Agric. Sci. 48: 323-329.

McSweeney, C.S., Palmer, B., Bunch, R. and Krause, D.O. (2001). Effect of the tropical forage calliandra on microbial protein synthesis and ecology in the rumen. J. Appl. Microbiol. 90, 78-88.

Nagaraja, T.G., C.J. Newbold, C.J. Van Nevel, & D.I. Demeyer, 1997. Manipulation of ruminal fermentation. In: The Rumen Microbial Ecosystem, P.N. Hobson & C.S. Stewart, 2nd edition, Chapman and Hall, London. 523-632.

Nollett, L., D. Demeyer, & W. Verstraete, 1997. Effect of 2-bromoethanesulfonic acid and *Peptoctreptococcus prodoctus* ATCC 35244 addition on stimulation of reductive acetogenesis in the ruminal ecosystem by selective inhibition of methanogenesis. Appl. Environ. Microbiol. 63: 194-200.

Nollett, L., L. Mbanzamihigo, D. Demeyer, W. Verstraete, 1998. Effect of addition of *Peptoctreptococcus prodoctus* ATCC 35244 on reductive acetogenesis in the ruminal ecosystem after inhibition of methanogenesis by cell-free supernatant of *Lactobacillus plantarum* 80. Anim. Feed Sci. Technol. 71: 49-66.

Pavlostathis, S.G., T.L. Miller, & M.J. Wolin, 1990. Cellulose fermentation by continuous cultures of *Ruminococcus albus* and *Methanobrevibacter smithii*. Appl. Microbiol. Biotechnol. 33: 109-116.

Rieu-Lesme, F., C. Dauga, G. Fonty, & J. Dore, 1998. Isolation from the rumen of a new acetogenic bacterium closely related to *Clostridium difficile*. Anaerobe. 4: 89-94.

Sawyer, M.A., W.J. Hoover, & C.J. Sniffen, 1971. Effects of a methane inhibitor on growth and energy metabolism in sheep. J. Dairy Sci. 54: 792-793 (Abstract).

Thauer, R.K., K. Jungermann, & K. Decker, 1977. Energy conservation in chemotrophic anaerobic bacteria. Bacteriological Reviews 41: 100-180.

Trei, J.E., R.C. Parish, Y.K. Singh, & G.C. Scott, 1971?. Effect of methane inhibitors on rumen metabolism and feedlot performance of sheep. J. Dairy Sci. 54: 536-540.

Trei, J.E., G.C. Scott, & R.C. Parish, 1972. Influence of methane inhibition on energetic efficiency of lambs. J Anim. Sci. 34: 510-515.

United States Environment Protection Agency, 1994. International Anthropogenic Methane Emissions: Estimates for 1990. Office of Policy Planning and Evaluation, Washington DC.

Wolin, M.J., 1975. Interactions between the bacterial species of the rumen. In: Digestion and Metabolism in the Ruminant, I.W. McDonald & A.C.I. Warner, University of New England Publishing Unit, Armidale, Australia. 134-148.

Wolin, M.J., T.L. Miller, & C.S. Stewart, 1997. Microbe-microbe interactions. In: The Rumen Microbial Ecosystem, P.N. Hobson & C.S. Stewart, 2nd edition, Chapman and Hall, London. 467-491

Greenhouse Gases and Animal Agriculture
J. Takahashi & B. A. Young (Editors)

Reducing ruminal methane production by chemical and biological manipulation

H. Itabashi

Dept. of Biological Production, Tokyo University of Agriculture and Technology, Fuchu-shi,Tokyo, 183-8509, Japan

The substantial loss of feed energy as methane in ruminants has prompted research on the use of feed additives to reduce methanogenesis. These additives aimed to manipulate rumen fermentation by chemical method which inhibit the growth of methanogens directly, or serve as electron acceptors that is utilized by hydrogen utilizing bacteria. There may be also biological methods to reduce methanogenesis that mainly act by controlling the protozoal population, because methanogens corelate closely with protozoa. The use of probiotics may become important in inhibiting methanogens. In all cases, methane can be decreased by promoting a shift in fermentation toward propionic acid production in the rumen.

For evaluation of a methane-inhibiting compound, it is very important to develop strategies to mitigate methane production, without causing negative impacts on ruminant nutrition, such as depress in food intake and fiber digestion, increase in hydrogen gas output, and a possible adaptation of the microbes to the compound.

The importance of molecular method to analyse methanogens will also be discussed.

Keywords: methane, propionic acid, fumaric acid, protozoa, fiber digestion

1. INTRODUCTION

Methane produced during ruminal fermentation represents a substantial loss of energy to the ruminants, and it is also a significant source of greenhouse gas emission from animal industry. Therefore, reducing methane production by ruminants would have significant economical and environmental benefits (Johnson & Johnson, 1995 ; McAllister *et al.*,1996).

Commitment made by many governments to reduce greenhouse gases in Kyoto in 1997 have made it necessary to re-examine ruminal methane production with the objective of finding means of reducing this emissions. According to Kyoto Protocol, Japan has agreed to reduce the greenhouse gases emissions by 6% of the emissions occurring in 1990 until 2008-2012.

There are a variety of dietary and microbial factors influencing ruminal methane production. Changes in methane production caused by altering the composition of the ration and feeding method, or processing ration components, will not be discussed here.

Considerable research efforts have been devoted to the manipulation of rumen fermentation including reducing methanogenesis with the final aim of improving the productivity of ruminants. Thus, many chemical compounds have been tested as potential feed additives for ruminants on the basis of their direct or indirect effects on reducing methane formation in the rumen.

In this paper, the author discuss the reducion of methane emissions by two ways developed by our laboratory. One is a 'chemical' method which inhibit methanogenic bacteria directory, or act as electron acceptor, and enhance a shift in fermentation pattern toward propionic acid production. The other is a 'biological method' that eliminates or decreases the activity of rumen protozoa, because protozoa have close relationship with methanogens.

2. CHEMICAL CONTROL OF METHANE PRODUCTION

2.1. Use of toxic compounds to methanogen

It has been known that several compounds, including chlorinated methane analogues, trichloroethyl adipate, bromoethanesulfonic acid and pyromellitic diimide, are toxic to methanogenic bacteria (McAllister, 1996). However, the commercial use of these compounds as feed additives has not been possible because microbes adapt to them or they are too volatile. Recently, McCrabb *et al.*(1997) studied the effect of a chemical complex of bromochloromethane (BCM) and cyclodextrin (CD), which is chemically stable when added to feed. They reported that this BCM-CD compound can reduce methane production remarkably without any negative effect on fiber digestion, and we also confirmed this (Itabashi *et al.*, 1999). This compound is believed to inhibit methanogenesis by reacting with reduced vitamin B_{12} which inhibits cobamide-dependent methanogenesis (Chalupa, 1977). Analysis of molecular diversity of rumen bacteria showed remarkable decrease in archae by BCM-CD. However, there remain some doubts for the use of these halogenated analogues, because these compounds may have ill effects on environmental problems and animal reproductive function.

We sought natural and safe substances to reduce methane production without any harmful effects on environment. One example is horse-radish. It was found that addition of horse-radish-CD can reduce methanogenesis (Fig.1), and increase ruminal propionic production in vitro and in vivo. Digestibility of dry matter and number of ruminal protozoa in cattle were not affected, suggesting that this compound can be recommended for practical use as inhibitors of methane production. Addition of allylisothiocyanate, which is main effective component of horse-dadish, also showed reducing effect on methane production. Thus the reduction in methane production observed when horse-radish is added in the diet results primarily from inhibitory effect of allylisothiocyanate on methanogens. The suitable level of horse-radish in the diet without increasing in hydrogen gas remains to be studied (Fig.1).

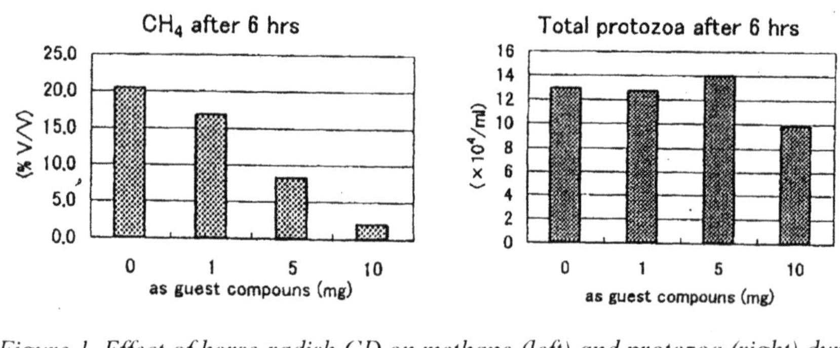

Figure 1. Effect of horse-radish-CD on methane (left) and protozoa (right) during 6-h.

2.2. Use of alternative electron acceptors

Methane production can be reduced by a decrease in the formation of H_2, which is the major substrates for methanogenesis. However, direct inhibition of this reaction may depress fermentation by cellulolytic bacteria that produce H_2, because this reaction is a means for the disposal of electrons that are liberated from energy-yielding substances (Hino *et al*., 1995). Accordingly, a reduction of H_2 formation by enhancement of reactions that accept electrons and a shift in the fermentation toward propionic acid production are desired (Asanuma *et al*., 1999).

It has been shown that many compounds, such as sulphate, nitrate, long chain free unsaturated fatty acids, can reduce methane production as alternate electron acceptors. These compounds direct electron flow toward microorganisms other than methane bacteria. This approach is further supported by the fact that reduction of nitrate and sulphate are thermodynamically favoured over carbon dioxide reduction. However, intake of nitrate often exceeds the reductive capacity of rumen microbes, and excess nitrate became a cause of nitrate poisoning in ruminants, and nitrate and sulphate may depress feed intake (Nakamura *et al*., 1981). Therefore, these compounds, even as alternate electron acceptors, cannot be used as practical methane inhibitors.

Recently, some dicarboxylic acids (e.g., aspartic acid, fumaric acid, malic acid) have been studied in vitro as feed additives for ruminants(Callaway & Martin, 1996, Lopez *et al*., 1999). Fumaric acid is an intermediate in the propionic acid pathway, in which it is reduced to succinic acid by fumaric acid reductase. In this reaction, reducing equivalents are needed and therefore fumaric acid may provide an alternative electron sink for hydrogen. As hydrogen is used to reduce fumaric acid, it may be possible to decrease the utilization of H_2 for methanogenesis.

It was found that the addition of fumaric acid in vitro can decrease methane production by 6% (Lopez *et al*., 1999), and 10% (Asanuma *et al*.,1999), depending on the amount of fumaric acid added. However, little information is available on the actual effects of fumaric acid on rumen digestion and animal performance in vivo. Isobe & Shibata (1993) observed that rumen fermentation in goats was enhanced by fumaric acid, increasing propionic acid formation as well as cellulose digestion.

We have firstly studied the effect of fumaric acid on energy metabolisim including methane production in vivo (Erudenbayaru *et al*., 2001). Methane production was reduced by about 20% when roughage alone was fed to cattle, and by about 5% when roughage-concentrate was fed. Methane output pattern in the experiment fed roughage alone are shown in Fig.2. In both diets, ruminal propionic acid formation was increased by the addition of fumaric acid, indicating that this organic acid act as an effective hydrogen sink, competing with methanogenesis.

These results also showed that more positive effects were related to the use of high-roughage rations which normally give rise to large quantities of methane and low proportions of propionic acid in the rumen. And, in the case of roughage-concentrate diets, the digestibility of fiber increased, which is in agreement with the result of Isobe & Shibata (1993).

We also tried to enhance the effect of fumaric acid on methane production by using ionophore (salinomycin) in cattle. When salinomycin alone was added, methane production decreased by about 15% and ruminal propionic acid production increased, whereas retained energy, digestion of fiber and number of ruminal protozoa were decreased. When salinomycin plus fumaric acid were added, more effect on reduction of methane production was not observed, however, digestion of fiber and number of protozoa were increased.

There were other benefical effects of fumaric acid. Fumaric acid also increased the concentration of plasma glucose and milk protein synthesis in dairy cows (unpublished data), which has to be attributed to incresed propionic acid production in the rumen.

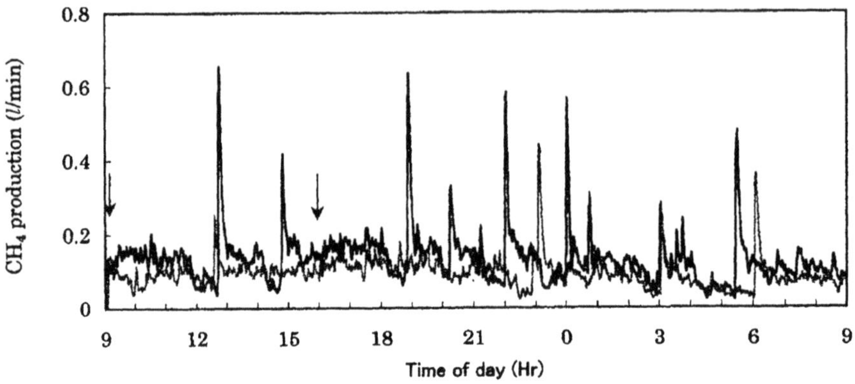

Figure 2. Effect of fumaric acid on methane production in cattle fed silage alone.
(Time of feeding, Control, Fumaric acid added)

It was concluded that fumaric acid may be put to practical use for ruminants diet, because the administration of fumaric acid have a benefical effects through decreased methane production, increased plasma glucose concentration and net energy retention.

3. BIOLOGICAL CONTROL OF METHANE PRODUCTION

Ruminal methanogenic bacteria corelate with hydrolytic microorganisms that produce hydrogen. There are such potent hydrogen producers as Ruminococci, Eubacteria, Clostridia, anaerobic fungi and ciliate protozoa. Among them protozoa are the most potent producers. They may produce up to 5 nmoles of hydrogen/cell/day, and it is estimated that about 50 liters/day of hydrogen were produced by protozoa of sheep (Ushida *et al.*, 1997). Methanogenesis from this electron donor may explain a large part of the source of methane in the rumen.

Hydrogen transfer to the methanogens from protozoa is defined as facultative interspecies hydrogen transfer. The cell surface of protozoa is often colonized by methanogens (Vogels *et al.*, 1980). Stumm *et al.*(1982) showed that up to 20% of the methanogens in the rumen may be associated with protozoa. Finlay *et al.*(1994) reported that the methanogens might occupy up to 1% of the internal volume of protozoa, and calculated that protozoa could account for 37% of the total methane production. Recently, Newbold *et al.*,(1995) showed that methanogens associated with rumen protozoa were apparently responsible for between 9 and 25% of methanogenesis in the rumen.

Therefore, removal of protozoa, defaunation, can reduce methane production in vivo (Itabashi *et al.*, 1984, ; Whitelaw *et al.*,1984). Defaunation markedly increased the total number of ruminal bacteria, but decreased the number of methanogens (Arakaki *et al.*, 1994). As the role of each protozoan genus on methanogenesis may not be uniform, ciliate protozoa may affect methanogens in a genus-dependent manner. Thus, the number of methanogens was lower in defaunated calves than in mono-faunated or mixed population-faunated ones. The effect of refaunation was significant except for the *Epidinium*-monofaunated calves (Takenaka & Itabashi. 1995). This might be related to the findings of Vogels *et al.*(1980), in which they reported that the rate of attachment of methanogen to the *Epidinium* sp. was lower than to those with other protozoan genera.

We have also studied the effect of each protozoan genus on methanogenesis of calves. Methane production was the lowest in defaunated followed by *Epidinium*+

Eudiplodinium-bifaunated, *Entodinium*-monofaunated, *Entodinium*+*Eudiplodinium*-bifaunated and *Eudiplodinium*-monofaunated. The highest methane production was recorded in the calves harbouring a mixed protozoan population (Fig.3). Thus, ruminal methane production appears to be influenced by the generic composition of the protozoa population (Itabashi *et al.* 1994).

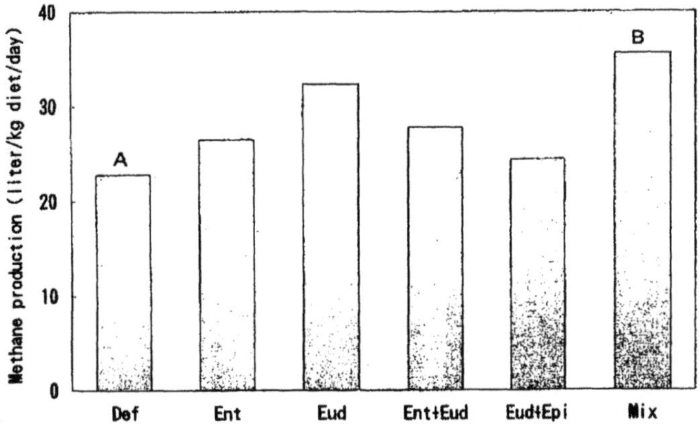

Figure 3. Effect of each genus of protozoa on methane production in calved fed hay and concentrate. (Def,defaunated ; Ent,Entodinium ;Eud,Eudiplodinium maggii; Epi, Epidinium ;Mix.Mixed protozoa)

Lastly it must be stressed that protozoa can be benefical to the ruminants because they stimulate fiber digestion, maintain ruminal homeostasis and so on (Jouany & Martin,1997). As defaunation may depress fiber digestion, complete elimination of protozoa is not recommended as a method for reducing methane. Therefore control of fauna composition or 'partial defaunation' (Ivan, private communication) rather than total removal of protozoa may have the potential to reduce methane production without depression fiber digestion in the rumen.

We are now seeking natural and safe substances to reduce the number of protozoa, such as medium chain fatty acid and some tropical feed. Further investigations are needed to establish the effects of probiotics or acetogenic bacteriaq on ruminal methane production.

REFERENCES

Arakaki,C.,M.Mitsumori,H.Itabashi,S.Shirasaka, & H.Minato,1994. Influence of the presence of protozoa on the rumen microbial population of cattle. J.Gen. Appl. Microbiol.,40:215- 226.

Asanuma, N., M.Iwamoto, & T.Hino, 1999. Effect of the addition of fumarate on methane production by ruminal microorganisms in vitro. J.Dairy Sci.,82:780-787.

Callaway,T.R. & S.A.Martin,1996. Effects of organic acid and monensin treatment on in vitro mixed ruminal micro-organisms fermentation of cracked corn. J.Anim.Sci.,74:1982-1989.

Chalupa,W., 1977. Manipulating rumen fermentation. J.Anim.Sci.,46:585-599.

Erudenbayaru,S.Kanda,T.Nishida,S.Ando,M.Ishida,T.Itoh,Y.Isobe,K. Nagara, & K.Hara, H.Itabashi, 1999. Effect of fumaric acid or bromochloromethane-CD on ruminal methanogenesis in cattle. S.Afr.J. Anim. Sci., 29:138.

Finlay,B.J.,G.Esteban,K.J.Clarke,A.G.Williams,T.M.Embley & R.P.Hirt, 1994. Some rumen cilaites have endosymbiotic methanogens. FEMS Microbiology Letters. 117:157-162.

Hino,T., T.Miwa, H.Saitoh, H.Esaki, S.Takagi, N.Asanuma & J.Umemori, 1995. Effect of methane production on plant fiber digestion by rumen microorganisms. Bull.Agric.Meiji Univ., 106:27-33.

Itabashi,H., T.Kobayashi & M.Matsumoto, 1984. The effects of rumen ciliate protozoa on energy metabolism and some constituents in rumen fluid and blood plasma of goats. Jpn.J.Zootech.Sci.,55:248-256.

Itabashi,H., Y.Washio, A.Takenaka, S.Oda & T.Ishibashi, 1994. Effects of a controlled fauna on methanogenesis, fibre digestion and rate of growth in calves. Proc.Soc.Nutr. Physiol., 3: 179.

Itabashi,H.,T.Takahashi,Erudenbayaru,S.Kanda,T.Nishida,S.Ando,M.Ishida,T.Itoh,Y.Isobe,K Nagara, & K.Hara, 1999. Effect os fumaric acid or bromochloromethane-CD on ruminal methanogenesis in cattle. S.Afr.J. Anim. Sci., 29:138.

Isobe,Y. & F.Shibata,1993. Rumen fermentation in goats administered fumaric acid. Anim. Sci.Technol.,64:1024-1030.

Johnson,K.A. & D.E.Johnson,1995. Methane emissions from cattle. J.Anim.Sci., 73:2483-2492.

Jouany,J.P. & C.Martin,1997. Effect of protozoa in plant cell wall and starch digestion in the rumen. In: Rumen microbes and digestive physiology in ruminants. Onodera,R. H.Itabashi, K.Ushida,H.Yano, & Y.Sasaki, Japan Sci. Soc. Press, Tokyo /S.Karger, Basel.11-24.

Kandylis,K.,1984. The role of sulphur in ruminant nutrition. A review. Liv. Prod. Sci.,11:611-624.

Lopez,S.,C.Valdes,C.J.Newbold, & R.J.Wallace, 1999. Influence of sodium fumarate addition on rumen fermentation in vitro, Br.J.Nutr.,81:59-64.

McAllister,T.A.,E.K.Okine,G.W.Mathison, & K.J.Cheng, 1996. Dietary, environmental and microbiological aspects of methane production in ruminants, Can.J.Anim.Sci.,76:231-243.

McCrabb,G.J.,K.T.Berger,T.Magner,C.May, & R.A.Hunter, 1997. Inhibiting methane production in Brahman cattle by dietary supplementation with a novel compound and the effects on growth. Aust.J.Agric.Res.,48:323-329.

Nakamura,Y.,Y.Tada,T.Saito,J.Yoshida, & R.Nakamura,1981. Nitrate metabolism of microorganisms in the rumen of sheep fed high nitrate Italian ryegrass silage. Jpn.J. Zootech. Sci.,52:512-518.

Newbold,C.J.,B.Lassalas,& J.P.Jouany,1995.The importance of methanogens associated with ciliate protozoa in ruminal methane production in vitro. Lett. Appl. Microbiol.,21:230-234.Stumm,C.K., H.J.Gijzen & G.D.Vogels, 1982. Association of methanogenic bacteria with ovine rumen ciliates. Br.J.Nutr.47:95-99.

Takenaka,A.& H.Itabashi,1996. Changes in the population of some functional groups of rumen bacteria including methane bacteria by changing the rumen ciliate in calves. J.Gen. Appl. Microbiol., 41:377-387.

Ushida,K., M.Tokura, A.Takenaka & H.Itabashi,1997. Ciliate protozoa and ruminal methanogenesis. In : Rumen microbes and digestive physiology in ruminants. Onodera,R., H.Itabashi,K.Ushida,H.Yano, & Y.Sasaki, Japan Sci. Soc. Press, Tokyo /S.Karger, Basel. 209-220.

Vogels,G.D.,W.F.Hoppe, & C.K.Stumm,1980. Association of methanogenic bacteria with rumen ciliate. Appl.Environ. Microbiol.,40 : 608-612.

Whitelaw,F.G.,J.M.Eadie,L.A.Bruce & W.J.Shand,1984. Methane formation in faunated and ciliate-free cattle and its relationship with rumen volatile-fatty acid proportions. Br.J.Nutr., 52:261-275.

Greenhouse Gases and Animal Agriculture
J. Takahashi & B. A. Young (Editors)

Animal and manure-derived methane emissions as affected by dietary fatty acids and manure storage system

M. Kreuzer[1], F. Dohme[1], D. R. Külling[2], F. Sutter[1], P. Lischer[3] & H. Menzi[2]

[1]Institute of Animal Sciences, Animal Nutrition, ETH Zurich, ETH Zentrum/LFW, CH-8092 Zurich, Switzerland; [2]Federal Research Station of Agroecology and Agriculture, Liebefeld, CH-3003 Bern, Switzerland; [3]Constat, CH-3095 Spiegel, Switzerland

Twelve dairy cows were fed forage-concentrate (3:2) based rations which either were supplemented with 40 g/kg stearic acid (C_{18}; considered to be ineffective against CH_4) or with the same amount of lauric acid (C_{12}; assumed to suppress ruminal methanogenesis). CH_4 release was recorded in cows (respiratory chambers), complete slurry and, separately, urine-rich slurry and farmyard manure stored for 14 weeks. CH_4 release of the cows was lower in the C_{12}-fed cows than with C_{18} by about 20%, but the corresponding CH_4 emission from their manure was higher. Manure-derived CH_4 on average accounted for 8.2% and 15.4% of total CH_4 production with 7 and 14 weeks of storage, respectively. Total system greenhouse gas potential (CH_4 from enteric fermentation and manure, and manure N_2O) accounted for 8.7 and 10.5 kg equivalents of CO_2/cow/d with C_{12} and C_{18}, respectively. In conclusion, successful dietary CH_4 suppression has to be effective both in the animals and in the manure.

Keywords: methane, diet, fatty acids, dairy cows, manure storage

1. INTRODUCTION

Domestic livestock is estimated to release 80 Tg/a of CH_4 with another 25 Tg/a being emitted from their manure (Chynoweth, 1996). In order to cope with the challenge to reduce CH_4 emission from ruminants, various dietary mitigation strategies are investigated (Moss et al., 2000). As one promising natural option the use of lipids was described, and medium-chain fatty acids and fats thereof evolved to be particularly efficient against ruminal CH_4 synthesis (cf. Dohme et al., 2001a). It is, however, not yet clear whether CH_4 mitigation strategies, effective in enteric fermentation, are subject to a compensatory higher CH_4 release from manure storage. The objective of the present study was to determine the effects of a medium-chain saturated fatty acid on direct animal-derived and manure-derived CH_4 emission and to assess the effect of the manure storage system (complete slurry vs urine-rich slurry/farmyard manure) and duration (7 vs 14 weeks). In order to be able to quantify the overall effects on global warming potential also the N_2O emission from manure was followed.

2. MATERIALS AND METHODS

At the ETH research station Chamau, 12 early-lactating Brown Swiss cows were subjected to two dietary treatments, the medium-chain fatty acid lauric acid (C_{12}) or the long-chain fatty acid stearic acid (C_{18}), supplemented at a level of 40 g/kg feed dry matter (DM) each. The complete diets were composed of forage (grass and maize silage, hay) and concentrate (including 3% Ca-rich mineral premix) in a ratio of 3:2. The diets were equivalent in supply of absorbable protein, crude protein and net energy. Each animal performed a 10-day control period on the C_{18} diet and a 15-day period receiving either the C_{12} or the C_{18} diet. During the last 6 days faeces and urine were quantitatively collected by the use of urinals attached with Velcro straps glued onto the clipped skin with special adhesive. Gaseous exchange of the cows was measured in open-circuit respiratory chambers on days three and four of the collection period. Further details on the cow trial are described in Dohme et al. (2001b).

Two typical farm manure collection and storage systems were simulated, the slurry system as applied in cowsheds with slatted floors collecting all excreta in the slurry, and the traditional urine-rich slurry/farmyard manure system. Urine-rich slurry contained 90% of urine and 10% of faeces, and the remainder formed the farmyard manure. The manure storage experiment was carried out as described in detail by Külling et al. (2001) on mixed faeces and urine collected from half of the animals per treatment. Both slurries were mixed and diluted with water (1:2) and farmyard manure was supplemented with chopped straw. The slurries were stored at 20°C and 70% humidity while the farmyard manure containing buckets were placed into a heated bath following temperatures declining as is typical for stacked manure. A closed chamber was used to monitor CH_4 and N_2O emission rates over 14 weeks. Methane in air samples was analysed with gas chromatography and N_2O with diode laser spectroscopy. Compositional analyses of feed and faeces were performed by standard procedures. Data were subjected to analysis of variance. Dietary treatment was considered as effect in the animal experiment, while in the manure storage experiment additionally manure type and the interaction of diet × manure type were regarded. Furthermore, diet effects within manure type were evaluated by monofactorial analysis of variance. Tables display means and standard errors of the means (SEM).

3. RESULTS AND DISCUSSION

Table 1 gives the effects of C_{12} compared to C_{18} in the cows. Although cows of both dietary treatments were offered the same amount of feed, cows of the C_{12} group refused about half of the concentrate due to palatability reasons and therefore consumed 3.3 kg/d DM less than the C_{18}-fed cows. Since only concentrate was refused, NDF intake was relatively less affected than total feed intake. The C_{12}-fed cows digested NDF slightly less efficiently than C_{18}-fed cows (0.536 vs 0.576) and their faeces had a significantly higher NDF content in DM. Considering certain initial differences in milk yield between the cows allocated to the C_{12} and C_{18} group (30.8 vs 28.7 kg milk/d), performance was also reduced by C_{12}. CH_4 release was lower by 21% with C_{12} than with C_{18} but was equal per unit of fibre digested. Relative to DM intake and milk yield, the decline in CH_4 release with C_{12} compared with C_{18} was 3% and 19%, respectively. Despite the recently described efficacy of medium chain fatty acids (Dohme et al., 2001a), C_{12} failed to exhibit a clear direct CH_4-suppressing effect in the present study. The lower CH_4 release of the C_{12}-fed cows mainly resulted from the suppression of concentrate intake. The reasons for the lack of the direct effect of C_{12} in methanogenesis remained unclear although interactions with carbohydrates and minerals (Ca), possibly binding and inactivating C_{12}, cannot be excluded.

Table 1. Treatment differences in cow-related traits variables (n=6)

Dietary treatment	C_{18}	C_{12}	SEM
Intake and performance			
Dry matter intake (DM, kg/d)	17.7^a	14.4^b	0.85
NDF intake (kg/d)	5.8	5.1	0.31
N intake (g/d)	449^a	354^b	26.8
Milk yield (kg/d; energy corrected)	28.0	27.0	1.51
Excretion			
Faeces & urine (kg/d)	50.1	46.5	3.81
Faecal NDF (kg/d)	2.4	2.4	0.13
NDF content (g/kg)			
in faecal DM	430^b	522^a	17.4
in complete faeces & urine	49	52	2.9
CH_4 release			
g/d	385^a	304^b	15.1
g/kg DM intake	21.9	21.3	1.08
g/kg digested NDF	118	112	7.81
g/kg milk	14.0^a	11.4^b	0.76

Over a storage period of 14 weeks, higher CH_4 emission rates in all manure types derived from the C_{12}-fed cows compared with the C_{18}-fed cows occurred. Furthermore, CH_4 release from C_{12} and C_{18} treatment manures showed a different evolution with storage time. It started earlier with C_{18} in the slurries and, less clear, in farmyard manure. This was later reversed. When related to the faeces and urine amount as daily excreted by the cows, the diet effects were clearly expressed in complete slurry after 14 weeks of storage but effects were much less pronounced after 7 weeks of storage (Table 2). In the urine-rich slurry/farmyard manure system, CH_4 emissions were lower during 14 weeks of storage than in slurry. Combining enteric and manure emissions over 14 weeks of storage, CH_4 release was lower by 26 g/cow/day (5.8%) with C_{12} than with C_{18} in the complete slurry system. The corresponding reduction was 90 g cow^{-1} day^{-1} (21.3%) in the two-component system. Accordingly, the reduction found in animal-derived CH_4 emission was mostly compensated by manure CH_4 release in complete slurry and partially in the two-component system. The proportions of total CH_4 emission originating from manure storage over 7 and 14 weeks were about 8% and 15%, respectively. CH_4 formation in ruminants is mainly the result of fibre fermentation. The increase in fibre content of the faeces with C_{12} in the present study could explain the diet effects on CH_4 release from manure. Also the type of manure storage had an effect on CH_4 release. Initially farmyard manure was a more important source of CH_4 than the two types of slurry similar to previous findings (Külling et al., 2001), with higher values in the two-component system than with complete slurry. However, the ranking changed around week 7.

Table 2. Total emission of greenhouse gases from enteric fermentation and manure stored for 7 and 14 weeks and related to the daily amount of faeces and urine per cow (n= 4)

Type of manure	Complete Slurry		Slurry & farm-yard manure		Effect of			
Type of diet	C_{18}	C_{12}	C_{18}	C_{12}	Diet	Manure	D×M	SEM
CH_4 (g/cow/d)[c]								
Manure (7 weeks)	27	31	11	20	*	ns	*	9.0
Manure (14 weeks)	42[b]	115[a]	19	28	***	***	***	14.2
Total (7 weeks)	430[a]	335[b]	413[a]	324[b]	***	ns	*	9.0
Total (14 weeks)	445	419	422	332	***	***	***	14.2
CH_4 originating from manure (% of total)								
7 weeks stored	6	9	5	12	**	ns	*	2.2
14 weeks stored	10[b]	27[a]	9[b]	16[a]	***	***	***	2.9
N_2O (g/cow/d)								
Manure (7 weeks)	0.1	0.1	1.5	1.0	ns	**	ns	1.21
Manure (14 weeks)	4.4	3.7	4.7	1.9	**	**	*	1.77
Greenhouse gas equivalents of CO_2 (kg/cow/d)[d]								
Total (7 weeks)	9.0[a]	7.1[b]	9.1[a]	7.1[b]	***	**	ns	0.49
Total (14 weeks)	10.7	9.9	10.3	7.6	***	ns	**	0.71

[a,b] Means with unequal superscripts within manure type are significantly different at $P < 0.05$.
[c] Total CH_4 comprises manure- & animal-derived methane (403 and 304 g methane/cow/d with C_{18} and C_{12}, respectively; only animals whose manure was subjected to storage).
[d] Considering animal and manure-derived CH_4 ($21 \times CO_2$ potential) as well as N_2O ($230 \times CO_2$ potential) from manure (IPCC, 1996).

N_2O emissions were low in urine-rich slurry and remained unaffected by the diet as also in complete slurry, however at a N_2O emission rate 43 times higher than in urine-rich slurry. The level of N_2O emission was highest in farmyard manure. When related to the daily amount of excreta per cow, N_2O emissions were similar in all treatments except in the urine-rich slurry/farmyard manure system with slurry of C_{18}-fed cows (Table 2). Overall, there was a certain compensation of differences between diets and between manure systems (after 14 weeks of storage) in CH_4 and in N_2O emissions in terms of overall global warming potential due to their mostly contrary variation with treatment. Greenhouse gas equivalents (GHGeq) were still lower in the C_{12} treatment than with C_{18}. GHGeq differences between types of manure were less clear. N_2O emissions made up about 50% of manure-derived GHGeq during 14 weeks of manure storage but only accounted for 10% of total GHGeq. At a storage time of 7 weeks, the N_2O contribution to total GHGeq was minor in the urine-rich slurry/farmyard manure system and negligible in complete slurry. This is contrary to Johnson et al. (2001)

who calculated that N_2O is 1.2fold more important for GHGeq than CH_4 in U.S. livestock production systems. However, their study assumed high N_2O losses during grazing.

In conclusion, data indicate that there could be a tendency for a compensatory increase in CH_4 formation in manure in cases of reduced enteric CH_4 formation when associated with a reduced fibre fermentation. It still remains open whether or not a compensation also takes place in situations with a CH_4 suppression independent from faecal fibre output. Irrespective of manure type and storage duration, enteric CH_4 release was obviously of greater importance for the global warming potential of the dairy husbandry systems investigated than CH_4 and N_2O emissions during manure storage. However, the contribution of these two trace gases for GHGeq increases with storage time and has to be considered in a more integrated and holistic way in assessments of the efficacy of dietary CH_4 mitigation strategies.

REFERENCES

Chynoweth, D. P., 1996. Environmental impact of biomethanogenesis. Environ. Monitor. Assess. 42: 3-18.

Dohme, F., A. Machmüller, A. Wasserfallen & M. Kreuzer, 2001a. Ruminal methanogenesis as influenced by individual fatty acids supplemented to complete ruminant diets. Letters Appl. Microbiol. 32: 47-51.

Dohme, F., F. Sutter, A. Machmüller & M. Kreuzer, 2001b. Methane formation and energy metabolism of lactating cows receiving individual medium-chain fatty acids. In: Energy Metabolism in Animals, A. Chwalibog & K. Jakobsen, eds., EAAP Publ. 103, Wageningen Pers, Wageningen, The Netherlands, 369-372.

Külling, D. R., H. Menzi, T. F. Kröber, A. Neftel, F. Sutter, P. Lischer & M. Kreuzer, 2001. Emissions of ammonia, nitrous oxide and methane from different types of dairy manure during storage as affected by dietary protein content. J. Agric. Sci., Camb., in press

IPCC, 1996. Climate Change 1995 - The Science of Climate Change, J. T. Houghton, G. J. Jenkins & J. J. Ephraums, eds., Cambridge University Press, UK.

Johnson, D. E., A. F. Seidl & H. W. Phetteplace, 2001. Methane, nitrous oxide and carbon dioxide emissions from U.S. beef production systems. In: Energy Metabolism in Animals, A. Chwalibog & K. Jakobsen, eds., EAAP Publ. 103, Wageningen Pers, Wageningen, The Netherlands, 161-164.

Moss, A. R., J.-P. Jouany & J. Newbold, 2000. Methane production in ruminants: its contribution to global warming. Ann. Zootech. 49: 231-253.

Propionate precursors as possible alternative electron acceptors to methane in ruminal fermentation

C. J. Newbold[1], J. O. Ouda[2,4], S. Lopez[3], N. Nelson[1], H. Omed[4], R. J. Wallace[1] & A .R. Moss[2]

[1]Rowett Research Institute, Bucksburn, Aberdeen, AB21 9SB, UK, [2]ADAS Feed Evaluation and Nutritional Sciences, Alcester Road, Stratford-upon-Avon, CV37 9RQ, UK, [3] Dept. Produccion Animal, University of León, E-24071, LEÓN, Spain, [4]University of Wales, Bangor, Gwynedd, LL57 2UD, UK.

Precursors of propionate in ruminal fermentation were tested for their ability to decrease methane production by ruminal fluid *in vitro*. Acrylate and fumarate had the most consistent effects with 50% of the added acids being fermented to propionate and methane production decreasing by between 8 and 14%.

Acrylate and fumarate were evaluated further in Rusitec. Again, both compounds weere fermented to propionate, but the ability of fumarate (44% capture of hydrogen based on the equation $4H_2 + CO_2 = CH_4 + 2H_2O$ and the reduction of fumarate to succinate by 2H) to divert metabolic hydrogen from methane to propionate was almost twice that of acrylate (22% capture of hydrogen).

Four diets comprising grass hay supplemented with fumaric acid at a rate of 0, 20, 40 and 80 g /kg DM were offered at the maintenance plane of nutrition to wether sheep in a 4 x 4 Latin square design. Methane production was measured in open-circuit respiration chambers after 17 d adaptation to the diet. There was no effect of fumaric acid on NDF digestibility. Methane production decreased significantly in a linear response ($P<0.01$) to increasing inclusion of fumaric acid. The reduction in methane production with inclusion of fumaric acid was 2.9, 4.2 and 12.3% for 20, 40 and 80 g fumaric acid/ kg DM. Hydrogen capture ranged from 65 to 97%.

Keywords: methane, fumarate, rumen, organic acids

1. INTRODUCTION

Methane formation in the rumen represents a substantial loss of energy to the animal and is also a major source of greenhouse gas emissions from agriculture. Methanogenesis is the main means of disposal of hydrogen during rumen fermentation. The formation of propionate represents an alternative route of hydrogen disposal, providing sufficient propionate precursors are available. Theoretically, adding propionate precursors should stimulate propionate production and decrease methane production. In the present study, the effects of potential precursors of propionate on rumen fermentation and methane production were investigated *in vitro* and *in vivo*.

2. MATERIALS AND METHODS

2.1. In vitro batch fermentations

Short-term *in vitro* incubations were carried out with rumen fluid withdrawn from three rumen-cannulated sheep fed 1 kg/d of a good quality hay. Strained rumen fluid was transferred anaerobically (20% vol/vol) to medium containing (per L): 475 ml distilled water, 1.25 ml trace elements solution (13.2 g $CaCl_2.2H_2O$, 10 g $MnCl_2.4H_2O$, 1 g $CoCl_2.6H_2O$ and 0.8 g $FeCl_2.6H_2O$ per L), 237.5 ml of buffer solution (35 g $NaHCO_3$ and 4 g NH_4HCO_3 per L), 237.5 ml of macrominerals solution (5.7 g Na_2HPO_4, 6.2 g KH_2PO_4 and 0.6 g $MgSO_4.7H_2O$ per L), 1.25 ml of 0.1% (w/v) resazurin solution and 47.5 ml of freshly prepared reducing solution (6.25 g cysteine hydrochloride, 40 ml 1 N NaOH and 6.25 g $Na_2S.9H_2O$ per L). After mixing, 50 ml of buffered ruminal fluid was dispensed anaerobically into 120-ml serum bottles containing 400 mg of a 80% forage and 20% concentrate diet, previously ground to pass through a 1 mm mesh screen. Solutions of the different additives were prepared in distilled water, and one ml of each solution (400 μmol of additive) was added to each of two incubation bottles. After 24 h of incubation, fermentation parameters and formation of fermentation products (total gas, methane and VFA) were determined in each bottle following the methods of analysis described in López *et al.* (1999).

2.2. Effect of fumarate and acrylate on fermentation in Rusitec

An artificial rumen fermentor (Rusitec) with twelve vessels was used as described by Czerkawski and Breckenridge (1977). Each vessel was fed 20 g/d of a mixed diet consisting of grass hay, barley, molasses, white fish-meal and a vitamins and minerals mixture (500, 299.5, 100, 91 and 9.5 g/ kg dry matter, respectively). In 8 vessels, either acrylate or fumarate (6.25 mmol/d, both added as the sodium salt, 4 replicates per treatment) was added to the basal diet. The remaining 4 vessels received no addition. The experiment lasted 3 weeks and all measurements were made in the last 7 d of the experiment. The volume of gas produced was measured with a wet meter and gases were analysed by gas chromatography. The volume of effluent collected was measured and samples taken for analysis of volatile fatty acids. Methods of analysis are described further in Newbold *et al.* (1996).

2.3. The effect of fumaric acid on methane production by sheep *in vivo*

Four diets comprising grass hay supplemented with fumaric acid at a rate of 0, 20, 40 and 80 g/kg DM were offered at the maintenance plane of nutrition to rumen cannulated wether sheep in two equal feeds in a 4 x 4 Latin square design. Fumaric acid was added to the hay with molasses 25 g (fresh weight) per sheep per day to improve the palatability. Fresh water was available at all times. The study comprised four 21-d periods with 10 d acclimatisation to the diet, 7 d collection of faeces and urine followed by four 24-h measurements of methane production in open-circuit respiration chambers. Rumen samples were withdrawn as described by Moss *et al.* (1995).

3. RESULTS

Methane production decreased in the presence of fumarate and acrylate, with the size of the reduction in stoichometric agreement with the increase in propionate production. 2-Oxoglutarate and to a lesser extent pyruvate also decreased methane production (Table 1), however the size of the decrease was larger than would have been predicted from the increase in propionate, suggesting a more complex effect on the rumen fermentation.

Acrylate and fumarate were further evaluated in Rusitec. Of the added acrylate and fumarate, 33% and 44% respectively was recovered in propionate (Table 2). However, fumarate appeared more efficient in decreasing methane production (28 vs. 14% decrease).

Sheep consumed 0, 21, 41.6 and 81 g/d, s.e.d. 1.54, fumaric acid on the 20, 40 and 80 g /kg DM fumaric acid diets, respectively. Fumaric acid stimulated intake but had no effect on digestibility (Table 3). Increasing the inclusion rate of fumaric acid supplementation decreased total VFA and rumen pH linearly (Table 3). The molar proportion of acetate was significantly lower with the addition of fumaric acid while the molar proportion of propionate increased. Methane production decreased significantly with the inclusion of fumaric acid in the diet.

Table 1. The effects of adding propionate precursors on in vitro fermentation

	Methane (μmol/d)	C2 (μmol/d)	C3 (μmol/d)	C4 (μmol/d)	H recovered (%)
Blank	636	1420	458	187	95.0
Acrylate	546 ab	1519	696 ab	113 b	88.2
Aspartate	596	1556 a	666 ab	205	89.9
Citrate	614	2162 ab	495 b	206	68.4
Fumarate	582 b	1515	730 ab	198	91.9
2-Oxoglutarate	552 ab	1841 ab	505 b	242 ab	72.0
Lactate	641	1579 ab	609 ab	270 ab	89.3
Malate	613	1600 ab	693 ab	227 ab	89.5
Pyruvate	586 b	1643 ab	506 b	202	81.9
Tartrate	654	1769 ab	436 b	222 ab	81.0
Propionate	663	1450	904 a	194	106.3
s.e.d.	23.6	41.2	28.4	7.8	

a and b indicate that treatment is significantly ($P<0.05$) different from control and propionate treatments, respectively.

Table 2. Influence of acrylate and fumarate addition (6.25 mmol/d) on pH and the production volatile fatty acids and gases (mmol/d) in Rusitec

	control	acrylate	fumarate	s.e.d
pH	6.71	6.71	6.78	0.26
Acetate	16.3	17.3	17.9	0.98
Propionate	7.6	9.7	10.4	0.58**
Butyrate	8.1	8.6	9.6	1.02
Hydrogen	0.34	0.26	0.27	0.085
Methane	2.46	2.12	1.78	0.161*

*$P<0.05$, ** $P<0.01$,

Table 3. The effect of fumaric acid supplementation on organic matter intake, digestibility, rumen pH, concentrations of volatile fatty acids (VFA) and methane production in sheep.

Measurement	Fumaric acid (g/kg DM)				SED
	0	20	40	80	
OM intake (g/d)	809	832	844	870	7.3***
OM digestibility	0.603	0.604	0.628	0.621	0.0155
pH	6.44	6.35	6.33	6.24	0.066*
Total VFA (mM)	70.7	68.6	68.4	64.5	1.48***
Molar proportions of VFA (mol/mol)					
Acetic acid	0.686	0.684	0.674	0.659	0.0038***
Propionic acid	0.219	0.226	0.237	0.259	0.0035***
Butyric acid	0.084	0.080	0.081	0.073	0.0011***
Methane output					
Methane (L/d)	26.7	26.5	26.4	24.2	0.54***
Methane (L/kg DM intake)	31.0	30.1	29.7	27.2	0.74***

*$P<0.05$, ** $P<0.01$, ***$P<0.001$

4. CONCLUSION

Based on the equation $4H_2 + CO_2 = CH_4 + 2H_2O$ and the reduction of fumarate to succinate by 2H, the efficiency of hydrogen trapping by fumarate, in the Rusitec, was 44%. In sheep, the data are confounded by changes in intake, but when expressed on litres per kg of DM intake the capture of hydrogen by fumarate ranged from 65 to 97%. In conclusion, fumaric acid may be a useful additive to decrease methane production by ruminants, but further work is required to investigate the effects of fumarate on methane production on wider range of diets.

ACKNOWLEDGMENTS

This work was funded by MAFF. The Rowett Research Institute receives financial support from the Scottish Office Agriculture, Environment and Fisheries Department. J. O. Ouda is grateful to the Kenya Agricultural Research Institute (KARI) for his scholarship.

REFERENCES

Czerkawski, J.W. & Breckenridge, G., 1977. Design and development of a long term rumen simulation technique (Rusitec). British Journal of Nutrition, 38: 371-384.

López, S., Valdes, C., Newbold, C.J., & Wallace, R.J., 1999. Influence of sodium fumarate on rumen fermentation *in vitro*. British Journal of Nutrition, 81: 59-64.

Moss, A.R., Givens, D.I. & Garnsworthy, P.C., 1995. The effect of supplementing grass silage with barley on digestibility, in sacco degradability, rumen fermentation and methane production in sheep at two levels of intake. Animal Feed Science and Technology, 55: 9-33.

Newbold, C.J., Wallace, R.J. & McIntosh, F.M., 1996. Mode of action of the yeast *Saccharomyces cerevisiae* as a feed additive for ruminants. British Journal of Nutrition, 76: 249-261.

Evaluating the dietary pre-conditions for a significant methane-suppressing effect of myristic acid in ruminants

A. Machmüller, C. R. Soliva, R. H. Razminowicz, S. J. Duleba & M. Kreuzer

Institute of Animal Sciences, Animal Nutrition, ETH Zurich, ETH Zentrum/LFW, CH-8092 Zurich, Switzerland

The objective of the present *in vivo* experiment was to determine the extent to which and under which dietary pre-conditions the methane-suppressing effect of myristic acid can be realised in feeding practice. Six sheep were subjected to a 6×6 Latin square arrangement. Pure myristic acid (0 and 5%) was supplemented to two basal diet types differing in forage to concentrate ratio (1:1.5 and 1:0.5). Levels of dietary calcium were 0.4 and 0.9% (the latter only in combination with the myristic acid supplemented diets). The results indicate that the extent of the suppressing effect of myristic acid on methane emission depends on the type of basal diet and on the level of dietary calcium. Supplementing 5% myristic acid to the basal diet with a high concentrate proportion significantly suppressed methane emission of the sheep by 47% and 58%, with and without an additional dietary calcium supply, respectively. With the basal diet characterised by a high forage proportion, the extent of the methane-suppressing effect of 5% dietary myristic acid was lower (22%) and became insignificant when the level of dietary calcium was also increased. The present experiment showed that the composition of the complete diet is decisive for the methane-suppressing effect of myristic acid.

Keywords: methane, myristic acid, forage proportion, calcium, sheep

1. INTRODUCTION

Saturated medium-chain fatty acids (MCFA), such as lauric ($C_{12:0}$) and myristic acid ($C_{14:0}$), have the potential to suppress methane production in the total digestive tract of ruminants as was first shown by Blaxter & Czerkawski (1966). The use of coconut oil, which is selectively rich in these two fatty acids, turned out to be effective to reduce methane release both *in vitro* (Dong *et al.* 1997; Machmüller *et al.* 1998; Dohme *et al.* 1999, 2000) and *in vivo* (Machmüller & Kreuzer 1999; Machmüller *et al.* 2000), even when used in proportions of only about 3% in feed, but particularly at levels of 6-7%. The decrease in methane production achieved *in vivo* was 20-70%, and no major adverse influence was observed on digestion and utilisation of energy and protein (Machmüller & Kreuzer 1999; Machmüller *et al.* 2000).

It has been shown that free MCFA could directly affect the growth of ruminal species of all three domains of life *Archaea*, *Bacteria* and *Eucarya*. In the pure culture study of Henderson (1973) the growth of *Methanobrevibacter ruminantium* was found to be inhibited by adding $C_{10:0}$, $C_{12:0}$ and $C_{14:0}$. Galbraith *et al.* (1971) demonstrated in an *in vitro* screening test that some species of Gram positive bacteria are also inhibited by these fatty acids, whereas Gram negative bacteria seem to be less sensitive. Also rumen ciliates are suppressed by the MCFA, particularly $C_{10:0}$ and $C_{12:0}$ (Matsumoto *et al.*, 1991). The extent of these positive as well as negative direct anti-microbial effects of the MCFA is determined by the concentration of non-esterified free fatty acids in the medium. Interactions of MCFA with other feed ingredients will influence the concentration of free fatty acids in rumen fluid and thus be decisive for the methane-suppressing effect. Possible important interactions may arise from the following mechanisms: (1) free fatty acids of MCFA and alkaline earth metals, especially calcium, could form insoluble and inactive soaps in rumen fluid (El Hag & Miller, 1972); (2) triglycerides or fatty acids could be adsorbed onto feed particles (Harfoot *et al.*, 1974). Therefore, the objective of the present *in vivo* experiment was to evaluate the effect of varying forage to concentrate ratio and of dietary calcium on the methane-suppressing effect of a free medium-chain fatty acid. Myristic acid was used as a model since lauric acid could result in a depression of feed intake (Blaxter & Czerkawski, 1966; Dohme *et al.*, 2001b).

2. MATERIAL AND METHODS

The study was carried out with six castrated male sheep (liveweight 41±2 kg) of the Swiss White Hill breed. They were subjected to a 6×6 Latin square arrangement. Each sheep received a different sequence of treatments. Two basal diet types differing in forage to concentrate ratio (1:1.5 and 1:0.5) were fed. Feed was calculated to be isoenergetically in terms of metabolisable energy, equivalent to 494 $kJ/kg^{0.75}$ liveweight (RAP, 1999). This meant either 320 g hay and 475 g concentrate or 618 g hay and 307 g concentrate per day. The concentrate consisted of 70 % barley and 30 % soybean meal. The two basal diet types were supplemented with pure non-esterified myristic acid (0 and 5%) and dietary calcium (0.42 and 0.90%, the latter only in combination with the myristic acid supplemented diets). Calcium carbonate was used as calcium source. The whole study comprised six subsequent 24 d measurement periods each consisting of 14 d of adaptation, 8 d of complete collection of faeces and urine, and 2 d of quantitative measurement of gaseous exchange in open-circuit respiratory chambers. Rumen fluid samples were taken 5 h after morning feeding on the last day of each measurement period, with a flexible stomach tube inserted through the oesophagus. The experiment was carried out in accordance to the Swiss guidelines for animal welfare. Data were analysed by the general linear model (GLM) procedure (SAS version 6.12, SAS Institute Inc.). Analysis of variance was carried out regarding diet type, supplementation, diet type × supplementation, animal and measurement period. Multiple comparisons among treatment means were performed by Tukey's studentised range test.

3. RESULTS AND DISCUSSION

Changing the basal diet type did not significantly affect the rumen fluid properties given in Table 1. At calcium level of 0.42% the average pH for both diet types was 7.0, the ammonia concentration was 9.4 mmol L^{-1}, the bacteria count was 12×10^9 mL^{-1}, and the ciliate count was 6.5×10^5 mL^{-1}. Calcium supplementation decreased pH and ammonia but did not significantly affect microbial counts although average ciliate counts tended to decrease. Interaction occurred between diet type and supplementation in ammonia concentration.

Table 1. Effects of diet type and supplementation on selected rumen fluid properties (n=6).

Diet type	Hay : concentrate = 1:1.5			Hay : concentrate = 1:0.5			
Myristic acid, %	0	5	5	0	5	5	
Calcium, %	0.42	0.42	0.90	0.42	0.42	0.90	SEM
pH^{A}	7.00^{a}	6.62^{ab}	6.37^{b}	7.04^{a}	6.56^{ab}	6.70^{ab}	0.111
Ammonia (mmol L^{-1})^{B,C}	10.99	5.36	4.90	7.87	5.85	7.12	1.430
Bacteria ($\times 10^9$ mL^{-1})	12.3	11.6	11.6	11.1	10.0	8.5	1.69
Ciliates ($\times 10^5$ mL^{-1})	5.87	5.03	1.32	7.10	3.63	2.33	2.029

^{A}Supplementation main effect, $P < 0.001$
^{B}Supplementation main effect, $P < 0.01$
^{C}Diet type × supplementation interaction, $P < 0.05$
^{a,b}Within rows, means followed by the same superscript are not significantly different at $P < 0.05$

Supplementation enhanced oxygen consumption (Table 2), but this was numerically an effect just of myristic acid. Carbon dioxide release was not significantly affected by supplementation but by basal diet type. Instead, a higher carbon dioxide production occurred when the diet had a higher proportion of hay. This affected the respiratory quotient (CO_2:O_2). Without supplementation the average respiratory quotients did not differ between the two basal diet types but the addition of myristic acid reduced CO_2:O_2 in both diet types. This effect was more pronounced when the concentrate proportion of the diet was high.

Table 2. Effects of diet type and supplementation on gas exchange (n=6).

Diet type	Hay : concentrate = 1:1.5			Hay : concentrate = 1:0.5			
Myristic acid, %	0	5	5	0	5	5	
Calcium, %	0.42	0.42	0.90	0.42	0.42	0.90	SEM
O_2 (L d^{-1})^{A}	321	354	354	341	353	355	7.7
CO_2 (L d^{-1})^{B}	337	342	348	359	358	360	6.0
CO_2:O_2 (L:L)^{A,B,C}	1.05^{a}	0.97^{c}	0.98^{bc}	1.06^{a}	1.01^{b}	1.02^{b}	0.008
CH_4 (L d^{-1})^{A,B,D}	25.1^{a}	10.5^{c}	13.2^{c}	25.5^{a}	19.8^{b}	21.8^{ab}	1.18
CH_4:CO_2 (mL:L)^{A,B,D}	74.9^{a}	30.8^{c}	38.1^{c}	71.2^{a}	55.5^{b}	60.7^{ab}	3.59

^{A}Supplementation main effect, $P < 0.001$
^{B}Diet type main effect, $P < 0.001$
^{C}Diet type × supplementation interaction, $P < 0.01$
^{D}Diet type × supplementation interaction, $P < 0.001$
^{a,b,c}Within rows, means followed by the same superscript are not significantly different at $P < 0.05$

The average daily methane release differed widely among the six treatment groups. Irrespective of the dietary calcium level, supplementing 5% myristic acid to the basal diet with a high concentrate proportion significantly reduced total tract methane emission of the sheep from 25 L d^{-1} to on average 12 L d^{-1}. With the basal diet including a high forage proportion, the extent of the methane-suppressing effect of 5% myristic acid was lower, since the average daily methane release was only reduced by 6 L d^{-1}. Furthermore, when increasing the level of dietary calcium, the supplementation of myristic acid had no significant effect anymore. Since these differences among the treatment groups also persisted when relating the methane release to carbon dioxide release it can be assumed that the fermentation processes in rumen were not depressed to a major extent by supplementing myristic acid as was also shown *in vitro* by Dohme *et al.* (2001a). More information about the effects within the rumen will be given by the data on volatile fatty acid concentration in rumen fluid, *Archaea* counts in rumen fluid and total tract nutrient digestibility, which all are still under investigation.

The data presented indicate that the extent of the methane-suppressing effect of myristic acid depends on basal diet type and on the level of dietary calcium. The extent of the adsorption of myristic acid onto feed particles is determined by the number and type of feed particles in the rumen. This will be affected by the amount of daily feed intake and the fermentation rate. Machmüller *et al.* (2001) already demonstrated *in vitro* that the methane-suppressing effect of medium-chain fatty acids (in this case coconut oil) decreased when changing the basal diet from intensive-type (low-structural carbohydrate content) to extensive-type (high-structural carbohydrate content). Depending on the diet type the diurnal development of pH and the average pH will also be affected and therefore will influence the formation of insoluble and inactive soaps (Palmquist *et al.*, 1986). This might explain the different response in methane to the two basal diets to calcium supplementation. In line with the present results, a low methane-suppressing effect of myristic acid was also reported in dairy cows fed a diet with a forage:concentrate ratio of 1:0.67 supplemented by a mineral premix containing calcium (Dohme *et al.*, 2001b). In conclusion, the present experiment showed that it is very important that the composition of the complete diet has to be considered in order to benefit mostly from the methane-suppressing effect of myristic acid.

REFERENCES

Blaxter, K. L. & J. Czerkawski, 1966. Modifications of the methane production of the sheep by supplementation of its diet. J. Sci. Food Agric. 17: 417-421.

Dohme, F., A. Machmüller, B. L. Estermann, P. Pfister, A. Wasserfallen & M. Kreuzer, 1999. The role of the rumen ciliate protozoa for the methane suppression caused by coconut oil. Letters Appl. Microbiol. 29: 187-192.

Dohme, F., A. Machmüller, A. Wasserfallen & M. Kreuzer, 2000. Comparative efficiency of various fats rich in medium-chain fatty acids to suppress methane release, methanogens and ciliates as measured with Rusitec. Can. J. Anim. Sci. 80: 473-482.

Dohme, F., A. Machmüller, A. Wasserfallen & M. Kreuzer, 2001a. Ruminal methanogenesis as influenced by individual fatty acids supplemented to complete ruminant diets. Letters Appl. Microbiol. 32: 47-51.

Dohme, F., F. Sutter, A. Machmüller & M. Kreuzer, 2001b. Methane formation and energy metabolism of lactating cows receiving individual medium-chain fatty acids. In: Energy Metabolism in Animals. A. Chwalibog & K. Jakobsen, Wageningen Pers, Wageningen, The Netherlands. 369-372.

Dong, Y., H. D. Bae, T. A. McAllister, G. W. Mathison & K.-J. Cheng, 1997. Lipid-induced depression of methane production and digestibility in the artificial rumen system (RUSITEC). Can. J. Anim. Sci. 77: 269-278.

El Hag, G. A. & T. B. Miller, 1972. Evaluation of whisky distillery by-products VI. The Reduction in digestibility of malt distiller's grains by fatty acids and the interaction with calcium and other reversal agents. J. Sci. Fd. Agric. 23: 247-258.

Galbraith, H., T. B. Miller, A. M. Paton & J. K. Thompson, 1971. Antibacterial activity of long chain fatty acids and the reversal with calcium, magnesium, ergocalciferol and cholesterol. J. Appl. Bact. 34: 803-813.

Harfoot, C. G., M. L. Crouchman, R. C. Noble & J. H. Moore, 1974. Competition between food particles and rumen bacteria in the uptake of long chain fatty acids and triglycerides. J. appl. Bact. 37: 663-641.

Henderson, C., 1973. The effects of fatty acids on pure cultures of rumen bacteria. J. Agric. Sci. (Camb.) 81: 107-112.

Machmüller, A. & M. Kreuzer, 1999. Methane suppression by coconut oil and associated effects on nutrient and energy balance in sheep. Can. J. Anim. Sci. 79: 65-72.

Machmüller, A., D. A. Ossowski, M. Wanner & M. Kreuzer, 1998. Potential of various fatty feeds to reduce methane release from rumen fermentation *in vitro* (Rusitec). Anim. Feed Sci. Technol. 71: 117-130.

Machmüller, A., D. A. Ossowski & M. Kreuzer, 2000. Comparative evaluation of the effects of coconut oil, oilseeds and crystalline fat on methane release, digestion and energy balance in lambs. Anim. Feed Sci. Technol. 85: 41-60.

Machmüller, A., F. Dohme, C. R. Soliva, M. Wanner & M. Kreuzer, 2001. Diet composition affects the level of ruminal methane suppression by medium-chain fatty acids. Aust. J. Agric. Res. 52: in press.

Matsumoto, M., T. Kobayashi, A. Takenaka & H. Itabashi, 1991. Defaunation effects of medium-chain fatty acids and their derivates on goat rumen protozoa. J. Gen. Appl. Microbiol. 37: 439-445.

Palmquist, D. L., T. C. Jenkins & A. E. Joyner, 1986. Effect of dietary fat and calcium source on insoluble soap formation in the rumen. J. Dairy Sci. 69: 1020-1025.

RAP, 1999. 'Fütterungsempfehlungen und Nährwerttabellen für Wiederkäuer.' Landwirtschaftliche Lehrmittelzentrale, Zollikofen, Switzerland.

Greenhouse Gases and Animal Agriculture
J. Takahashi & B. A. Young (Editors)

Effect of a stepwise increase of medium-chain fatty acid supply on ruminal methanogenesis *in vitro*

C.R. Soliva[1], I.K. Hindrichsen[1], C. Gerig[1], L. Meile[2], M. Kreuzer[1] & A. Machmüller[1]

[1]Institute of Animal Sciences, Animal Nutrition, [2]Institute of Food Science, Laboratory of Microbiology, ETH Zurich, ETH Zentrum, CH-8092 Zurich, Switzerland

The Hohenheim gas test apparatus was used to investigate the dose effects of non-esterified $C_{12:0}$ and $C_{14:0}$ fatty acids as well as a potential synergetic effect of a mixture of $C_{12:0}$ and $C_{14:0}$ on methanogenesis in rumen fluid. Additionally, the relationship between methane release and methanogen count, analysed by the fluorescence *in situ* hybridization, was followed. In experiment 1, $C_{12:0}$ and $C_{14:0}$ were compared with $C_{18:0}$, which served as an ineffective control. Each of these fatty acids was added at 14 different levels from 0 to 32.5 mg in steps of 2.5 mg to a rumen fluid/buffer mixture and incubated for 24 h. In experiment 2, $C_{12:0}$ was supplied at 13 different levels up to 30 mg. Each concentration was supplemented by either 0, 10, 20 or 30 mg $C_{14:0}$. The only substrates were H_2 and CO_2. When provided alone, only $C_{12:0}$ decreased methanogenesis, increasingly with the $C_{12:0}$ level. $C_{12:0}$ and $C_{14:0}$ suppressed methanogen count by the same extent as methane was reduced by $C_{12:0}$. $C_{14:0}$ added to $C_{12:0}$ enhanced the methane inhibitory effect of $C_{12:0}$, and also the effects on methanogen count were enforced. This illustrates the advantage of using $C_{12:0}$ and $C_{14:0}$ together for instance in common diets.

Keywords: methane, lauric acid, myristic acid, methanogens, FISH

1. INTRODUCTION

Similar to long-chain polyunsaturated fatty acids, certain medium-chain saturated fatty acids are effective in suppressing methane release from ruminal fermentation (Dohme et al., 2001). This explains why coconut oil, a lipid rich in the medium chain fatty acids lauric ($C_{12:0}$) and myristic ($C_{14:0}$) acid, depresses methanogenesis more effectively than fats characterised by other fatty acids (Dong et al., 1997; Machmüller et al., 1998). Dong et al. (1997) gathered, that $C_{12:0}$ and $C_{14:0}$ are responsible for the low titre observed for methanogens, but also for that of cellulolytes and amylolytes; coconut oil seemed to have a direct toxic effect on these microbes. Such a direct effect on *Methanobacterium ruminantium* had been demonstrated for polyunsaturated fatty acids in pure culture studies by Prins et al. (1972). Additionally, fatty acids might have an additive and even a synergetic methane suppressing effect. Accordingly, Koster and Cramer (1987) demonstrated that methanogenesis in batch cultures with granular sludge was more inhibited by adding a mixture of $C_{12:0}$ and $C_{14:0}$ than by adding $C_{12:0}$ only.

The objectives of the present *in vitro* study were (i) to compare the dose effects of non-esterified $C_{12:0}$ and $C_{14:0}$ on methanogenesis, (ii) to explore a potential synergetic effect of a mixture of $C_{12:0}$ and $C_{14:0}$ in terms of methane inhibition, and (iii) to investigate the relationship of variations in methane release and in methanogen count.

2. MATERIAL AND METHODS

Two *in vitro* experiments were carried out using the Hohenheim gas test apparatus (Menke *et al.*, 1979). This system was developed to determine the energetic value of ruminant feeds. The equipment used consisted of an incubator with a rotor designed for 57 glass pistons (incubation units) of 100 ml volume. This *in vitro* system is a static system which means that, after the start of the incubation, no food can be added and metabolic products, apart from gases, cannot be removed. Therefore, the incubation period is limited to 1–2 days. The pistons were modified as suggested by Steingass (personal communication) by mounting a second exit closed with a (what is the unabbreviated description)PTFE-layered septum. This offered the opportunity to remove gas samples with a gas tight 0.15-ml syringe for compositional analysis. The complete gas volume produced was obtained directly from the calibrated scale printed on the pistons. Rumen fluid was collected before the morning feeding from a Brown Swiss cow receiving a diet composed of 60% hay and 40% concentrate. After adjusting the pH to 7.0, 30 ml of rumen fluid/buffer mixture (1:2) were filled anaerobically into the pre-heated pistons and incubated for 24 h at 39°C. Exclusively H_2 and CO_2 (4:1) were provided as substrates to evaluate direct effects of medium-chain fatty acids only on the methanogens. This approach excluded interactions of fat with feed particles and suppressed the growth of other microbes.

In a first experiment, the response of methanogenesis to increasing amounts of non-esterified $C_{12:0}$ and $C_{14:0}$, as potentially active fatty acids, was investigated. Stearic acid ($C_{18:0}$), which is assumed to be ineffective in terms of inhibiting methanogenesis, served as control. Each of the three fatty acids was added in 14 different levels increasing from 0 to 32.5 mg in steps of 2.5 mg (one incubation per level). In a second experiment, $C_{12:0}$ was supplied to rumen fluid at 13 different levels increasing from 0 to 30 mg in steps of 2.5 mg. Each single concentration of $C_{12:0}$ was supplemented either by 0, 10, 20 or 30 mg of $C_{14:0}$.

After incubation for 24 h, rumen fluid and gas samples were obtained. Rumen fluid was analysed for pH, volatile fatty acids (gas chromatography) and total protozoa counts (how were protozoa enumerated). For the quantification of the methanogens, the fluorescence *in situ* hybridization technique (FISH) was applied as outlined in the manual of Stahl et al. (1995). Briefly, cell material was fixed for 3 h in 4% (w/v) formaldehyde dissolved in phosphate buffered saline (pH 7.2, PBS), then pelleted and resuspended in PBS and ethanol. Additionally to the manual, the samples were homogenized by vortexing and ultrasonification (TEC-25, Telsonic, CH-Bronschhofen) for 2 × 30 s to avoid cell clumping. Afterwards 3 µl of fixed cells were applied to gelatine-coated slides. After dehydration with 50, 80 and 100% ethanol for 3 min each, followed by air-drying, the slides were hybridized in 8 µl hybridization buffer and 1 µl (25 ng) of the 16S rRNA probe Arch 915 targeting methanogens (Stahl et al., 1995). The probe was fluorescein-labelled with Cy3. The hybridized slides were covered with an antifade reagent (Fluoro Guard™, Bio Rad, Hercules, California, USA) and examined with a microscope equipped for epifluorescence measurements (BX-60, Olympus Optical AG, CH-Volketswil). Images of the fluorescein-labelled methanogens were taken with a camera mounted onto the microscope, and methanogens were automatically counted by the analySIS software programme (analySIS, version 3.1, Soft Imagine System GmbH, CH-Uster).

3. RESULTS AND DISCUSSION

In both experiments pH was not affected by any of the supplementations of medium-chain fatty acids and averaged at 7.3 to 7.5. Although in several former studies a protozoa suppressing effect of feeds rich in $C_{12:0}$ and $C_{14:0}$ was found (e.g., Machmüller and Kreuzer, 1999), this did not occur in the present approach leading to an average count of 4.3×10^4 mL^{-1} in experiment 1 and 1.9×10^4 mL^{-1} in experiment 2. The lack of a specific fatty acid effect on protozoa may have been due to the short incubation period and the lack of feed.

In experiment 1, both $C_{18:0}$ and $C_{14:0}$ did not affect the methane formation, not even at the highest concentration of 32.5 mg per incubation unit (Fig. 1). The average methane release from the $C_{14:0}$ and $C_{18:0}$ treated rumen fluid was 18.8 mL/24 h per incubation unit. In contrast, with $C_{12:0}$ a distinct polynomially shaped dose effect occurred. From a concentration of 10 mg $C_{12:0}$ onwards methane inhibition was noted which, at a level of 32.5 mg, descreased to 16% of zero supplementation (Fig. 2). Similarly, methanogens were suppressed by $C_{12:0}$ to 13% of controls. However, the latter was also the case with $C_{14:0}$ although methane formation had not responded to $C_{14:0}$. Applying an extended *in vitro* incubation period using the Rusitec system both $C_{12:0}$ and $C_{14:0}$ suppressed methane and methanogen count (determined by cultivation technique) to the same extent (Dohme et al., 2001). Also from the present results the basic efficacy of $C_{14:0}$ is obvious from the clear decrease in the population density of methanogens with increasing supply of this fatty acid. However, it seems that in the case of $C_{12:0}$, other than with $C_{14:0}$, a shorter period of time is required until not only the number but also the methanogenic activity of the remaining microbes is reduced to the original level. With $C_{14:0}$ the methanogenic activity per cell must be drastically increased thus utilising the hydrogen which is available in excess due to the reduced number of methanogens.

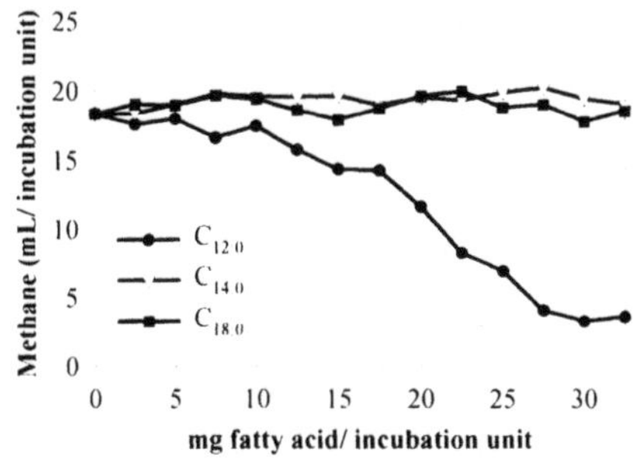

Fig. 1: Methane release in experiment 1

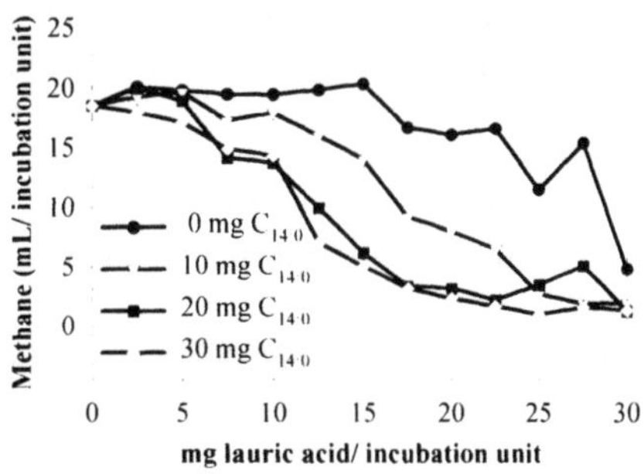

Fig. 3: Methane release in experiment 2

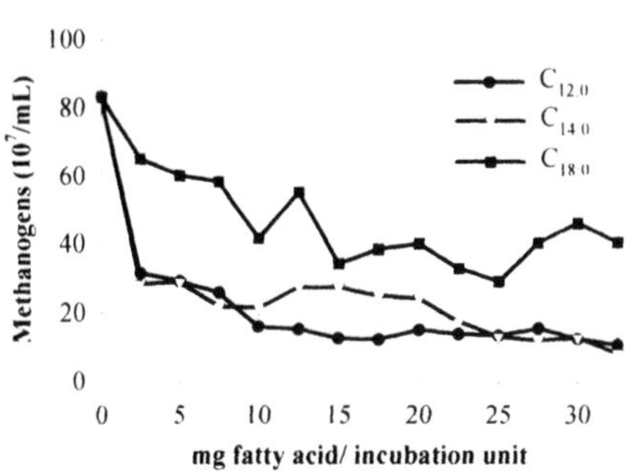

Fig. 2: Methanogens in experiment 1

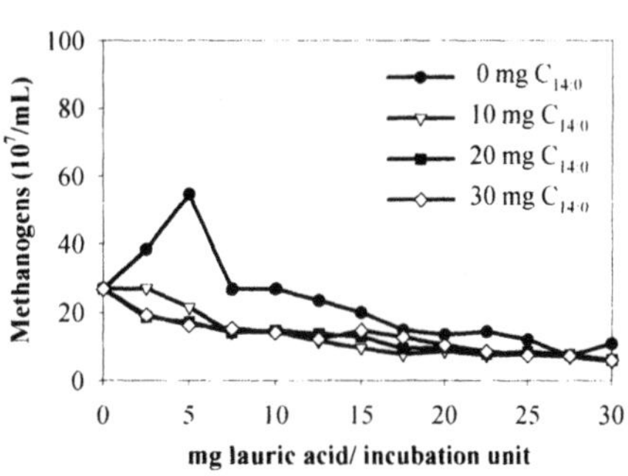

Fig. 4: Methanogens in experiment 2

Hydrogen available in case of a reduced methane release was almost completely recovered as free hydrogen, whereas almost the complete amount of hydrogen supplied was incorporated into methane in the $C_{14:0}$ and $C_{18:0}$ treatments. There was neither a major change in the acetate : propionate ratio nor an effect on the fatty acid concentration in rumen fluid with increasing levels of any of the fatty acids. This seems unexpected as Whitelaw *et al.* (1984) described a close inverse relationship between propionic acid percentage and the quantity of methane produced based on the assumption that propionic acid formation directly competes with methane formation for the available hydrogen. Furthermore, an increase in reductive acetogenesis with excessive hydrogen amounts could have been expected (Le Van et al., 1998). The assumed major reason for the lack of effect on volatile fatty acid synthesis in the present experiment is that deliberately no feed was supplied, which leaves the methanogens as the only species properly supplied with substrates (H_2 and CO_2). Furthermore, the short incubation period might have played a role.

In experiment 2, additional $C_{14:0}$ clearly enhanced the inhibitory effect of $C_{12:0}$ in methanogenesis (Fig. 3). A supply of 10 mg of $C_{14:0}$ added to different levels of $C_{12:0}$ was already sufficient to support methane reduction by $C_{12:0}$ thus decreasing methane release at far lower concentrations of $C_{12:0}$ with than without $C_{14:0}$. Another 10 mg of $C_{14:0}$ further enhanced this effect while an addition of totally 30 mg $C_{14:0}$ gave no increase in $C_{12:0}$ efficacy any more. Similar to experiment 1, although at a lower initial population density, increasing levels of $C_{12:0}$ decreased the methanogen count (Fig. 4). At low additions, this effect was similarly enhanced by 10, 20 and 30 mg additional $C_{14:0}$. Overall, methanogen count then was reduced by 76% with 30 mg $C_{12:0}$ relative to the unsupplemented treatments whereas without additional $C_{14:0}$ the maximum reduction was 60%. These results clearly indicate the presence of additivity and, particularly with respect to methane release, synergism of $C_{12:0}$ and $C_{14:0}$ in the suppression of methanogens and methanogenesis. Like in experiment 1, there are indications that reductions in methanogen count might (transiently) result in an increased methanogenic activity per cell (e.g., low $C_{12:0}$ levels supplied with 10 mg additional $C_{14:0}$). As in experiment 1, amount and pattern of volatile fatty acids did not change with varying concentrations of $C_{12:0}$ and $C_{14:0}$. The hydrogen recovery rate was calculated to account for 97% in all treatments and the hydrogen added was exclusively recovered in methane and free hydrogen.

In conclusion, both $C_{12:0}$ and $C_{14:0}$ affect methanogens although in a different way. Both are able to decrease methanogen population, but responses in methane formation during 24 h of exposure to these fatty acids differ. There is a synergetic effect of $C_{12:0}$ and $C_{14:0}$ illustrating that dietary fats such as coconut oil, containing both fatty acids, might be more efficient to suppress methane in ruminants compared to fatty acids given isolated. It remains to be explored to which extent the variation in methane release per methanogen cell found is due to a generally different cell activity at higher hydrogen supply or to shifts in methanogen species with a differing inherent activity. (It would be worth speculating why methanogens are still present at 10^7/ml when methane production has nearly fallen to zero.)

REFERENCES

Dohme, F., A. Machmüller, A. Wasserfallen & M. Kreuzer, 2001. Ruminal methanogenesis as influenced by individual fatty acids supplemented to complete ruminant diets. Letters Appl. Microbiol. 32: 47-51.

Dong, Y., H. D. Bae, T. A. McAllister, G. W. Mathison & K.-J. Cheng, 1997. Lipid-induced depression of methane production and digestibility in the artificial rumen system (RUSITEC). Can. J. Anim. Sci. 77: 269-278.

Koster, I. W. & A. Cramer, 1987. Inhibition of methanogenesis from acetate in granular sludge by long-chain fatty acids. Appl. Environ. Microbiol. 52: 403-409.

Le Van, T. D., J. A. Robinson, J. Ralph, R. C. Greening, W. J. Smolenski, J. A. Z. Leedle & D. M. Schaefer, 1998. Assessment of reductive acetogenesis with indigenous ruminal bacterium populations and *Acetitomaculum ruminis*. Appl. Environ. Microbiol. 64: 3429-3436.

Machmüller, A. & M. Kreuzer, 1999. Methane suppression by coconut oil and associated effects on nutrient and energy balance in sheep. Can. J. Anim. Sci. 79: 65-72.

Machmüller, A., D. A. Ossowski, M. Wanner & M. Kreuzer, 1998. Potential of various fatty feeds to reduce methane release from rumen fermentation *in vitro* (Rusitec). Anim. Feed Sci. Technol. 71: 117-130.

Menke, K. H., L. Raab, A. Salewski, H. Steingass, D. Fritz & W. Schneider, 1979. The estimation of the digestibility and metabolizable energy content of ruminant feeding-stuff from the gas production when they are incubated with rumen liquor *in vitro*. J. Agric. Sci. (Camb) 93: 217-222.

Prins, R. A., C. J. Van Nevel & D. I. Demeyer, 1972. Pure culture studies of inhibitors for methanogenic bacteria. Antonie van Leeuwenhoek 38: 281-287.

Stahl, D. A., R. I. Amann, L. K. Poulsen, L. Raskin & W. C. Capman, 1995. Use of fluorescent probes for determinative microscopy of methanogenic archaea. In: Archaea: Methanogens: A laboratory manual. Eds: Sowers, K. R. & H. J. Schreier. Cold Spring Harbor Laboratory Press, New York. pp. 111-121.

Whitelaw, F. G., J. M. Eadie, I. A. Bruce & W. J. Shand, 1984. Methane formation in faunated and ciliate-free cattle and its relationship with rumen volatile fatty acid proportions. Br. J. Nutr. 52: 261-275.

Greenhouse Gases and Animal Agriculture
J. Takahashi & B. A. Young (Editors)

Treatment effects on methane emissions from grazing cattle

L. A. Harper[1], F. M. Byers[2] & R. R. Sharpe[1]

[1]USDA-ARS, 1420 Experiment Station Road, Watkinsville, GA 30677 USA,
[2]USDA-ARS, Building 200, Room 100, BARC-E, Beltsville, MD 20705 USA

Methane is believed to contribute about 18% of the global integrated radiative force with emissions from enteric fermentation representing about 16 % of anthropogenic emissions. A micrometeorological technique was developed to measure methane production by cattle in pasture and feedlot conditions. The technique can be used continuously for extended periods, allows for short-term observations (eight-minute samples) to assess activity and diurnal effects, permits a relatively large number of animals to be sampled, and is virtually non-intrusive to the animals being evaluated. Thirty-six animals receiving one of three supplement strategies and grazing sorghum haygrazer were used in this research. Three strategic grazing nutritional treatments were examined: 1. Control (with a mineral supplement only); 2. A specifically blended-variable escape protein plus mineral supplement; and 3. An ionophore in the specifically blended-variable escape protein plus mineral supplement. The protein-mineral supplement somewhat increased growth rate ($P<.10$), which also trended to increase methane emissions. The strategic ionophore protein-mineral supplement decreased methane emissions by over 50% from those of the control cattle, while maintaining growth rate similar to controls. Use of the micrometerological technique provided a non-interference means of testing methane emissions from animals under natural conditions and verified that management options can have a significant effect on methane emissions from animal production systems.

Keywords: Enteric, micrometeorological, mass-difference, ionophores, grass quality

1. INTRODUCTION

As a greenhouse gas, methane (CH_4) is second in global importance to carbon dioxide (CO_2). Methane is the most abundant organic gas in the earth's atmosphere and evidence has shown that CH_4 concentrations have increased globally at a rate of about 1% or 15 ppb yr^{-1} during the decade preceding 1991 (Watson et al, 1990). It has increased 2.5 times in the last two centuries due to anthropogenic activity (IPCC, 1995). In the U.S., enteric emissions is the fourth largest source of greenhouse gas emissions and is a potential source for reduction (USEPA, 2001).

Anthropogenic sources account for about 70% of the total annual production or release. Enteric fermentation emissions estimates range from 7% (Byers & Turner, 1991) to 20% (Lerner et al., 1988) of total emissions. It also represents an animal production loss of 2 to 12% of the gross energy consumed by ruminants (Johnson & Johnson, 1995). Most of our current knowledge of methane production by animals comes from measurements in respiration-chamber experiments to assess the nutritional value of feeds (e.g. Blaxter and Clapperton, 1965). In fact, the Blaxter and Clapperton data set is the basis of a number of current methodologies for estimating national global production of methane from ruminant animals. However, recent studies under field conditions have given more field-realistic results (Denmead et al., 1998; Harper et al., 1999; Johnson & Johnson, 1995; Lassey et al., 1997). Harper et al. (1999) showed a larger influence of feed quality on CH_4 production than was allowed for with current global and national inventory procedures; with significantly lower CH_4 production for high-grain diets. The purpose of this experiment is to assess the methanogenesis response to management and treatment effects on strategies to control emissions of this greenhouse gas and increase feed conversion efficiency.

2. MATERIALS AND METHODS

Field studies were conducted at the Texas A&M Experiment Station on the Brazos River floodplain near College Station, TX, during the summer of 1995. Six separate experiments, by treatments, were conducted to compare CH_4 emissions by cattle grazing the same forage and but different forage supplement treatments. Mostly dry, sunny weather prevailed during the experiments with maximum daytime temperatures ranging from 27 to 34°C during the study with maximum solar radiation intensities slightly over 1000 W m^{-2}.

Forage and Treatments. A summer annual forage called Haygrazer (sudan sorghum) was planted, fertilized, and irrigated to provide continuous grazing for cattle in this study. Thirty-six animals receiving one of three supplement strategies, while grazing sorghum haygrazer, were used in this research, with a parallel group (n=36) concurrently grazed for growth rate assessments. Forage was allowed to grow to sufficient height and rotationally grazed at approximately 2- 3 week intervals over the course of the 72 day experimental period. Forage was sampled at weekly intervals over the study and averaged 69.6% NDF and 39.7% ADF. Three strategic grazing nutritional treatments were examined: 1. control (with a mineral supplement continuously available); 2. 0.45 kg d^{-1} of a specifically blended 24% variable escape protein and mineral supplement; and 3. 0.45 kg d^{-1} of the specifically blended-variable escape protein and mineral supplement including an ionophore (monensin at 0.07 kg per 0.454 kg this does not seem correct). Measurements of CH_4 production were made by a non-interference, micrometeorological mass difference method during all measurement periods. Animals were placed on the grazing study and adapted to supplements for 3 weeks prior to initiating the first measurements of methane emissions. Methane emissions were measured on 6 animals at a time from each treatment group.

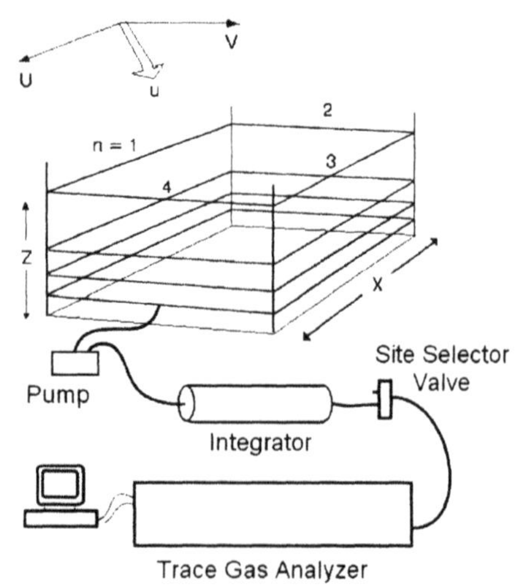

Fig 1. Measurement and calculation scheme for the micrometeorological mass balance technique.

Initially, the measurement equipment was set up in a pre-selected area with fresh, ungrazed forage for the first measurement run. At the end of each measurement period, the animals were removed and the measurement equipment was again set up in a selected area with fresh, ungrazed forage for the next run. There was sufficient forage in the test area for 4 days of grazing by the six test animals. Experimental animal groups and treatments were selected at random order for the first 3 and again for the subsequent 3 measurement periods. The emissions measurement procedure was thus conducted 6 times to measure both groups of 6 animals from each supplement strategy once. The measurement phase lasted 6 weeks.

Flux measurement procedure. The MMD technique developed for this type of study is a modification of the integrated horizontal flux (IHF) technique described by Harper et al. (1999). Methane budgets were calculated for the field in which the animals were feeding from measurements of windspeeds and atmospheric CH_4 concentrations on the upwind and downwind boundaries (Fig. 1) at 0.5, 1, 2, and 3.5 m above the ground. Air was drawn into the sampling tubes equally along their length through capillary tubes inserted at 1 m intervals, and pumped by separate air lines to a mobile laboratory where measurements of CH_4 concentration were made using tunable diode laser spectroscopy. Coupled with measurements of windspeed and wind direction at the same heights, the concentration measurements permitted calculation of the fluxes of CH_4 across each boundary. Measurements were made continuously allowing preparation of a new CH_4 budget every 8 min.

The method was tested by constructing CH_4 budgets for the test field when no cattle were present, and by releasing CH_4 inside the field at known rates. Denmead et al. (1998) discussed constraints and possible problematic meteorological situations when using the technique.

3. RESULTS AND DISCUSSION

The 36 animals managed in the parallel 72 day study to establish baseline growth response to the nutritional strategies averaged 0.59, 0.73 and 0.60 kg d^{-1} gains for control mineral, protein, and protein- ionophore supplemented groups, respectively, with the protein supplemented group trending ($P<.10$) to increase rate of gain. The response to this blended protein was as expected, based on expected ruminal response to protein with a range of degradation rates. Typically, however, a response in rate of gain is also expected with the ionophore-protein combination, which was not observed in this study. Fig. 2 shows the results of methane emissions for each of the three treatment groups, including 2 measurement periods per supplement group. There was no significant difference in methane emissions between

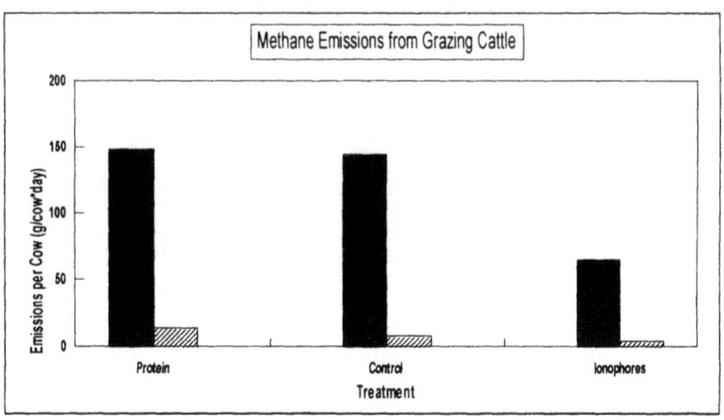

Fig. 2. Methane emissions from grazing cattle expressed by individual measurements and by treatments.

the 2 groups of animals measured in different periods, for any of the treatment groups, and thus overall treatment means are presented for each treatment group. The dark bar designates the mean emissions for treatment and the hatched bar is the standard error.

The protein supplement tended to increase (NS) methane emissions over levels for the animals receiving the mineral supplement. Protein most likely enhanced ruminal cellulose digestion, and consequently, methane production (Protein may have also resulted in an increase in forage intake). Addition of the ionophore to the protein supplement significantly reduced CH_4 emissions by 55% below emissions for the cattle on the control treatment, and by 62% compared to the animals receiving only the protein supplement. The reduction in methane production with monensin has been documented previously, in animals in short term studies in respiration chamber facilities. However, others have noted that the reduction measured in these confined animal facilities appeared to be transitory, with a return to near control levels in 3- 4 weeks. In contrast, in this current study conducted under typical grazing conditions for grazing animals with continuous exposure to a natural environment, there was no indication for any change in the impact of the ionophore in reducing methane production, as the emissions measured during either of the 2 periods where the protein ionophore animals were measured were similar, and much lower than for the other treatment groups.

Use of the micrometerological technique provided a non-interference means of testing methane emissions from animals under natural conditions and verified that currently available management options can have a significant effect on methane emissions from animal production systems.

REFERENCES

Blaxter, K. L. and J. L. Clapperton. 1965. Prediction of the amount of methane produced by ruminants. Br. J. Nutr. 19:511-522.

Byers, F. M. and N. D. Turner. 1991. The role of methane from beef cattle in global warming. Beef Cattle Research in Texas, PR-4838, pp. 69-74.

Denmead, O. T., Harper, L. A., Freney, J. R., Griffith, D. W., Leuning, R., and Sharpe, R. R. 1998. A mass balance method for non-intrusive measurements of surface-air trace gas exchange. Atmos. Environ. 32: 3679-3688.

Harper, L. A., O. T. Denmead, J. R. Freney, and F. M. Byers. 1999. Direct measurements of methane emissions from grazing and feed lot cattle. J. Anim. Sci. 77:1392-1401.

Johnson, K.A. and D.E. Johnson. 1995. Methane emissions from cattle. J. Animal Sci. 73:2483- 2492.

Lassey, K.R., M.J. Ulyatt, R.J. Martin, C.F. Walker, and I.D. Shelton. 1997. Methane emissions measured directly from grazing livestock in New Zealand. Atmos. Environ. 31:2905- 2914.

Lerner, J., E. Matthews, and I. Fung. 1988. Methane emission from animals: A global high-resolution data base. Global biogeochem. Cycles 2:139-156.

USEPA. 2001. Inventory of U.S. Greehnouse Gas Emissions and Sinks: 1990-1999. Office of Atmospheric Programs (6204N), EPA 236-R-01-001, April 2001.

Watson, R. T., H. Rohde, H. Oeschger, and U. Siegenthaler. 1990. Greenhouse gases and aerosols. pp. 1-40. In: J. T. Houghton, G. J. Jenkins, and J. J. Ephraums (eds.) Climate Change: The IPCC Scientific Assessment, Cambridge University Press, Cambridge.

The prediction of methane conversion rate from dietary factors

M. Kurihara[1], T. Nishida[1], A. Purnomoadi[2], M.Shibata[3] & F. Terada[1]

[1]National Institute of Livestock and Grassland Science, Tsukuba Norin-kenkyu-danchi P.O. Box 5, Ibaraki, 305-0901 Japan, [2]Diponegoro University, Kampus Baru Tembalang, Semarang, Indonesia, [3]National Agricultural Research Center for Kyushu Okinawa Region, Nishigoshi-machi, Kumamoto, 861-1192 Japan

CH_4 conversion rate (MCR; MJ CH_4 energy per 100 MJ gross energy intake) in ruminants was influenced by energy intake and other nutrients in the diet. At maintenance levels of energy intake (M), MCR increased when the proportion of concentrate in the diet increased, and with an increase in crude protein (CP) content of the diet from 4 to 9 %. In contrast, at higher levels of energy intake (about 2.5*M), MCR was lower when the proportion of concentrate in the diet was higher. Supplementations with calcium soap of fatty acids, a known inhibitor of CH_4 production, reduced MCR by 0.5 to1.7 % units. MCR of diets fed to dairy cows could be predicted from CP, crude fiber (CF) and nitrogen free extracts (NFE) content (%DM), and DM digestibility (DMD: %) of feed consumed as follows:
(1) MCR=3.37-0.272*CP+0.119*DMD for energy intake level of cows < 1.5*M
(2) MCR=6.34-0.427*CP+0.095*DMD for 1.5*M < energy intake level of cows < 2.5*M
(3) MCR=13.81-0.668*CP-0.195*NFE+0.203*DMD for 2.5M <energy intake level of cows
Multiple correlation coefficients of the equation (1), (2) and (3) were 0.7 ($p<0.001$), 0.57 ($p<0.001$) and 0.76 ($p<0.001$), respectively.

Keywords: methane conversion rate, prediction, dairy cow

1. INTRODUCTION

Shibata *et al.* (1992) demonstrated the import influence that dietary factors have on methane (CH_4) production (liter/day) using dairy cattle, beef cattle, sheep and goats, and developed a predictive relationship for CH_4 production based on dry matter intake. CH_4 conversion rate (MCR; MJ CH_4 energy per 100 MJ gross energy intake), termed by the Intergovernmental Panel on Climate Change (IPCC, 1996), is one of the main input parameters used to predict CH_4 production from ruminant populations in greenhouse gas inventories. Blaxter & Clapperton (1965) reported that MCR in sheep and cattle was related both to dietary energy digestibility and feeding level. Johnson *et al.* (1991) reported that the equation of Blaxter and Clapperton (1965) predicted a considerably narrower range of MCRs than those observed for beef cattle. The objective of this study was to clarify the effect of dietary factors on MCR and to develop a relationship to predict MCR of diets fed to dairy cows from dietary factors. The data used for this analysis were obtained from 185 and 16 total energy balance trials with dairy cows and goats, respectively. CH_4 production was measured for 2 to 4 consecutive days in an open circuit respiration chamber during each trial.

2. EFFECT OF DIETARY FACTORS ON METHANE CONVERSION RATE (MCR)

2.1. Effect of Roughage-concentrate ratio

The influence of roughage-concentrate ratio on MCR depended on the energy intake level of dairy cows (Table 1). While at maintenance levels of energy intake MCR was higher at lower roughage intake, at higher levels of energy intake (about 2.5 x maintenance) it was lower at lower roughage intake. Blaxter & Clapperton (1965) reported that as the apparent digestibility of feed increases, MCR increases markedly at maintenance level of feeding, but at feeding levels of 3 x maintenance it falls. The present data agreed with their result. On the other hand, while CH_4 production per 100 MJ metabolizable energy intakes for dry cows fed on roughage less than 80% of diets was similar, the value for lactating cows had the tendency, which declined with the decline of roughage consumed. These results suggest that daily CH_4 production from lactating cows could be reduced by increasing concentrate up to 70% of diets consumed, but feeding concentrate more than 20% of diets is not effective for lowering daily CH_4 production from dry cows.

Table 1. Effect of roughage-concentrate ratios on methane (CH_4) production.

	Dry cows				Lactating cows		
Roughage component of diet (%)	100	80	60	40	70	50	30
DM[1] digestibility, %	57.5^{D}	67.2^{C}	71.2^{B}	75.7^{A}	62.8^{B}	66.2^{A}	66.5^{A}
CH_4, MJ/100MJ GEI[1]	7.8^{B}	7.8^{B}	8.9^{A}	9.0^{A}	7.3	7.1	5.8
CH_4, MJ/100MJ MEI[1]	16.4^{A}	13.7^{B}	14.6^{B}	13.9^{B}	14.3	12.9	10.5

1: DM, dry matter; GEI, gross energy intake; MEI, metabolizable energy intake. ^{ABCD}Means in the same row within the same animal with different superscripts differ ($p<0.05$) by Tukey's multiple range test.

2.2. Effect of crude protein (CP) Content of Diets

MCR significantly increased from CP 4% diet to CP 9% and CP 12% diets in goats (Table 2). As the CP content in diets increased from 4 to 9 %, MCR tended to increase. This tendency agreed with the results in dry cows. However when ruminants were fed sufficient CP diets above maintenance energy intake, daily CH_4 production was negatively related to daily CP intake (Shibata *et al.*, 1992) or protein digested (Neergaard, 1973). Furthermore in dairy cows fed on diets of which CP content was above 9 %, MCR was negatively related to CP content over all energy intake level (Table 4). Demeyer *et al.* (1972) showed that hydrogen was recovered in microbial cell material as well as CH_4, propionate and butyrate in fermentation balances. Therefore it is hypothesized that more hydrogen derived from fermentation is recovered in microbial cell material when microbial cell synthesis increases with the rise in rumen degradable nitrogen. On the other hand it is thought that the decrease in CH_4 production from ruminants fed on lower CP diets is related to the depression of rumen fermentation as microbial protein synthesis is limited by the supply of nitrogen, a lower limit of 90 g CP per kg dry matter of the diet (ARC, 1980).

Table 2. Effect of dietary crude protein (CP) content on methane (CH_4) production.

	Dry cows		Goats			
CP content of the diet	6.5	8.5	4	7	9	12
DM digestibility, %	49.1	51.3	55.0	57.7	58.5	57.0
CH_4, MJ/100MJ GEI	3.8	4.4	7.3^{B}	7.9^{AB}	8.8^{A}	8.3^{A}
CH_4, MJ/100MJ MEI	10.0	11.2	14.9	15.7	17.7	17.0

Abbreviations, ^{AB}: same as in *Table 1*.

2.3. Effect of Calcium soap of Fatty Acid (CaFA) feeding to diets

Two types of CaFA were used; C18:1 (C18:0 17%, C18:1 65%; Fine feed I, Miyoshi Oils and Fat Co., Ltd.) and C18:2 (C18:0 20%, C18:1 40%, C18:2 35%; Fine feed B). The supplementation with CaFA to basal diet lowered MCR by 1.7 and 0.5 % units for dry and lactating cows, respectively. Chalupa *et al.* (1986) observed a reduction in CH_4 production per mole of unsaturated fatty acid increased with increasing unsaturation of the acid but no effects of degree of saturation and amount of CaFA fed were observed in our experiment, in which Fine feed B contained only 35% unsaturated fatty acid (C18:2) so the effects of degree of unsaturation might not be observed clearly.

Table 3. Effect of calcium soap of fatty acid (CaFA) on methane (CH_4) production.

	Dry cows				Lactating cows	
	Control	C18:1L	C18:1H	C18:2H	Control	C18:2
CaFA intake, g/day		300	600	600		500
DM digestibility, %	71.2	71.6	69.7	70.5	68.1	68.0
CH_4, MJ/100MJ GEI	8.9^{A}	7.5^{B}	7.1^{B}	7.0^{B}	6.7^{a}	6.2^{b}
CH_4, MJ/100MJ MEI	14.6^{A}	12.6^{B}	11.3^{B}	11.2^{B}	11.3^{a}	10.3^{b}

Abbreviations, ^{AB}: same as in *Table 1*.

^{ab}Means in the same row within the same animal with different superscripts differ ($p<0.1$).

3. PREDICTION OF MCR FROM DIETARY FACTORS

Although several equations for CH_4 prediction have been published, most of them are for prediction of daily CH_4 production. Blaxter & Clapperton (1965) and Johnson et al. (1991) derived the equations for the prediction of MCR from digestibility of dietary energy and the level of intake relative to maintenance. Holter & Young (1992) derived the equations for the prediction of MCR from many dietary factors and animal performance. We tried to derive simple equations for the prediction of MCR of diets fed to dairy cows from dependent variables, which can be easily measured, in different levels of energy intake (Table 4). For the data used in the present analysis, MCR ranged from 3.4 to 10.4 %. CP contents of all diets were more than 9%. All regression equations were significant ($p<0.001$). Each standard error of regression estimate (Se) was less than 15% of each mean MCR. The difficulty in accurately predicting MCR increases as the level of energy intake decreases. The accuracy of these equations was similar to the results of Johnson et al. (1991).

Table 4. Relationship between MCR (MJ/100MJ GE intake) and some dietary factors.

Intake level	Mean	Range	Prediction equation for MCR	R^2	Se
< 1.5*M	7.4	4.5[1]	(1) = 15.82-0.351*CP-0.155*CF	0.27	1.1
(n=53)		10.4[2]	(2) = 3.37-0.272*CP+0.119*DMD	0.49	0.9
1.5 – 2.5*M	6.6	3.8[1]	(3) = 1.48-0.202*CP+0.079*NFE+0.206*CF	0.33	1.0
(n=73)		8.9[2]	(4) = 6.34-0.427*CP+0.095*DMD	0.33	1.0
> 2.5*M	5.7	3.3[1]	(5) = 6.10-0.289*CP+0.269*CF	0.36	0.8
(n=39)		7.9[2]	(6) = 13.81-0.668*CP-0.195*NFE+0.203*DMD	0.59	0.7

1, minimum MCR; 2, maximum MCR; Se, Standard error of regression estimate; M, Maintenance energy intake; CP, Crude protein; CF, Crude fiber; NFE, Nitrogen free extracts; DMD, Dry matter digestibility;

From these equations and under the Japanese feeding standard for dairy cattle (1999) MCR of typical diets fed to dairy cows in Japanese practical feeding conditions, it was estimated to be around 6 % other than high fat diet. This MCR, 6 %, was coincident with the recommended value for cattle in developed countries, except for feedlot cattle consuming diets with a large quantity of grain by IPCC (1996).

REFERENCES

AFFRC Secretariat, MAFF. 1999. Japanese Feeding Standard for Dairy Cattle. Central Association of Livestock Industry, Tokyo, Japan. pp189.

ARC, 1980. Requirements for protein. In The nutrient requirements of ruminant livestock, Commonwealth Agricultural Bureaux, Slough. U.K. 120-181.

Blaxter, K.L. & J.L. Clapperton, 1965. Prediction of the amount of methane produced by ruminants. *Br. J. Nutr.* 19: 511-522.

Chalupa, W., B. Vecchiarelli, A.E. Elser, D.S. Kronfeld, D. Sklan & D.L. Palmquist, 1986. Ruminal fermentation in vivo as influenced by long chain fatty acids. *J. Dairy Sci.* 69: 1293-1301.

Demeyer, D., H. Henderickx & C. Van Nevel, 1972. Evaluation of rumen fermentation balance, corrected for cell synthesis. *Proc. Nutr. Soc.* 31: 54A.

Holter, J.B. & A.J. Young, 1992. Methane prediction in dry and lactating cows. J. Dairy Sci. 75: 2165-2175.

IPCC, 1996. Agriculture. In Revised 1996 IPCC Guidelines for National Greenhouse Gas Inventories, vol. 3, IPCC, Bracknell. 4.1-4.140.

Johnson, D.E., T.M. Hill, B.R. Carmean, M.E. Branine, D.W. Lodman & G.W. Ward. 1991. New perspectives on ruminant methane emissions. In Energy metabolism of farm animals. C. Wenk & M. Boessinger, EAAP publication No.58. Switzerland. 376-379.

Neergaard, Lise, 1973. Methane production in growing calves in relation to the composition of the food. In Energy metabolism of farm animals. K.H. Menke, H.-J. Lantzsch & J.R. Reichl, EAAP publication No.14. Stuttgart, B.R.D. 103-106.

Shibata, M., F. Terada, K. Iwasaki, M. Kurihara & T. Nishida, 1992. Methane production in heifers, sheep and goats consuming diets of various hay-concentrate ratios. *Anim. Sci. Technol. (Jpn.).* 63: 1221-1227.

Greenhouse Gases and Animal Agriculture
J. Takahashi & B. A. Young (Editors)

Effect of heat exposure on the methane emission from expiratory gas in sheep fed with high concentrate diets

T. Takahashi, Y. Suzuki and K. Horiguchi

Faculty of Agriculture, Yamagata University, Tsuruoka, Yamagata, 997-8555 Japan

Four sheep (37kg in average body weight) were used in this experiment and the environmental temperature was set at 20℃ (control temperature) and 35℃ (heat exposure). Experimental animals were fed high concentrate diets (roughage : concentrate=2 : 8). Diets used were Italian ryegrass as roughage and commercial formula feed as concentrate. The amount of diets was set up at two levels of maintenance (2% of body weight as dry matter) and 80% of *ad libitum* Feeding. Feeding trials were conducted for 17 days by the cross back design of each environmental temperature. Methane was collected eight times (6 minutes per one time) at 3h intervals from expiratory gas with the method of vacuum pressure. A digestion trial was practiced by total collection of feces. The amount of methane emission per dry matter unit intake increased significantly in the case of heat exposure at both of the two feeding levels. Increasing rate of heat exposure to control temperature was 30% at maintenance and 53% at *ad libitum* feeding, respectively. In heat exposure at the two levels of diets, eating time was remarkably prolonged and digestibility of fibrous components tended to be higher. The decreasing rate of ruminal passage was observed in heat exposure. It is thought that in heat exposure with high concentrate feeding, the methane emission from expiratory gas by sheep increase because the digestibility of fibrous components is improved owing to prolongation of diets in the rumen, regardless of the amount of diets.

Keywords: methane emission, heat exposure, high concentrate, digestibility, sheep

1. INTRODUCTION

At present, methane emitted from ruminants has being watched as a gas with greenhouse effect. By the way, there is very little research on the relationship between methane emission from ruminants and environmental temperature. In dairy cattle, methane increases with the rise in the temperature under high roughage feeding (Shibata, 2001). The reason why the amount of methane emitted from ruminant increased at high temperature (32℃) is thought to be because the retention time in the digestive tract was lengthened in steers fed with grasses, and the digestibility was improved (Warren *et al.*, 1974). In high concentrate feeding under high temperature (32℃), however, there was no effect on methane emission in comparison with 18℃ and 26℃ at the ambient temperature (Shibata, 2001). Furthermore, Shibata *et al.*(1993) reported that though the amount of the methane emission increased with the increase of a dry matter intake, methane emission per dry matter unit decreased. But, the knowledge on methane emission from expiratory gas of ruminants at the ambient temperature over 32℃ was not obtained.

This experiment was carried out to examine the effect of heat exposure (35℃) on the methane emission from expiratory gas in sheep with different levels of high concentrate, and the factors which could be related to the methane emission.

2. MATERIALS AND METHODS

Four sheep cut at less than 2cm in wool length (37kg in average body weight) were used in this experiment. Environmental temperature was set at 20℃ (normal temperature (control)) and 35℃ (heat exposure), and relative humidity was kept to about 70% in both environmental temperatures. Experimental animals were fed with high concentrate diets of which the ratio was 8 : 2 in dry matter. Diets used were Italian ryegrass cut at the length of 4cm as roughage and commercial formula feed as a concentrate, respectively. The amount of mixed diets of concentrate and roughage was set up at two levels of 2% body weight per day in dry matter (maintenance) and 80% of *ad libitum* feeding (*ad libitum*). The amount of diets in 80% of *ad libitum* feeding was determined to be 67.2g/Body Weight $(kg)^{0.75}$ per day in dry matter from the result of pre-feeding tests. The diets were given once daily at 0900 and water was available freely during this experiment. Feeding trials were conducted by the cross back design at each environmental temperature. Each feeding period lasted for 17 days with a preliminary period of 12 days. On the fifth day of each experiment, methane emission was determined from expiratory gas collected with the method of vacuum pressure; i.e., expiratory gas was sampled eight times (for 6minutes per one time) at 3h intervals during a 24 hour period and methane concentration was analyzed by gas chromatography. A digestion trial was practiced by total collection of feces for 5 days. The indigestible particles of mixture polycarbonate and polyethylene were dosed into the reticulo-rumen (300 in numbers per head). The size of one particle (1.8 in specific gravity) is 2mm diameter and 4mm length. Feces were collected every 4 hours for 4 days after reticulo-rumen dosing. All data were evaluated by Student's t-test of SAS.

3. RESULTS AND DISCUSSION

Methane emission (L/kg(DMI)) from expiratory gas in experimental trials is shown in Figure 1. Methane emission was significantly larger at heat exposure (35℃) than in normal temperature (20℃) with both feeding levels: i. e., 17.1 ± 1.6; control and 26.2 ± 3.1; heat exposure at maintenance feeding level, and 19.0 ± 1.9; normal temperature and 24.8 ± 3.5; heat exposure at *ad libitum* feeding level, respectively. Increasing rate of heat exposure to normal temperature was 53.2% at maintenance level and 30.5% at *ad libitum* level, respectively. Comparing with the same feeding levels, in case of maintenance feeding, methane emission tended to be larger at heat exposure than in normal temperature. In case of the *ad libitum* feeding level, however, the difference between both ambient temperatures was not observed. As a result, the difference between same feeding levels tended to be greater at normal temperature than at heat exposure.

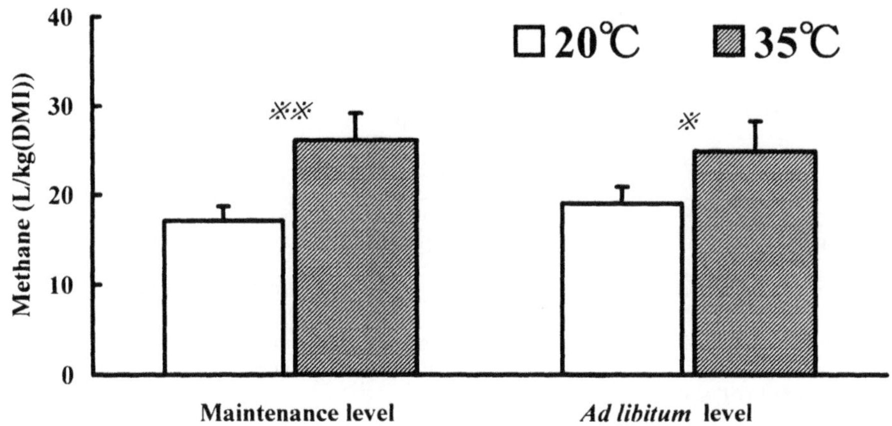

Figure 1. Effect of heat exposure (35 ℃) on methane emission from expiratory gas in sheep.
※※; P<0.01, ※; P<0.05.

Effects of heat exposure on digestibility and on passage rate of diet in sheep are shown in Table 1 and Table 2, respectively. At both feeding levels, digestibilities of fibrous components tended to be higher at heat exposure than at normal temperature. Especially, in *ad libitum* feeding, digestibilities of crude fiber and organic cell wall were higher (P<0.05) at heat exposure than at normal temperature. Digestibilities of components such as crude protein, crude fat and OCC except for fibrous component did not differ between the temperatures regardless of feeding level. Ruminal passage rate of diet (K1) was significantly (P<0.10) lower in heat exposure than in normal temperature in both feeding levels. At heat exposure, the transit time (TT) that the particle dosed in diet appears in feces tended to be delayed more than at normal temperature. However, there was no effect on the rate of post ruminal passage rate (K2) and the total mean retention time (TMRT) of the alimentary tract between two ambient temperatures in both feeding levels.

The time consuming diets lengthened (P<0.05) by heat exposure; i.e., for being about 60 minutes at normal temperature without relating to the feeding levels, they were 146 minutes at maintenance level and 181 minutes at *ad libitum* level by heat exposure.

Table 1. Effect of heat exposure (35 ℃) on digestibility of diets in sheep

Feeding level	Maintenance		*Ad libitum*	
Temperature	20℃	35℃	20℃	35℃
Crude protein	60.2	64.4	62.3	63.7
Crude fat	86.2	86.3	84.3	89.2
Crude fiber	26.4	32.4	29.7^{b}	35.2^{a}
Organic cell wall	57.4	66.4	59.4^{b}	70.5^{a}
Organic cell content	80.4	74.4	78.2	76.3

Means with different superscripts are significantly different at the 0.05 probability level between same feeding levels.

Table 2. Effect of heat exposure (35 ℃) on passage rate of diets in sheep

Feeding level	Maintenance		*Ad libitum*	
Temperature	20℃	35℃	20℃	35℃
K1 (%/hour)	4.2[A]	2.2[B]	4.7[a]	3.2[b]
K2 (%/hour)	5.1	12.8	7.5	6.5
TT (hour)	25.3	57.1	21.3[b]	33.6[a]
TMRT (hour)	74.1	101.6	71.7	86.6

Means with different superscripts are significantly different at the 0.10 probability level between same feeding levels.
K1= rate of ruminal passage, K2=rate of post ruminal passage, TT=transit time (the time from dose until first appearance of particle in feces), TMRT=total mean retention time of alimentary tract, calculated from an equation of Grovum and Williams (1973).

Hirayama *et al.* (2000) reported that in goats fed with grass hay by heat exposure (33℃), the reason why neutral detergent fiber digestibility rose would be due to the decreasing passage rate of feeds through the digestive tract and the lowered contraction frequency of rumen. From the expiratory gas in ruminants fed with high concentrate diets under heat exposure, methane emission did not increase (Kurihara, 1995 & Shibata *et al.*, 2001). However, the methane emission from expiratory gas in sheep fed with high concentrate under severe heat exposure seemed to increase as the result of an active digestion system in the rumen of fibrous component of diet, because the time when the animal takes diet and the retention time of diet in the rumen are lengthened.

The above result indicates that in sheep fed with high concentrate under severe heat exposure, the amount of methane emission would increase without relation to the feeding level.

REFERENCES

Grovum, W. L. & V. J. Williams, 1973. Rate of passage of digesta in sheep. 4. Passage of marker through the alimentary tract and the biological relevance of rate-constants derived from the changes in concentration of marker in faeces. British Journal of Nutrition. 30: 313-329.

Hirayama, T., S. Oshiro, K. Katoh & M. Ohta, 2000. Effects of heat exposure on the rumination and passage rate of feeds through the digestive tract of goats. Animal Science Journal. 71: J258-J267.

Kurihara, M., 1995. Digestion of fibrous components and methane emission. Animal Husbandry. (Japanese) 49: 1213-1218.

Shibata, M., F. Terada, M. Kurihara, T. Nishida & K. Iwasaki, 1993. Estimation of methane production in ruminants. Animal Science and Technology. 64: 790-796.

Shibata, M., 2001. Control of greenhouse gas in animal husbandry 6. Japan Livestock Technical Association. (Japanese) Tokyo. 35-69.

Warren, W. P., F. A. Marts, K. H. Asay, E. S. Hilderbrand, C. G. Payne & J. R. Vogt, 1974. Digestibility and rate of passage by steers fed tall fescue, alfalfa and orchardgrass hay in 18 ℃ and 32 ℃ ambient temperatures. Journal of Animal Science. 39: 93-96.

Greenhouse Gases and Animal Agriculture
J. Takahashi & B. A. Young (Editors)

Effects of ß 1-4 galacto-oligosaccharide (GOS) and *Candida kefyr* on nitrate-induced methemoglobinemia and methane emission in sheep

C. Sar, B. Santoso, X. G. Zhou, Y. Gamo, A. Koyama, T. Kobayashi, S. Shiozaki & J. Takahashi

Depart. of Animal Science, Obihiro University of Agriculture and Veterinary Medicine, Inada-cho, Obihiro, Hokkaido, 080-8555 Japan

4 × 4 latin square trial was carried out to assess the effects of ß 1-4 galacto-oligosaccharide (GOS) and yeast (*Candida kefyr* Strain: Y8A) culture on methemoglobinemia and alteration in gas exchange and metabolic rate in sheep treated with nitrate. Nitrate (1.3 g $NaNO_3$ $kg^{-0.75}$ body weight) with or without GOS and Y8A were administered into the rumen through fistulae as a single dose 30 min after the morning meal. GOS and Y8A were supplemented by sprinkling onto the feed and through rumen fistulae, respectively. Physiological saline was used as the control treatment. In each period, gaseous exchange and metabolic rate were monitored by an open-circuit respiratory system. The administration of nitrate alone induced almost 20% methaemoglobin in total haemoglobin. Consequently, drops in oxygen consumption, carbon dioxide production and metabolic rate were demonstrated as well as a marked reduction in methane production. The simultaneous administration of GOS and GOS mixed with Y8A with nitrate suppressed methaemoglobin formation, and consequently higher rates of the gas exchange and metabolic rate were observed. Thus, the results of this study indicate that GOS and GOS supplemented with Y8A are very likely to suppress methaemoglobinemia and depression of gas exchange and metabolic rate induced by nitrate, respectively.

Keywords: ß 1-4 galacto-oligosaccharide, Candida kefyr, methemoglobinemia, methane

1. INTRODUCTION

High level of nitrate in forages, water and food products could exert serious effects on animal health because of the acute toxicity syndrome known as nitrate poisoning (Jones, 1972). Ruminants are particularly vulnerable, as nitrate is reduced to ammonia by certain rumen microbes with nitrite as an intermediary product (Marais, *et al.*, 1988). Nitrite produced from ruminal reduction of nitrate is responsible for methaemoglobin arising from the reversible oxidation of ferrous iron in haemoglobin to the ferric valence (Takahashi and Young, 1992). Advanced methemoglobinemia results in severe anoxia. A subclinical nitrate-nitrite toxicity decreased oxygen consumption, carbon dioxide and methane production and metabolic rate as 20 % methaemoglobin formed (Takahashi and Young, 1994).

Oligosaccharides are naturally occurring carbohydrates with a low degree of polymerisation and consequently low molecular weight, being commonly found to perform in the various plant and animal sources. However, the function of Oligosaccharide in the rumen ecosystem is so far not known.

Meanwhile, Yeast culture (Diamond V Yeast culture, Diamond V Mills, Inc, Cedar Rapids, IA 52407) has received considerable attention as a nitrate intoxication preventive among livestock producers in Oklahoma (Stone, 1974 and 1976). As reported by Gerald *et al.* (1984), however, it has been indicated that the yeast culture doesn't decrease the methemoglobinemia produced by consumption of high-nitrate forages by sheep or cattle.

This experiment was, therefore, conducted on the effect of ß 1-4 galacto-oligosaccharide (GOS) and yeast (*Candida kefyr* Strain: Y8A) culture on methemoglobinemia and alteration in gas exchange and metabolic rate in sheep dosed with nitrate at sufficient levels to induce methaemoglobin.

2. MATERIALS AND METHODS

Four rumen-fistulated wethers with an initial weight of 43.8-59.3 kg were allocated to four dietary treatments in a 4 × 4 latin square design and fed on a basal diet of chopped orchard grass hay and timothy hay cube (50:50,w/w) at a maintenance level (55 g DM $kg^{-0.75}$ body weight) offered twice a day (8:00h and 16:00h meal). An interval of a week was allowed between each of the four trials to make sure there were no carry-over (lag) effects from a previous treatment. To examine the effect of ß 1-4 galacto-oligosaccharide (GOS) and Y8A culture on the nitrate-induced poisoning, nitrate (1.3 g $NaNO_3$ $Kg^{-0.75}$ body weight) with and without Y8A and GOS was directly administered into the rumen via fistulae as single dose 30 min after the morning feeding, and GOS (10 g a day) and Y8A were supplemented by sprinkling onto the feed and through rumen fistulae, respectively. Y8A culture used in this trial, had been extracted from naturally-fermented milk "Laban" produced from sheep milk in Yemen. Physiological saline (0.9% NaCL) was given as the control treatment. All animals were individually maintained in metabolic cages each equipped with a ventilated respiratory gas collection hood, with free access to water. All animals were weighed weekly before the beginning of each period to adjust daily allowance of feed and dosages of nitrate. Respiratory gaseous exchanges were monitored from 1 hour before to 8 hours after administration of the chemicals. To evaluate the advance of nitrate poisoning in relation to cardinal symptoms, blood and rumen fluid were taken at intervals during trial.

Data were analysed by repeated measurement analysis of variance according to the General Linear Models (GLM) Procedures of the Statistical Analysis Systems Institute (Cary, NC, 1994) for latin square arrangement of treatments.

3. RESULTS AND DISCUSSION

It is generally recognized that the toxicity of nitrate for ruminants depends upon its enzymatic reduction to nitrite by rumen microbes and the combination of nitrite with haemoglobin in blood to induce methemoglobinemia. A methaemoglobin content of about 20% of total haemoglobin was considered subclinical (Bondansky, 1951; Takahashi, 1992). In this experiment, a methaemoglobin level of just 20 % was achieved in the nitrate treatment as evident in Fig.1, which is comparable with these studies.

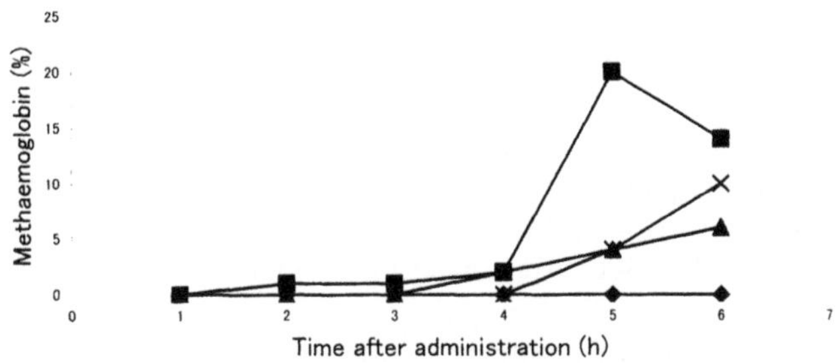

Fig.1. *Methaemoglobin accumulation in sheep after administration of saline control (◆), Nitrate (■), Nitrate+ GOS (▲) and Nitrate+ GOS+ Y8A (X). Each point indicates a mean of four animals.*

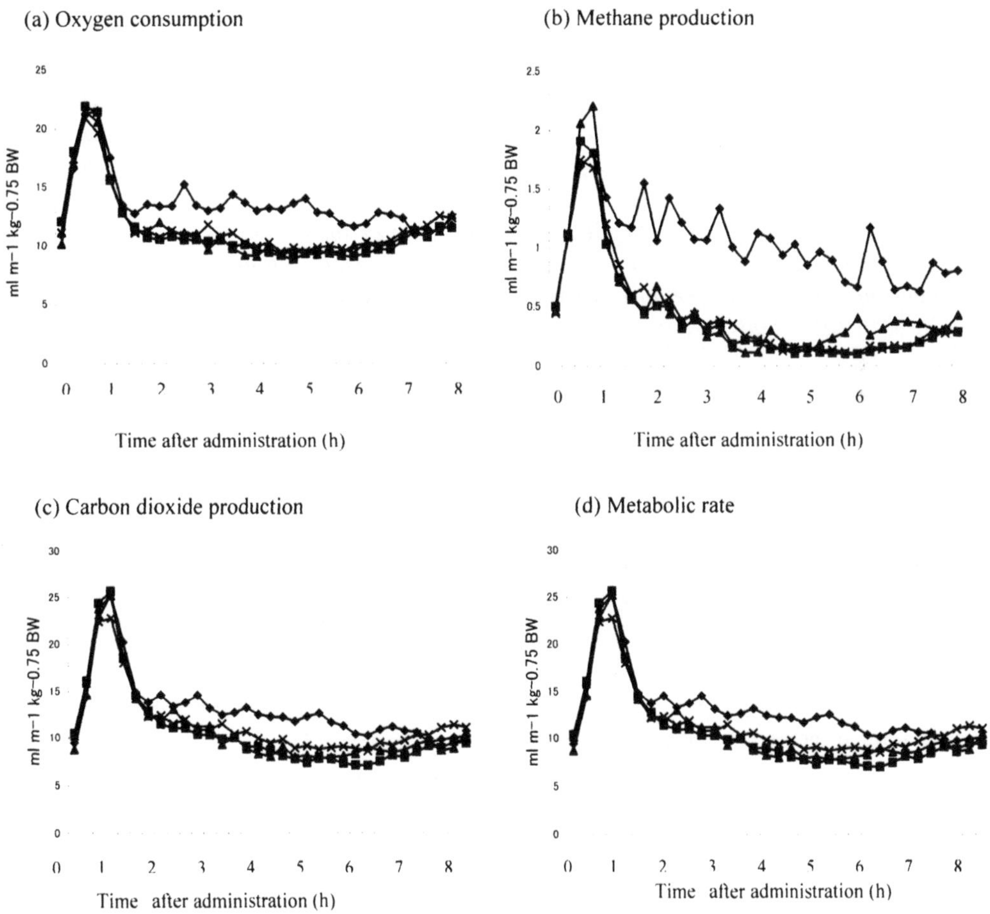

Fig.2. *Gas exchange and metabolic rate in sheep after administration of Saline control (◆), Nitrate (■), Nitrate+ GOS (▲) and Nitrate+ GOS+ Y8A (X). Each point indicates a mean of four animals.*

The result in Fig.2 show that oxygen consumption, carbon dioxide and metabolic rate were markedly decreased in sheep treated nitrate compared with the control treatment, being consistent with the report made by Takahashi *et al.* (1983). Moreover, the experimental observation indicated here is in agreement with the established fact that methanogenesis by rumen microbes has been inhibited by nitrate in *in vitro* and *in vivo* trials (Jones, 1972; Allison et al., 1981; Takahashi et al., 1983). Thus, the respiratory gas exchange and metabolic rate of the animal can be used to evaluate physiologically the extent of nitrate-induced methemoglobinemia.

Fig.2 (a), (b) and (d) also show the incomplete recovery of oxygen consumption, carbon dioxide production and metabolic rate to control level in the simultaneous administration of GOS and GOS plus Y8A with nitrate while GOS with or without Y8A suppressing nitrate-induced methaemoglobin formation, respectively. Meanwhile, Fig.2 (c) shows that methanogenesis by ruminal microbes was not observed to be lower in the simultaneous administration of nitrate with GOS and GOS plus Y8A than nitrate treatment alone given, respectively.

The results summarized in Table 1 show the suppressing effect of GOS and Y8A on nitrate-induced methemoglobinemia and disorders in respiratory gas and energy metabolism in sheep. Therefore, it is concluded that the appropriate amount of β 1-4 Oligosaccharide and *Candida kefyr* with nitrate is very likely to suppress rumen methanogenesis without poisoning.

Table1. *The suppressing effect of GOS and Y8A culture on nitrate-induced formation of methaemoglobin (expressed as a percentage of total haemoglobin) and respiratory gas exchange (ml^{-1} $kg^{-0.75}$ body weight) in sheep 5h after intra-ruminal administration of nitrate (1.3g $NaNO_3$/ $kg^{-0.75}$ body weight)*

	Treatment			
	Control	Nitrate	Nitrate plus GOS	Nitrate plus GOS plus Y8A
Methaemoglobin	0.0±0.0[1,a]	19.8±10.6[a]	4±2.3[a]	4.3±3.2[a]
O_2 consumption	13.5±0.8[a]	9.2±0.3[b]	9.8±0.8[b]	9.8±0.8[b]
CO_2 production	12.2±0.8[a]	7.8±0.3[b]	8.2±0.7[b]	9.1±0.9[b]
CH_4 production	1.0±0.1[a]	0.1±0.0[b]	0.2±0.1[b]	0.2±0.0[b]
Metabolic rate (W)	4.9±0.0[a]	3.1±0.1[b]	3.2±0.3[b]	3.4±0.3[b]

[1] Standard error of means from four animals.

Means within rows with different superscripts differ significantly ($P<0.05$).

REFERENCES

Allison, M.J., Reddy, C.A., and Cook, H.M., 1981. The effects of nitrate and nitrite on VFA and CH_4 production by ruminal microbes. J. Anim. Sci., (Suppl. 1): 391 (Abstract).

Bodansky, O., 1951. Methaemoglobinemia and methaemoglobin-producing compounds. Pharmacol. Rev., 3: 144-196.

Gerald W. H. and Keith S. L., 1984. Preliminary studies on the effect of yeast culture supplementation on nitrate/nitrite induced methemoglobinemia in lambs and steers. Vet. Hum. Toxicolo., 26(4): 309-313.

Jones, G.A., 1972. Dissimilatory metabolism of nitrate by the rumen microbiota. Can. J. Microbiol., 18: 1783-1987.

Marais, J.P., Therion, J., Mackie, I., Kistner, A., Dennisoson, C.D., 1988. Effect of nitrate and its reduction products on the growth and activity of the rumen microbial population. Br. J. Nutr., 59: 301-313.

Stone, C.W., 1974 and 1976. Yeast culture research briefs, Diamond V Mills, Inc, Cedar Rapids, IA.

Takahashi, J., Chiba T., Encho, T., Ikariga, J. and Fujita, H., 1983. The acute and chronic effect of nitrate on heat production and methanogenesis in ruminants. Proceedings of the 5th World Conference on Animal Production, 14-19 August 1983, Tokyo, Japan. Japanese Society of Zootechnical Science, Tokyo, Japan, 2: 351-352.

Takahashi, J. & Young, B.A., 1992. The modulation of nitrate-enhanced hypothermia by sulphur compounds in cold-exposed sheep. Anim. Feed Sci. and Technol., 39: 347-355.

Takahashi, J. & Young, B.A., 1994. The regulation of energy metabolism in sheep by nitrate and L-cysteine. In: Aguilera, J.F. (Ed), Energy Metabolism and Farm Animals. EAAP publication No. 76, Granada, pp: 387-390.

Greenhouse Gases and Animal Agriculture
J. Takahashi & B. A. Young (Editors)

Methane emission and nitrogen excretion by goats fed tropical byproducts based pelleted diet

M. Islam[1], O. Enishi[2], K. Higuchi[2], A. Purnomoadi[2], N. Takusari[2] & F. Terada[2]

[1]Presently at: Laboratory of Nutritional Science, Faculty of Agriculture, Kyoto University, Oiwakecho, Kitashirakawa, Sakyo-ku, Kyoto 606-8502, Japan, [2]Energy Metabolism Laboratory, National Institute of Livestock and Grassland Science, Tsukuba, Norin-danchi, Ibaraki 305-0901, Japan

Effects of pelleting and levels of intake of tropical byproducts based diet on methane (CH_4) and nitrogen (N) excretion by goats were investigated by a series of metabolism trials in a 2x3 split plot factorial experiment using 4 goats. The diet consisted of 50% sugarcane tops, 10% soybean husk, 15.5 % wheat bran, 10% molasses, 10% rice bran, 2% fishmeal, 1% urea, 1% mineral mixture and 0.5% common salt. The feeding systems were pellet and conventional forms, and intake levels were high (2.0M), medium (1.5M) and low (1.0 M). Goats consumed all the allotted feed for up to 2.0 M level in pellet form while in conventional feeding the goats refused some of the roughage portion at the 2.0 M level. DM and OM digestibilities were similar for the pelleted and conventional forms, while higher ($P<0.001$) digestibilities were found for lower intake levels. Methane emission was slightly higher in pellet feeding but heat production (HP), urinary energy and N losses were higher in conventional form. Methane emission, HP rate, total excretion (fecal and urinary nutrient) losses were lower at the high and medium levels of intake. Thus, pellet feeding at higher levels of intake increased retained energy and retained N through reducing the rate of energy and N losses. The lower HP per DMI and nutrient excretion through urine in pellet feeding at high intake levels resulted in lower CH_4 and N emission from the diet.

Key words: pelleting, intake level, tropical byproducts, methane, goats

1. INTRODUCTION

Maximum energy intake can be achieved by maximizing dry matter intake (DMI) and by increased energy concentration of the diet. Increase in energy concentration of diet in tropical countries is costly, but increase in DMI using agro-industrial by-products could be comparatively easier and more cost effective. Nutrient qualities of the byproducts are low and therefore, in conventional long-chopped feeding systems, resulted in low DMI, consequently under-nutrition of the tropical ruminants. In addition, with the low quality of roughage, energy requirement for eating and rumination can increase from 10% to 30% of the metabolizable energy (ME) provided by the feed-stuff, due to heat production (HP) associated with eating and rumination (Susenbeth et al., 1998). Processing of low quality roughage could increase the DMI and digestible nutrient intake (Minson, 1990; Islam, 1999). Since decreased

CH_4 production is associated with increased rate of passage and pelleting increased the rate of passage, therefore, CH_4 emission may decrease when feed is pelleted. In addition, if pelleting reduces urinary nutrient loss, the combined reduction may increase the ME. Consequently, it would increase overall feed efficiency so that fewer animals are required for the same production resulting in a reduction of total CH_4 emission. Higher levels of feed intake also increase the rate of passage and a reduction in the proportion of the ration digested in the rumen. Faster passage rate of feed through rumen results in a greater post-ruminal digestion and more energy utilization of pelleted compared chopped hay (Black and Tribe, 1973) even equal proportions of feed per unit body weight (Shaw et al., 1960). Since ruminal digestion produces more CH_4 and accounts for a significant loss of digestible energy (DE), increase level of feed thus reduces the CH_4. Therefore, this study was conducted to examine the effects of pelleting of diet prepared from agricultural by-products and use at higher levels of intake on energy and N metabolism in goats.

2. MATERIALS AND METHODS

The diet consisted of 50% dried sugar cane tops (imported), 15% soybean meal, 10% wheat bran, 10% rice bran, 10% molasses, 2% fishmeal, 1% urea, 1.5% mineral mixture and 0.5% common salt. Three intake levels based on TDN requirement (Itoh et al., 1978) were, low (1.0), medium (1.5) and high (2.0), times were employed with four adult male Japanese goats in 6 digestion and metabolism trials. The goats were randomly assigned to two groups, with mean body weights of 26.8 and 24.9 kg for group A and B, respectively. In the first trial conventional (cane tops mixed with molasses and urea immediately before feeding with supplementation of concentrate) and pellet forms of the diet at 1.0M level were fed to the animals of group A and B, respectively. In subsequent trials, the two forms of the diet were given to the animal group B and A, respectively. Similarly, the trials for the levels of 1.5M and 2.0M were conducted. The detailed methods of metabolism trials, respiration chambers management, nutrient analyses, collection and analyses of rumen liquor, energy balance measurements were discussed in our earlier report (Islam et al., 2000). Statistical significance was determined by analysis of variances with the dietary intake levels and form of diet as factors in a split plot design where the factors at levels of intake are the main plot and the form of feeding is the sub plot. The analyses were carried out using a general linear model procedure of SAS (1994).

3. RESULTS AND DISCUSSION

Diet composition is shown in Table 1. Dry matter and organic matter (OM) digestibilities were higher ($p<0.01$) at lower intake levels (Figures 1), but the values were similar for pellet and conventional feeding. The DM and OM digestibilities were higher on conventional feeding at low intake level but at the medium and high levels of intake the digestibilities were similar or slightly higher on pellet feeding (Figure 1. a. and b). Total energy losses through feces and urine were also higher ($p<0.001$) on conventional feeding than that for the pellet feeding. The rate of the losses was lower ($p<0.001$) at medium and high levels of intake. The CH_4 emission (CH_4 Kg^{-1}DM, or methane conversion ratio, MCR, CH4/GE) were higher ($P<0.05$) in pellet feeding. However, the rate of heat production (HP) per unit feed was lower at the high level of intake and pellet feeding (Table2).

Table 1. Nutrient composition (g kg^{-1}DM, unless otherwise stated) of ingredients and diets

Feeds	DM (g kg^{-1})	OM	CP	GE (Kcal kg^{-1}DM)
Concentrate mixture	916	916	198	4501
Molasses-urea mixed sugar cane tops	643	892	89	4123
Complete feed of cane tops and conc.	939	899	120	4243

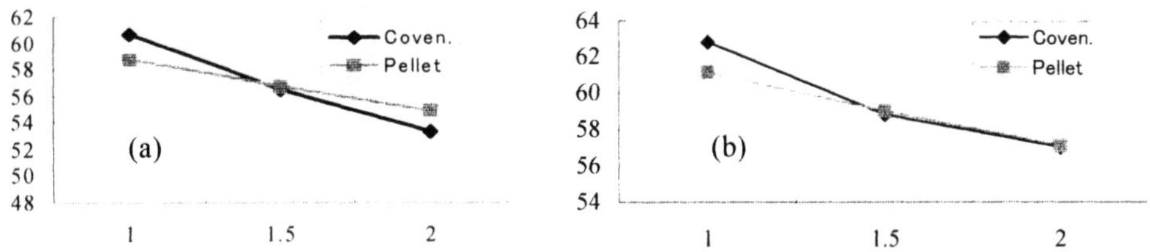

Figure 1. DM (a) and OM (b) digestibilities (%) on pellet and conventional feeding at three intake levels

Nitrogen intakes ($P<0.001$), and urinary nitrogen (UN) ($P<0.05$) output rates were higher for conventional feeding (Table 3). The proportional values for fecal N/N intake and UN/N intake were lower ($p<0.001$) for the higher levels of intake. The proportional nutrient losses as urine were lower ($P<0.001$) at medium and high levels of intake supports that high intake levels could reduce nutrient excretion. Therefore, using pelleted feed at a high level of intake could reduce urinary nutrient losses. The slightly higher ($P>0.05$) CH_4 emission (g day^{-1}), and CH_4 conversion ratio (MCR, %GE intake) ($P<0.001$) on pellet feeding was due to the small particle size in pellet and more area was allowed for microbial degradation in rumen, which was also evidenced by the higher digestibility values of the nutrients on pellet feeding at higher levels of intake (Table 2, Figure 1). The higher N losses (Table 3) as feces/unit N intake in pellet might be due to the increased rate of passage. However, the lower N loss as urine and lower total N losses at the higher level of intake showed a higher absorption of N that resulted in a higher retained N.

The MCR value ranged from 6-7%, which was similar to the values reported for sheep and goat on a range of roughage concentrate based diets at different intake levels (Terada et al., 1985: Islam et al, 2000). Of the N intake, the diet's form and intake levels affected the proportions lost as urine and feces. The high UN loss at low and medium intake levels resulted in the leaching NO_3-N from animal source to surface, ground and drinking water as well and thus, the excess N causes environmental pollution as ammonia and nitrate in the manure (Yano and Nakajima, 1996). The trend of N losses and differences of N losses as feces were small among the intake levels whilst the differences were higher in N losses as urine. About 23% of urinary N reduced at high level of intake compared to the low intake level. Pelleted feed slightly reduced the energy losses as HP per DMI. Goats refused some of the roughages at 2.0 level of intake, suggesting that in conventional feeding it is hard to achieve a high level of intake using agro-industrial byproducts based diet. Pelleting of tropical fibrous feed-based diet ensures a high intake level which maximizes net nutrient supply to the animal and, consequently maximizes animal productivity. Using pellet feed there was no refusal. However, in conventional or supplementation form of feeding, the waste of feed during feeding generally encountered was approximately 10-15%.

Table 2. Energy balance (Kcal.day^{-1}W$^{-0.75}$, unless otherwise stated) and partition of energy goats fed sugar cane tops based diet in different forms and levels of intake

Items[2]	Form		SEM[1]	Level of intake			SEM[1]
	Conv.	Pellet		High	Medium	Low	
Gross energy intake*	273	258	1.48***	339	274	185	4.02***
Fecal energy	121	114	1.32**	156	120	76	1.63***
Urinary energy*	10.6	10.1	0.25ns	12.1	10.8	8.1	0.20***
Methane (g/d/goat)	14.2	14.4	0.27ns	17.5	14.7	10.7	0.52***
Methane (g /KgDMI)	19.71	20.83	0.30*	18.23	20.45	22.13	0.45***
MCR (Methane/GE)	0.06	0.07	0.001*	0.06	0.06	0.07	0.002***
Heat production	106	99	1.24**	114	103	91	1.63***
HP per kgDMI (Mcal)	1.73	1.70	0.25ns	1.45	1.59	2.08	0.04***
Digestibility (DE/GE)	0.57	0.56	0.004ns	0.54	0.56	0.59	0.004***
Partition of digestible energy							
Urine**	0.07	0.07	0.001ns	0.07	0.07	0.08	0.001*
Methane	0.11	0.12	0.002*	0.10	0.11	0.12	0.004*
Metabolizable energy*	0.82	0.81	0.002*	0.83	0.82	0.81	0.004**
ME per kg DM (Kcal)	1997	1937	2.2ns	1922	1951	2028	3.5*
Partition of metabolizable energy							
Heat production	0.86	0.87	0.01ns	0.76	0.82	1.03	0.02***
Retained energy	0.13	0.12	0.01ns	0.24	0.18	-0.03	0.02***

[1] SEM–standard error of least square means. ns-non-significant; *-$P<0.05$; **-$P<0.01$; ***- $P<0.001$.
[2] Significant levels of the interaction effects of level of feeding and forms

Table 3. Nitrogen balance (g.day^{-1}W$^{-0.75}$) of goats fed sugar cane tops based diet in different forms and levels of intake

Items[2]	Diets		SEM[1]	Levels of intake			SEM[1]
	Conv.	Pellet		High	Medium	Low	
Intake**	1.29	1.17	0.01***	1.58	1.26	0.85	0.02***
Feces	0.41	0.42	0.009ns	0.55	0.43	0.27	0.01***
Urine**	0.73	0.68	0.02*	0.83	0.72	0.57	0.02***
Digested*	0.88	0.75	0.01***	1.03	0.83	0.58	0.01***
Retained*	0.14	0.07	0.015**	0.21	0.10	0.01	0.02***
Fecal N/ N intake	0.31	0.36	0.01***	0.35	0.34	0.31	0.01*
Urinary N/ N intake*	0.58	0.60	0.02ns	0.52	0.57	0.68	0.01***

[1] SEM–standard error of least square means. ns-non-significant; *-$P<0.05$; **-$P<0.01$; ***- $P<0.001$.
[2] Significant levels of the interaction effects of level of feeding and forms

The tropical sugar cane tops, which are very low in nutrients and generally left in the field, recently have been used as ruminant feed to meet the roughage shortage in Asia. Pelleting of the cane tops mixing with crop's byproducts and 1.0% urea as NPN could give a diet with an energy digestibility of 56%. The dried and imported sugar cane tops (3.5% CP) showed a very low palatability. Therefore, the tops was mixed with molasses urea solution (as a conventional practice in Asia used with other low quality roughage like straw) to improve palatability. The daily mixing is very cumbersome and is difficult for commercial large farms. Therefore, pelleting could be a better way to transfer the fibrous feeds to animal protein. A high level of intake could be used to reduce the rate of CH_4 and N excretion. However, the total input-output cost in terms of quality environment and productivity of ruminants are needed to study thoroughly.

ACKNOWLEDGEMENTS

The first author acknowledges The Ministry of Education, Culture, Sports, Science and Technology and Japan International Science and Technology Exchange Center (JISTEC) for awarding the STA/JSPS Fellowship.

REFERENCES

Black, J.L. & D.E. Tribe, 1973. Comparison of ruminal and abomasal administration of feed on the growth and body composition of lambs. Aust. J. Agric. Res., 24: 763-777.

Islam, M. 1999. Nutritional Evaluation and Utilization of Oil Palm (Elaeis guineensis) Frond as Feed for Ruminants. Ph.D. Thesis, Universiti Putra Malaysia, Malaysia.

Islam, M., H. Abe, Y. Hayashi, & F. Terada, 2000. Effects of feeding Italian ryegrass with corn on rumen environment, nutrient digestibility, methane emission, and energy and nitrogen utilization at two intake levels by goats. Small Rum. Res. 38, 165-174.

Itoh, M., T. Haryu, R. Tano, & K. Iwasaki, 1978. Maintenance requirement of energy and protein for castrated Japanese goats. Bull. Nat. Inst. Anim. Ind., Japan. No. 3, 41-50.

Minson, D.J., 1990. Forage in ruminant nutrition. Academic Press. San Diego, CA.

SAS (Statistical Analysis System), 1994. User's guide. Fourth edition. Version 6. SAS Institute Inc. Box 8000. Carry NC. 27511-800 USA.

Susenbeth, A. R. Mayer, B. Koehler, & O. Neumann, 1998. Energy requirement for eating in cattle. J. Anim. Sci. 76: 2701-2705.

Terada, F., K. Iwasaki, R. Tano, M. Itoh, & K. Kameoka, 1985. Comparison of metabolizibility and metabolizable energy contents among cattle, sheep, and goats fed the same diets. European Assoc. Anim. Prod. Pub. No. 32. pp. 130-133.

Yano, F. & T. Nakajima, 1996. Nutritional approaches for reducing nitrogen and phosphorus in Livestock Waste. Proc. Anim. Sci. Cong. 8th AAAP, Tokyo, Japan, Vol. 1. pp. 83-88.

Greenhouse Gases and Animal Agriculture
J. Takahashi & B. A. Young (Editors)

Development of nutritional management for controlling methane emissions from ruminants in Southeast Asia

S. Shioya, M. Tanaka, Y. Iwama & M. Kamiya

Department of Animal and Grassland Science, National Agricultural Research Center for Kyushu Okinawa Region, 2421 Suya Nishigoshi, Kikuchi, Kumamoto, 861-1192 Japan

To develop nutritional management for controlling methane emissions from Southeast Asian ruminants, the following two experiments were done on Holstein cattle.

1) Experiment 1 examined the influence which the nutritive value of tropical grass exerted on methane emission from ruminants. Four Holstein heifers, or dry cattle, were fed sudangrass silage, bahiagrass hay, or rice straw to measure nutritive value and amount of methane formation. Respective TDN contents in sudangrass, bahiagrass, and rice straw were 62, 55, and 54%, while methane emissions per dry matter intake were 31.3, 37.7, and 36.6L/day/kg, respectively. The relationship between methane emission per dry matter intake (y) and TDN contents (x) could be shown with the equation: $y=91.588-1.01x$. Therefore, methane emission per dry matter intake showed a tendency of decreasing concomitant with TDN content increase.

2) Experiment 2 examined the influence of feeding sweet potatoes for milk production and methane emission from cattle. Cattle were fed bahiagrass hay or bahiagrass hay and sweet potatoes. When potatoes were fed, milk production, milk element, and dry matter intake showed an increasing tendency. Also, cattle methane emission decreased by adding potatoes from the 146L/day to the 260L/day. Furthermore, ratios of methane emission to milk production decreased to 25.5L/day/kg by adding potatoes from 48L/day/kg. From the above results, it was considered effective to add some sort of potato to roughage using high nutrition to control cattle methane emissions in Southeast Asia by nutritional management.

Keywords: cattle, nutrition, methane, Southeast Asia, milk production

1. INTRODUCTION

It is known that methane emission from ruminants changes by feed quality. Especially, it increases greatly with intake of low quality forage. Also, methane emission from ruminants shows a tendency to increase in hot environments. Therefore, ruminant methane emission from Southeast Asia, where low quality tropical grass and agricultural by-products are the main feed is thought to be greater than in advanced countries. Still, it is considered that methane emissions can be decreased through improvement of feed quality and improved nutritional balance. Methane released by livestock in warm Southeast Asia is thought to constitute about 20% of methane released globally by ruminants. Furthermore, decreased methane emissions is an important subject in this area due to increased livestock populations in the future.

To develop nutritional management technologies for controlling ruminant methane emissions, methane must first be measured when various tropical grasses are digested by cattle. The influence exerted on methane emissions of the nutritive value of tropical grass is explained by these experiments. Also, technological development aims at mitigating methane formation by raising microorganism activity in the rumen. Therefore, influence of additional potato starch, which can be obtained at a low price in this area, on methane formation is explained.

2. MATERIALS AND METHODS

2.1. Experiment 1

Sudangrass silage, bahiagrass hay, and rice straw were used for the experiment. Sudanese grass was mown three times and made into roll-wrapped silage. Two other grass species were used after the first cut. Four Holstein heifers or dry cows were fed the above forage and nutritive values of those grasses were measured. The amount of methane formation was measured by respiration tests in an environmentally controlled room.

2.2. Experiment 2

Four Holstein milking cows (less than 10kg of milk production) with milk productivity similar to Southeast Asian dairy cows were fed bahiagrss hay, or bahiagrass hay and sweet potato. Methane formation amounts were measured by respiration tests in an environmentally controlled room. The amount of TDN of feed, dry matter intake, milk production, and milk element were measured.

3. RESULTS

3.1. Influence of feeding tropical grass on methane emissions from ruminants

Table 1 shows the effects on methane emissions of feeding tropical grass to ruminants. There were high amounts of methane emissions from cattle with bahiagrass hay and rice straw which had dry matter intake (DMI) amounts larger than sudangrass. Relationship between DMI (y) and methane emission (x) is shown with the equation $y=-46.508+41.179x$ (Fig. 1). Shibata et al (1992) reported the equation $y=-0.127+34.360x$. The DMI at this time is in the range of 1.5 times from one maintenance time, and the inclination is large in both equations. Furthermore, they reported that inclination of the equation becomes large concomitant with DMI decrease. The nutritive value of grass used in these experiments is low; moreover, the cattle appetite is thought to decrease according to high temperature. Therefore, it was thought that this inclination became greater from DMI being about equal to the maintenance amount.

Methane emissions varied by grass species, cutting, and so on. There were fewest methane emissions per DMI in sudangrass, in which TDN content was highest of the three species. Methane emissions per DMI decreased according to the rate of TDN content of grass was high when DMI was about equal as shown in Figure 1. Kurihara et al (1996) also reported observing a tendency of equivalence, as in this experiment, in comparison to feed when the rate of concentrate / roughage differs. Furthermore, they reported that this tendency becomes stronger as temperature increases. On the other hand, cattle energy requirements increase and DMI also increases under such a hot environment (Shioya and Terada, 1997). Therefore, methane emissions increase remarkably. In tropical areas, it is considered that it is very effective to feed cattle high TDN content diets which can fulfill energy requirements even if DMI is restricted to reduce cattle methane emissions.

Table 1. Influence of feeding tropical grass on ruminant methane emissions.

Forage	Cutting	Hay/Silage	TDN (%)	DMI (kg/d/head)	CH_4 (L/d/head)	CH_4 /DMI (L/kg)
Sudangrass	1	Silage	59.2	6.45	206.0	31.9
Sudangrass	3	Silage	61.0	5.15	155.0	30.1
Sudangrass	1	Silage	57.4	5.85	186.1	31.8
Bahiagrass	1	Hay	54.8	6.85	258.1	37.7
Rice straw	1	Hay	53.6	8.08	295.5	36.6

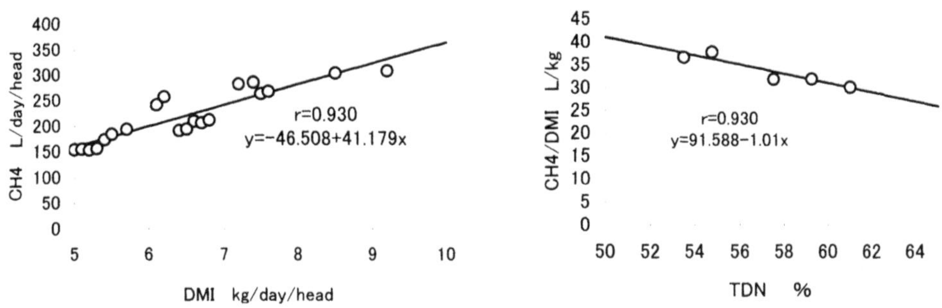

Figure 1. Relationship between DMI and CH_4 emissions (left). Relationship between TDN content and CH_4 emissions (right).

3.2. Influence of additive sweet potato for milk production and methane emission

Feed compositions and digestibility of bahiagrass hay and sweet potato are shown in Table 2. Sweet potatoes contain much starch and have high digestibility, so they are an effective energy source.

Table 3 shows the influence of sweet potato additives for milk production, milk elements, and methane emissions from dairy cows. The DMI and milk production showed a tendency to increase with added sweet potatoes. Methane emissions decreased in the opposite direction by adding sweet potatoes. Methane emissions per 4% fat-corrected-milk (FCM) decreased by about half by adding sweet potatoes. Also, methane emissions per FCM production decreased as much as FCM production increased. Shibata (2000) reported that it appears more remarkably at low level milk productivities. Therefore, it is important to enhance the ability of dairy cows in areas where cattle are fed low quality feed in developing countries. When FCM production was small, it was shown that it was very effective toward increasing FCM making while decreasing methane emissions for this experiment as well.

From the above results, to reduce CH_4 emissions from ruminants in Southeast Asia, it was effective to feed high TDN content grass and add an energy source like sweet potatoes.

Table 2. Feed composition and digestibility of bahiagrass hay and sweet potatoes (DM %).

Feed	CP	EE	CF	NFE	Ash	ADF	NDF	Digestibility
Bahiagrass hay	6.5	1.4	32.8	50.5	8.8	41.8	74.8	51.4
Sweet potato	2.9	0.3	2.5	91.4	2.9	3.8	5.4	80.8

Table 3. Influence of feeding sweet potatoes on milk productivity and methane emissions.

Treatment		Hay	Sweet potato
Body Weight	kg	573.5	569.3
Dry matter intake	kg/day	6.9 ^{A}	8.6^{B}
Hay	kg/day	6.9	3.1
Sweet Potatoes	kg/day	-	5.5
Milk Production	kg/day	4.67	4.92
FCM Production	kg/day	5.39	5.73
Milk Fat	%	5.07	5.12
Milk Protein	%	3.46	3.85
Milk Sugar	%	3.83	3.38
Solids-not-fat	%	8.28	8.23
Methane Emissions	L/day	259.1^{A}	146.3^{B}
Methane emissions / FCM	L/kg	48.1^{A}	25.5^{B}

A,B : $P<0.01$

REFERENCES

Kurihara, M. & T. Nishida, 1996. Global Environment Research of Japan in 1995(II). Research and Information Office Global Environment Department Environment Agency Government of Japan. 624-627.

Shibata, M., F. Terada, K. Iwasaki, M. Kurihara & T. Nishida, 1992. Methane production in heifers, sheep and goats consuming diets of various hay-concentrate rations. Anim. Sci. Technol. (Jpn), 63:1221-1227.

Shibata, M, 2000. Controlling Greenhouse Gases in Animal Production. (5). Society for Livestock Technology. 32-62.

Shioya, S., & F. Terada, 1997. The physiological response of milking cow under hot environment. Proc. Jpn. Soc. Anim. Nutr. Metab. 41:61-68.

Greenhouse Gases and Animal Agriculture
J. Takahashi & B. A. Young (Editors)

In vitro inhibition of microbial methane production by chemicals

N. Mohammed[1,2], N. Ajisaka[1], Z. A. Lila[2], Koji Hara[1] K. Mikuni[1], Kozo Hara[1], S. Kanda[2] & H. Itabashi[2]

[1]Bioresearch Corporation of Yokohama, Tsurumi-ku, Yokohama 230-0053, Japan
[2]Laboratory of Agricultural Production Technology, Tokyo University of Agriculture and Technology, Tokyo 183-8509, Japan

The objective of this study was to evaluate the effects of different concentrations of L-malate, 2-iodopropane- α -cyclodextrin complex (CD-IP) and their mixture on in vitro methane production by mixed rumen microorganism. Ruminal fluid was collected from dairy cows, fed hay and concentrate, and incubated at 39°C for 6 h with or without addition of different concentrations of malate, IP-CD or their mixture. At the end of the incubation period, methane and hydrogen were detected by gas chromatography and samples were taken for VFA and NH_3-N analysis. Fermentation of corn starch in the presence of 10 or 20 mM L-malate resulted in a decrease in methane production by 15 and 20%, respectively. Total VFA, total gas production and molar proportion of propionate tended to increase. Molar proportion of acetate, NH_3-N concentration and pH of the medium were unchanged. In the case of 20 or 40 mg CD-IP (0.2 or 0.4 mM IP), CH_4 production was decreased by 48 and 97%, respectively and H_2 production was 21.2 and 57.1 times higher as compared to the control, respectively. Total VFA and proportion of propionate tended to increase. NH_3-N concentration of the medium slightly decreased. Molar proportion of acetate, total gas production, and pH of the medium were unchanged. Addition of L-malate (10 or 20 mM) to IP-CD (0.2 or 0.4 mM IP) resulted in a decrease in methane production by 59 to 79% and 97 to 98%, respectively. On the other hand, hydrogen production was decreased by 61 to 69% and 63 to 68%, respectively as compared to IP-CD only. Total VFA, total gas production and molar proportion of propionate tended to increase. Molar proportion of acetate, NH_3-N concentration and pH of the medium were unchanged. In conclusion, these results suggested that addition of L-malate to CD-IP decreased methane production significantly as well as to decrease hydrogen production from CD-IP.

Keywords: methane, corrinoid inhibitor, α-cyclodextrin, volatile fatty acid, propionate

1. INTRODUCTION

Each year 30 more Tg (1 Tg = 10^{12}g) of methane is emitted into the atmosphere (Houghton et al., 1992). Of them, 20 – 30% of methane emissions originate from older and fossil sources including natural gas, mining and wetland emissions with the remaining 70 – 80% arising from contemporary sources including enteric fermentation (animal and insects), biomass burning, oceans and lakes, rice production and waste treatments (landfills, sewage etc.)

(Johnson and Johnson, 1995).

Methane is one of the products of digestive fermentation in ruminants and is released into the environment by eructation. Ruminants typically lose 2 – 12% metabolisable energy from feed by this pathway (Czerkawski, 1969). It is estimated that the world's population of ruminants produces about 15% of total atmospheric methane emissions (Moss, 1993). Greenhouse gases are a major concern for the environment and the effect on raising the atmospheric temperature. Methane is a highly potent greenhouse gas due to its high absorption ability of infrared radiation from the sun. On the other hand, methanogenesis represents a 5 to 15% energy loss to the ruminant animal.

Malate is a four-carbon dicarboxylic acid that is commonly found in biological tissues because it is an intermediate of citric acid cycle (Lehninger, 1975). It is a key intermediate in the succinate-propionate pathway and the predominant ruminal bacterium *Selenomonas ruminantium* uses this pathway (Gottschalk, 1986). More recent research showed that L-lactate uptake by S. ruminantium HD4 is increased in the presence of 10 mM L-aspartate, fumarate or L-malate; L-malate elicits the greatest response (Nisbet and Martin, 1990). Martin and Streeter (1995) reported that DL-malate caused a decrease in methane concentration, increase in final pH and propionate concentration. On the other hand, methanogens were shown to contain corrinoids at levels equal to or greater than those reported in *Clostridium thermoaceticum* and *Acetobacterium woodii* (Tanner et al., 1978). In these homoacetate-fermenting bacteria corrinoids function as methyl carriers during acetate synthesis via reductive carboxylation (Ghambeer et al., 1971). *M. barkeri* contain more corrinoids than do the other methanogens examined, but the exact level depends on the specific carbon and energy sources during growth (Krzycki and Zeikus, 1980). The role of corrinoids in methanogen methyl transfer reaction was explained previously (Wassenaar et al., 1998). The methyl group is transferred to coenzyme M (HS-CoM) resulting in the formation of the common substrate of the terminal methane-forming reaction, methyl-coenzyme M (CH_3-S-CoM). This transfer is catalyzed by the concerted action of two methyltransferases. A substrate-specific enzyme accepts the methyl group by binding it to the corrinoid prosthetic group which, in methanogenic enzymes, is typically 5-hydroxy-benzimidazolyl-cobamide (B_{12}-HbI). The methyl function is subsequently handed to coenzyme M by Co-methyl-B_{12}:HS-CoM methyltransferase. Iodopropane is a corrinoid inhibitor and it inhibited cell growth and methane production when *Methanobacterium thermoautotrophicum*, *Methanobacterium formicicum* and *Methanosarcina barkeri* were cultured on H_2-CO_2 (Kenealy and Zeikus, 1981).

The present paper investigated the effects of iodopropane, malate and their combination on ruminal microbial methane production and rumen fermentation in vitro.

2. MATERIALS AND METHODS

Ruminal contents were obtained from dairy cows fed 5 kg hay cube, 5 kg hay (sudan grass) and 6 kg concentrate in the morning, and 5 kg silage and 6 kg concentrate in the evening daily. The ruminal contents were obtained before morning feeding and squeezed through four layers of surgical gauge into an Erlenmeyer flask and purged O_2-free carbon dioxide. The rumen fluid was anaerobically transferred (1:2) to a medium containing 292 mg of K_2HPO_4, 292 mg of KH_2PO_4, 480 mg of $(NH_4)_2SO_4$, 480 mg of NaCl, 100 mg of $MgSO_4 \cdot 7H_2O$, 64 mg of $CaCl_2 \cdot 2H_2O$, 4000 mg of Na_2CO_3 and 600 mg cysteine hydrochloride per liter into separating funnel and incubated at 39° C for 30 min. Sixty milliliters of microorganisms and medium was anaerobically transferred to 120-ml serum bottles that contained 120 mg of corn starch as a substrate plus with or without malate, CD-iodopropane and their mixture. The bottles were capped with butyl rubber stoppers and aluminium seals and placed in a 39° C water bath for 6 h. At the end of the incubation period, 0.5 ml of gas was removed from each of the bottles, and methane and hydrogen were detected by gas chromatography (Model GC-8A, Shimadzu Co. Ltd., Kyoto). Total gases were measured by plunger method. The bottles were opened and samples were taken for VFA and NH_3-N analysis by gas chromatography (Model GC-14A, Shimadzu Co. Ltd., Kyoto) and Conway's micro-diffusion method, respectively.

3. RESULTS AND DISCUSSION

Addition of L-malate tended to decrease methane production gradually from corn starch without increasing hydrogen (Fig. 1 & Fig. 2). About 15 and 20% of methane production was decreased by 10 and 20 mM of L-malate, respectively. 8, 10 or 20 mM of L-malate tended to increase propionate and these corresponded to a decrease in acetate : propionate ratio. Total VFA and gas production were increased, and NH_3-N and pH of the medium were unchanged. Fermentation of corn starch in the presence of CD-IP (CD content 10% IP) resulted in a drastic decrease in methane production (Fig. 3) and hydrogen production was increased (Fig. 4). About 48 and 97% of methane production were inhibited by 20 and 40 mg (0.2 or 0.4 mM IP) of CD-IP, respectively. All three concentrations of CD-IP tended to increase total VFA, propionate and acetate : propionate ratio was decreased.

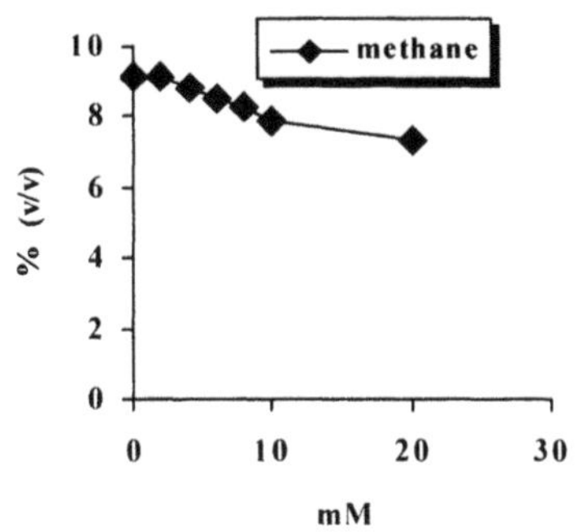

Fig. 1. Effect of malate on CH_4

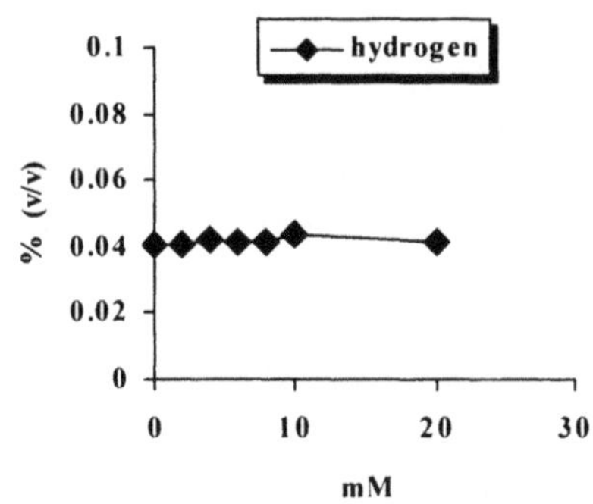

Fig. 2. Effect of malate on H_2

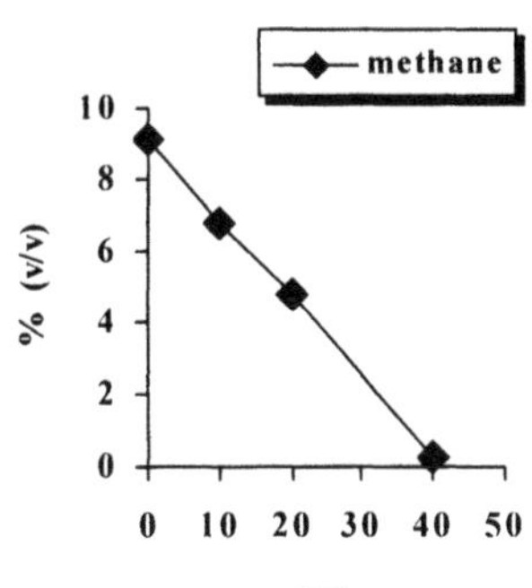

Fig. 3. Effect of CD-IP on CH_4

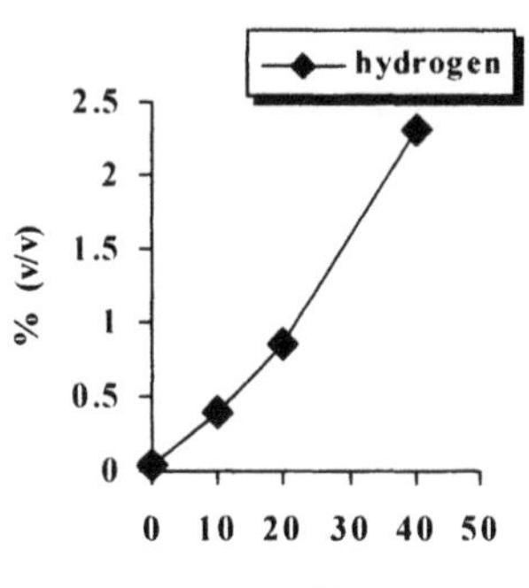

Fig. 4. Effect of CD-IP on H_2

NH_3-N concentration of the medium slightly decreased. Molar proportion of acetate, total gas production, and pH of the medium were unchanged. Addition of L-malate (10 or 20 mM) to IP-CD (0.2 or 0.4 mM IP) resulted in a decrease in methane production by 59 to 79% and 97 to 98%, respectively (Fig. 5). Hydrogen production was decreased by 61 to 69% and 63 to 68%, respectively as compared to IP-CD only (Fig. 6). Total VFA, total gas production and molar proportion of propionate tended to increase. Molar proportion of acetate, NH_3-N concentration and pH of the medium were unchanged.

Therefore, incorporation of malate to IP-CD into our incubations stimulated mixed ruminal microorganism fermentation, change in fermentation products and decreased hydrogen production.

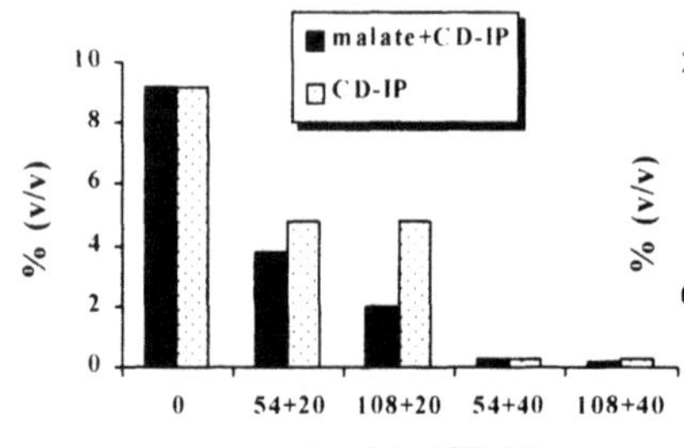

Fig. 5. Addition of malate to CD-IP on CH_4

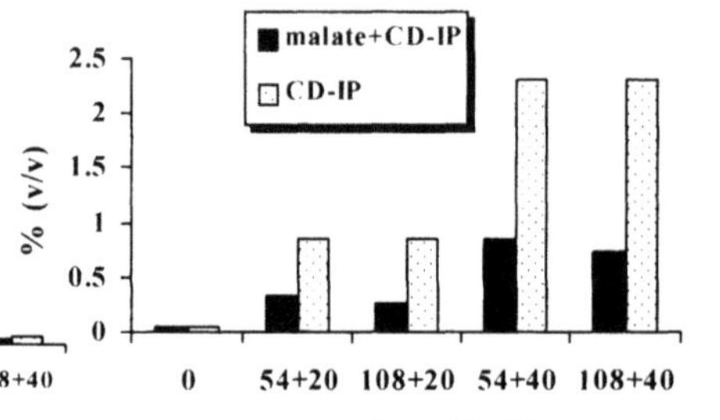

Fig. 6. Addition of malate to CD-IP on H_2

REFERENCES

Czerkawski, J. W., 1969. Methane production in ruminants and its significance. World review of ruminants and Dietetics. 11: 240-282.

Ghambeer, R. K., H. G. Wood, M. Schulman & L. Ljungdahl, 1971. Total synthesis of acetate from CO_2. III. Inhibition by alkylhalides of the synthesis from CO_2, methyltetrahydrofolate methyl-B_{12} by *Clostridium thermoacetocum*. Archieves of Biochemistry and Biophysics. 143: 471-484.

Gottschalk, G., 1986. Bacterial Metabolism (2nd Ed.). Springer-Verlag, New York. pp 242.

Houghton J. T., B. A. Callander & S. K. Varney, 1992. Climate change 1992. The supplementary report to the IPCC scientific assessment. University press, Cambridge UK.

Johnson, K. A. & D. E. Johnson, 1995. Methane emissions from cattle. Journal of Animal Science. 73: 2483-2491.

Kenealy, W. & J. G. Zeikus, 1981. Influence of corrinoid antaginists on methanogen metabolism. Journal of Bacteriology. 146: 133-140.

Krzycki, S., & J. G. Zeikus, 1980. Quantification of corrinoids in methanogenic bacteria. Current Microbiology. 3: 243-245.

Lehninger, A. L., 1975. Biochemistry (2nd Ed.). Worth Publishers, New York.

Martin, S. A. & M. N. Streeter, 1995. Effect of malate on in vitro mixed ruminal microorganism fermentation. Journal of Animal Science. 73: 2141-2145.

Moss, A. R., 1993. Methane: global warming and production by animals. Chalcombe Publications, Kingston, UK. pp 105.

Nisbet, D. J. & S. A. Martin, 1990. Effect of dicarboxylic acids and *Aspergillus oryzae* fermentation extract on D-lactate uptake by the ruminal bacterium *Selenomonas ruminantium*. Applied Environmental Microbiology. 56: 3515.

Tanner, R. S., R. S. Wolfe & L. G. Ljungdahl, 1978. Tetrahydrofolate enzyme levels in *Acetobacterium woodii* and their implication in the synthesis of acetate from CO_2. Journal of Bacteriology. 134: 668-670.

Wassenaar, R. W., J. T. Keltjens, C. Van Der Drift & G. D. Vogels, 1998. Purification and Characterization of dimethylamine:5-hydroxybenzimidazolylcobamide methyltransferase from Methanosarcina barkeri Fusaro. European Journal of Biochemistry. 253: 692-697.

Greenhouse Gases and Animal Agriculture
J. Takahashi & B. A. Young (Editors)

Effects of lactic acid bacteria, yeasts and galactooligosaccharide supplementation on *in vitro* rumen methane production

Y. Gamo[1], M. Mii[2], X. G. Zhou[1], C. Sar[1], B. Santoso[1], I. Arai[1], K. Kimura[3] & J. Takahashi[1]

[1]Obihiro University of Agriculture and Veterinary Medicine, Obihiro, Hokkaido, 080-8555 Japan, [2]Meiji Feed Company, Takasaki, Gunma, 370-0047 Japan, [3]Yakult Central Institute for Microbiological Research, Kunitachi, Tokyo, 186-0011 Japan

The manipulating effects of galactooligosaccharide and/or lactic acid bacteria and yeasts as probiotics on rumen methanogenesis were determined using *in vitro* gas production system. The treatments were control (CTL) without manipulator, β1-4galactooligosaccharide (GOS), lactic acid bacteria (*Leuconostoc mesenteroides subsp. mesenteroides* (L5), *Leuconostoc lactis* (L12), *Lactococcus lactis subsp. lactis* (L23), *Lactococcus lactis subsp. lactis* (L42)), lactic acid bacteria with galactooligosaccharide (L5+GOS, L12+GOS, L23+GOS, L42+GOS), yeasts (*Candida kefyr* (Y8A), *Candida kefyr* (Y4D), *Saccharomyces pastorianus* (Y3A), *Trichosporon sericeum* (Y7A)) and yeasts with β1-4 galactooligosaccharide (Y8A+GOS, Y4D+GOS, Y3A+GOS, Y7A+GOS). Methane production was reduced by GOS, L5, L12, L42, Y8A and Y7A. The amount of methane production in L5+GOS and Y8A+GOS was lower than L5 and Y8A, respectively. In addition, supplementation of GOS in L23, L42, Y8A, Y3A and Y7A tended to increase total VFA concentration. Degradation of β1-4 galactooligosaccharide by lactic acid bacteria and yeasts after1 hour incubation was about 80% and 90%, respectively.

Keywords: methane, lactic acid bacteria, yeast, probiotics

1. INTRODUCTION

Methane is produced from H_2 and CO_2 by methanogens in the rumen. Methane production in the rumen represents loss of feed energy for the host animals. The energy loss of methane is estimated to be 7-10% of gross energy and 12% of digestible energy. In addition, methane emission from ruminants contributes to global warming as greenhouse gas. Thus, the rumen methane production in ruminants must be reduced due to both economical and environmental issues.

Ionophores such as monensin, salinomycine or lasalocid have been used as feed additives to improve feed efficiency (Goodrich et. al., 1984; Jonson et. al., 1985). These polyether antibiotics can reduce rumen methanogenesis due to enhancing propionate production. But, they adversely affected the digestibility of fiber or the number of protozoa (Chen et. al., 1979; Hino, 1981). Furthermore, possible problems could be induced by the remaining antibiotics in the products and the appearance of resistant cells. It is important that rumen manipulators for methane suppression are safe for humans, animals and the environment. Thus, natural alternatives to ionophores must be sought for this purpose. Dawson (1988) reported that addition of lactic acid bacteria and yeast could decrease 12% of A/P ratio and 4% of methane production *in vitro* incubation trial. Lactic acid bacteria supplements also have been used for health and preventing disease in swine production (Pollman, et. al., 1980). Therefore, if lactic acid bacteria can compete with methanogens, methane production in rumen may be reduced. Yeast as supplements could stimulate digestion of cellulose and utilization of lactate in rumen (Callaway, et. al., 1997; Nisbet, et. al., 1991). On the other hand, methane production was increased because yeast stimulated microorganisms related to fiber (Sullivan, et. al., 1999). So far, effects of probiotics on rumen methanogenesis have not been consistently evaluated.

The objective of this study is to determine the effects of lactic acid bacteria, yeasts, and/or β1-4 galactooligosaccharide as feed supplements on *in vitro* rumen methane production.

2. MATERIALS AND METHODS

A four-separable flask was used as the fermentation flask of gas production system. The rumen fluid was collected from two rumen fistulated cows as the inoculum for incubation. The collected 750 ml of rumen fluid was mixed with 750 ml of McDougall's artificial saliva (McDougall, 1948) and then 5g of ground air-dried orchard grass silage was added to the mixed fluid as substrate. For the manipulators β1-4 galactooligosaccharide (GOS), four lactic acid bacterias (*Leuconostoc mesenteroides subsp. mesenteroides* (L5), *Leuconostoc lactis* (L12), *Lactococcus lactis subsp. lactis* (L23), *Lactococcus lactis subsp. lactis* (L42)), four lactic acid bacteria with galactooligosaccharide (L5+GOS, L12+GOS, L23+GOS, L42+GOS), four yeasts (*Candida kefyr* (Y8A), *Candida kefyr* (Y4D), *Saccharomyces pastorianus* (Y3A), *Trichosporon sericeum* (Y7A)), and four yeasts with β1-4 galactooligosaccharide (Y8A+GOS, Y4D+GOS, Y3A+GOS, Y7A+GOS). β1-4 galactooligosaccharide was produced from cow milk (Yakult Central Institute). For control (CTL) incubation was performed without manipulators. Lactic acid bacteria and yeasts were isolated from "Laban" which was a traditional fermented milk product in Yemen. In lactic acid bacteria examined L23 and L42 were identified as the same species but they had different activities. The flask was placed in a 39°C water bath and incubated anaerobically for 12 hours. Each treatment was performed in triplicate. Cumulative methane production was determined by a fully automated infrared gas analyzing system. The fluid samples collected every hour were taken by the peristaltic pump to determine pH, ORP, and VFA, ammonia nitrogen and lactic acid concentrations. β1-4 galactooligosaccharide were determined in 1 and 12 hour incubation fluids by using the Alliance HPLC System (2690) Waters (Yakult).

3. RESULTS AND DISCUSSION

In vitro methane production rate and cumulative methane production were extrapolated by non-liner regression analyses from the time course of methane production for 12hour incubation. Fig.1 shows effects of lactic acid bacteria, yeasts and/or β1-4 galactooligosaccharide on *in vitro* methane production rate. The rate of methane production without supplement (CTR) peaked after about 30 hours. With β1-4 galactooligosaccharide (GOS), it peaked after about 20 hours. Addition of lactic acid bacteria and yeasts was earlier than CTR and GOS. The rate of methane production with lactic acid bacteria or yeasts peaked after about 10 hours.

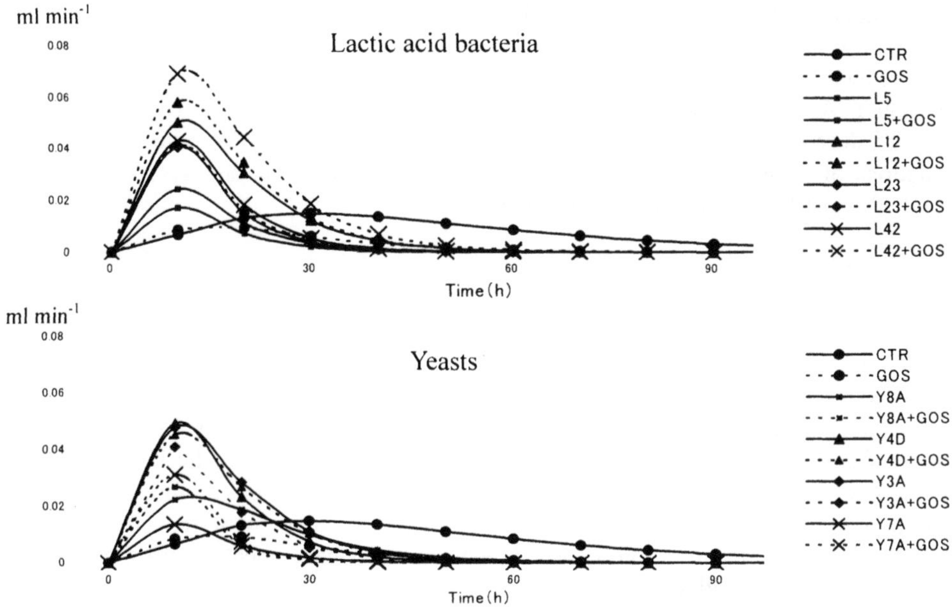

Fig. 1. Effects of lactic acid bacteria and yeasts from "Laban" and/or β1-4 oligosaccharide on the rate of methane production

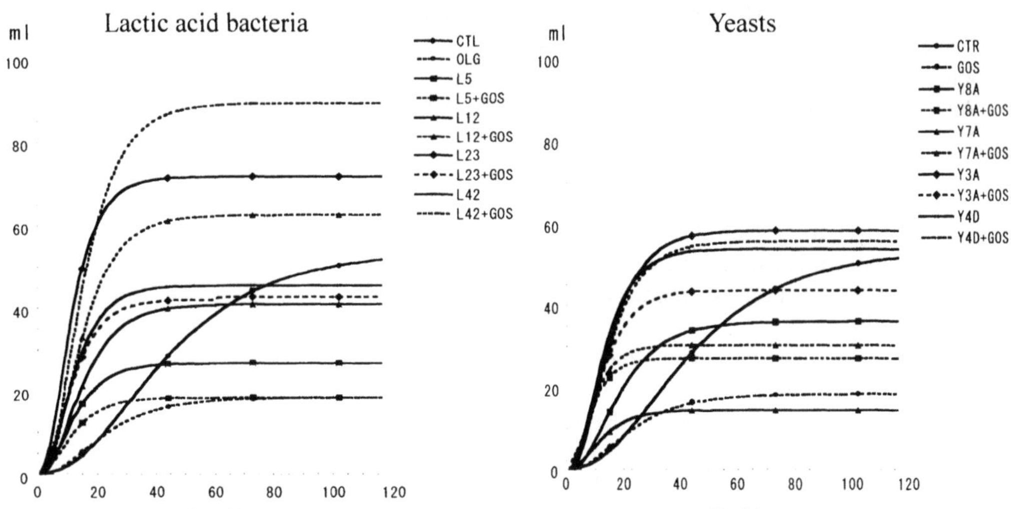

Fig. 2. Effects of lactic acid bacteria and yeasts from "Laban" and/or β1-4 galactooligosaccharide on cumulative methane production
Bertalanffy model : $CH_4 (ml) = a + b(1 - e^{-ct})^3$ Where, t= minutes

Fig. 2 shows the effects of lactic acid bacteria and yeasts on the cumulative methane production. The lowest methane production was obtained in β1-4 galactooligosaccharide (GOS) in treatments of lactic acid bacteria. L5, L12 and L42 also reduced cumulative methane production to less than the control level. In addition, cumulative methane production in L5+GOS was much less than L5. In treatments of yeast, Y7A and Y8A reduced cumulative methane production. However, cumulative methane production in Y7A+GOS was higher than Y7A, though Y8A+GOS reduced it much lower than Y8A. For VFA production GOS or L5 decreased acetate/propionate ratio. All yeasts tended to decrease the A/P ratio. Degradations of β1-4 galactooligosaccharide in lactic acid bacteria and yeasts were about 80% and 90% at 1 hour incubation, respectively.

4. CONCLUSION

Methane production was reduced by addition of β1-4 galactooligosaccharide. Among lactic acid bacteria and yeasts examined *Leuconostoc mesenteroides subsp. mesenteroides*, *Leuconostoc lactis*, *Lactococcus lactis subsp. Lactis*, *Candida kefyr* and *Trichosporon sericeum*. The simultaneous addition of β1-4 galactooligosaccharide in *Leuconostoc mesenteroides subsp. mesenteroides* or *Candida kefyr* was lower than the sole addition of these microbes. The inhibition of methane production by these alien microbes and/or β1-4 galactooligosaccharide might have occurred due to the stimulation of reduction reactions consuming hydrogen. In some probiotics, however, the enhancement of fiber degradation induced the increase in methane production.

REFERENCES

Callaway E. S, S. A. Martin, 1997. Effect of a Saccharomyces cerevisiae culture on ruminal bacteria that utilize lactate and digest cellulose. J. Dairy Sci. 81:1591-1598.

Chen M, M. J. Wolin, 1979. Effect of monensin and lasalocid-sodium on the growth of methanogenic and rumen saccharolytic bacteria. Appl. Environ. Microbiol. 38:72-77.

Dawson K. A., 1988. Manipulating ruminal fermentations: Are there natural alternatives to ionophores for beef production? Biotechnology in the Feed Industry. 101-112. Alletech Technical Publications.

Goodrich R. D, J. E. Garrett, D. R. Gast, M. A. Kirick, D. A. Larson, J. C. Mieske, 1984. Influence of monensin on the performance of cattle. J. Anim. Sci. 58:1484-1498.

Hino T, 1981. Action of monensin on rumen protozoa. Jap. J. Anim. Sci. Technol. 53:171-179.

Jonsson E, I. Olsson, 1985. The effect on performance, health, and fecal microflora of feeding Lactobacillus strains to neonatal calves. Swed. J. Agrc. Res., 15:71-76.

McDougall E. I., 1948. Studies on Ruminant Saliva , 1 The compositionand output of sheep saliva. Biochem. J. 43:99-109.

Nisbet D. J., S. A. Martin, 1991. Effect of a Saccharomyces cerevisiae culture on lactate utilization by the ruminal bacterium Selenomonas ruminantium. J. Anim. Sci. 69:4628-4633.

Pollman D. S., D. M. Danielson, E. R. Wren, E. R. Peo, K. M. Shahani, 1980. Influence of Lactobacillus acidophilus inoculum on gnotobiotic and conventional pigs. J. Anim. Sci. 51:629-637.

Sullivan H. M., S. A. Martin, 1999. Effects of Saccharomyces cerevisiae culture on in vitro mixed ruminal microorganism fermentation. J. Dairy Sci. 82:2011-2016.

Part 3

Bio-energy from agricultural bio-mass

Greenhouse Gases and Animal Agriculture
J. Takahashi & B. A. Young (Editors)

The current status of anaerobic digestion in China

R. Dong[1], G. Liu[1], Y. Hou[1], Y. Zeng[2], L. Yan[2], Y. Kitamura[3] & T. Yamaguchi[4]

[1]KLAB, China Agricultural University, Beijing 100083, P. R. China; [2]Chengdu Anaerobic Institute, Chengdu, P. R. China; [3]Shimane University, Japan; [4]Institute of Agricultural and Forest Engineering, the University of Tsukuba, Tsukuba City, 305-8572, Japan

China is a leading state in the research and application of Anaerobic Digestion Technologies (ADT) on the world specially in rural areas. As early as in 1890s the methane was known in South China. For the current application, there are four main fields including the treatment of Agricultural Waste (AWT), Industrial Organic Waste Waster (IOWT), Urban Waste Water and the Mud from Waste Waster (UWMT), and Urban Solid Organic Disposals (USODT). Higher gas production, optimal design and combination of components for the AD system, and better integration in a District Ecological Circle for sustainable development have been achieved. The problems remain for ADT in China and the potentialities have also been analyzed. Suggestions for efficient application of ADT in China and the developing countries have been proposed.

Keywords: anaerobic digestion, wasted resource, ecological benefit, comprehensive utilization

1. INTRODUCTION

China is one of the leading countries conducting AD researches and large-scale applications. Early on it was one of the counties with the most developed techniques in methane production. Simple digestion tanks appeared along the coastal areas of Guangdong Province in early 19th Century. For the protection of local economy, Mr. Guorui Luo, from Taiwan Province, developed the first artificial Anaerobic Digester with water-based gas pressure balance technology. The second ADT extension was initiated in Wuchang, while the "three-parts combination" technology was revised the Chinese Style AD was formed. At the beginning of 1980's, the special national AD propagation system came into practice. Series of AD technologies were further developed and popularized, the comprehensive utilization of methane has been strengthen combined with the ecological construction.

The main materials for Anaerobic Digester are agricultural residues, industrial waste water, household waste water and the sludge from the waste water treatment, and household solid residues. Large amount of organic content is in those wastes (Fig. 1, Fig. 2 and Table 1) (Bian Y., 2000; Wang K.,2000). Clean energy, high efficiency organic fertilizer as well as additives for animal feeds could be expected from the organic wastes after anaerobic treatment.

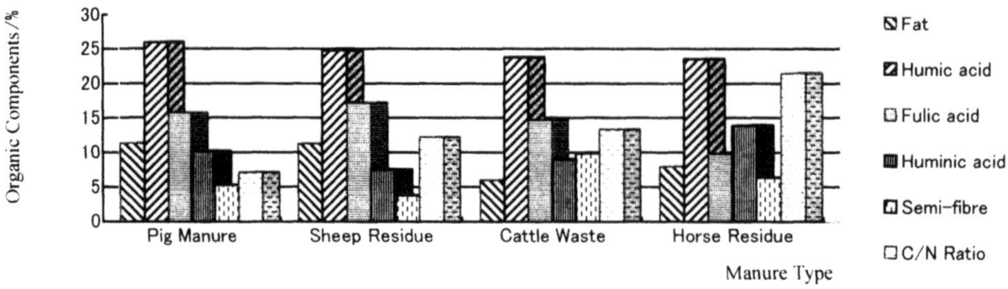

Fig.1 Organic components in Manure/%

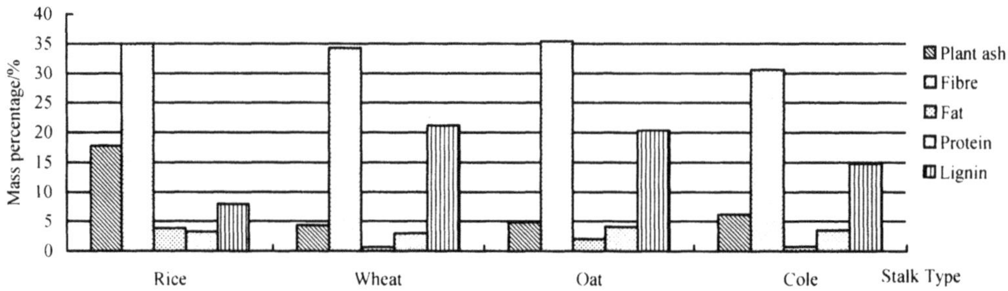

Fig.2 Mass percentage of organic components in several plant stalks/%

Table 1 Main components of lees from some food and drink products (mg/l) (Wang K., 2000)

Item \ Waste	Sweet potato lees	Corn lees	Molasses lees	Monosodium Glutamate lees	Lemon waste	Alcohol lees	Beer lees	Beet lees
PH	5.4	5.2	3.2-3.5	5.0-5.5	3-4	---	---	---
W(deoxidized sugar)	0.22	---	---	0.75	0.4	---	---	---
W(total sugar)	0.68	0.83	2.2	---	1.0	---	1.8	---
W(total solid)	5.2	0.57	11.5	11.54	4.0	---	---	---
W(suspend solid)	4.2	4.2	1.5	1.54	---	40	20-25	8.4
W(plant ash)	0.6	0.22	3.1	6.03	---	5.84		
W(N)	0.13	---	0.31		---	8.1	5.1	0.9
W(P)	0.02	---	0.05	0.16	---	0.11	---	0.05
W(glutamicacid)	---	---	---	1.74	---	---	---	---
W(amylum)	---	---	---	1.03	---	---	---	---
COD(g/L)	52.6	70	130	100-200	20-30	30-53	---	---
BOD(g/L)	23.3	---	50-60	---	---	---	---	---

The AD scale has developed from several square meters to as high as several thousands square meters. The 1999 Yearbook said that the household AD with 5-8m^3 has been 7.63 million, while the number of large AD is 959, among which the largest one has a volume of

8000m^3 with a capacity to treat 32.18 million tons of waste and to produce 0.42 billion m^3 of methane. There have been 79 thousand of household waste water anaerobic treatment stations with a capacity to treat 237.69 billion tons of waste water. There are 115 Methane Power stations with a capacity of 2342 kW and a yearly generation of 3.01 million kW-h electricity. The utilization of methane has been centralized and large scale, and comprehensive development (Table 2, Fig. 3 and Fig. 4) (Xu J., 2000; Zhang B., 1999).

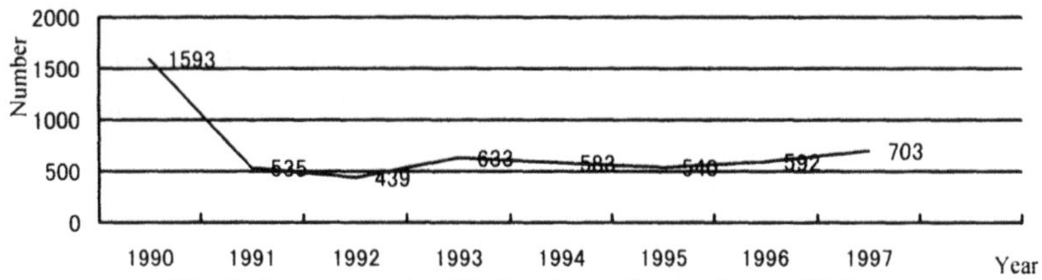

Fig.3 Large scale AD Stations through out China

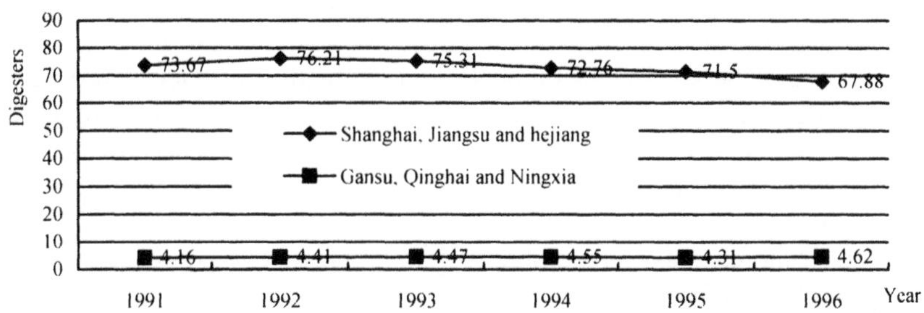

Fig.4 Household Digesters per ten thousand capita

Table 2 Distribution of livestock AD engineering through China (Xu J., 2000)

Province	AD stations /unit	Generating gas /10^4 m^3	User /10^{4}households	Electricity /kW	Fertilzer /10^4 ton	Feed /10^4 ton
Beijing	13	13.38	1.63	0	0	0
Hebei	17	9.14	0.025	0	0	0
Liaoning	10	0.7	0	0	0.5	0
Shanghai	41	991.34	1.72	0	1.07	0.01
Jiangsu	100	105.53	0.254	64	0.05	0
Zhejiang	33	351.13	0.88	3	0	0
AnHui	13	12.20	0.079	0	0.03	0
FuJian	6	37062	0.02	1.5	0	0
JiangXi	51	33.14	0.09	0	0	0
HuBei	63	72.96	0.26	118	0.08	0.57
HuNan	59	91.78	0.43	235	0.16	0
GuangDong	18	42.24	0.09	173	0.50	0
Total	460	1999.42	5.59	866	2.49	0.7

2. CHARACTERISTICS AND FEATURES OF AD TECHNOLOGIES IN CHINA

China that is called "The Hometown of Methane" has experienced deep research and exploitation on anaerobic digestion for a long time and acquired integrated techniques. At present, there are such advanced processes as AF (anaerobic filter), UASB (up-flow anaerobic sludge bed), AFBR (anaerobic fluid bed reactor), EGSB (expanded granular sludge bed), UASB+AF, MBS (two-phase anaerobic membrane biosystem) and AAFEB (anaerobic adhering film expanding bed), etc.

Up-flow Anaerobic Sludge Bed (UASB) (Zhang B., 1999): This is the most widely used high efficiency AD reactor (ADR), in which granular sludge can be formed and the blocking problem is solved. It has a simple inner structure, with only a water-collocating facility and a three-phase separator. It is adapted to a large range of organic concentration. The treating efficiency is more than one or two times than the ordinary ADR even COD is 1000-3000mg/l high. The COD removal rate is improved by 10-20%. The Hydraulic Retaining Time (HRT) is reduces by 50%. Most importantly, the cost is low.

Anaerobic Filter (AF) (He Y., 1998): This is a high speed reactor that use microorganism as the medium. It use soft or semi-soft stuff to lessen the volume and use microorganism fixed technique to prolong the sludge retaining time. The treating efficiency has been improved greatly. It can treat waste of various scales and can treat dissolved wastewater with load of 10-15kg COD/m^3·d. However, it is easily blocked and the cost rises because of a great number of stuff. Down-flow anaerobic filter is propitious to solve the problem, but the membrane is formed slowly and reactor volume load is low.

Two-step or Comprehensive Anaerobic Digestion: It is composed of two kinds of AD technology. It works with high efficiency and is resistant to heavy impact.

Inner Circulating Reactor: It depends on the density difference caused by ascending pipe and returning pipe to give flowing lift for natural circulation. It can treat successfully wastewater from potatoes, beer, foods and lemon acid.

Except for popularized in rural areas, ADR is widely used to treat wastewater because it has advantages as cost- and land-effective, high load capacity, various scale, and less sludge (table 3).

Table 3 Comparison of some AD process (Zhang B., 1999)

Technological index \ Type	Common digester	AF	UASB
Max COD load /kg·m^{-3}·d^{-1}	2~3	10~30	10~30
COD removal rate /%	70~80	90	90
Min effluent COD /mg·L^{-1}	5000	1000	1000~1500
Max effluent SS /mg·L^{-1}	50000	200	4000
HRT /d	10~20	0.5~3	0.5~5
Efficiency	low	high	high
Resistance to impact	low	medium	higher
Requirement for temperature	higher	lower	lower
Power requirement	high	low	low
Control difficulty	medium	easy	difficulty
Total cost	high	lower	low

2.1. The typical AD processes for different wastes

2.1.1. Agriculture waste

AD engineering techniques for livestock manure treatment is internationally important and China has got the most advanced technologies on the world (table 4).

The HRT would be much shorter using Down-flow Still Fix-Membrane ADR to treat pigfarm wastewater. Moderate concentrated wastewater passing the 1mm meshwork, under the temperature of 35°C, the HRT is about three days, and gas generating rate is $4.5m^3/m^3 \cdot d$.

UASB is applied to treat swine wastewater under low temperature (13-19°C) with passing meshwork, the gas generating rate is $0.3\text{-}0.5m^3/m^3 \cdot d$. This technique is currently taking the lead around the world, and will fill the vacancy of treating wastewater under low temperature and solve the remaining problem in North China.

A new technique "dynamic digesting under constant temperature" has been developed and used to treat the agricultural plant residues. The outstanding feature is the high solid organic material content of more than 90%. Liquid is circulated and heated outside the Steel tank digester. The storage is much efficient with a compressor, compared with the current typical storage. The total investment decreases by 60% and it is the first equipment that could be popularized in the rural areas under the market-driven condition.

2.1.2. Industrial organic wastewater (Chen H., 2000; Liu C., 2000; Wang K., 2000; Huang X., 2000; Zhang G., 1996)

Lees from alcohol companies are usually treated by separation-anaerobic digestion method. The separated liquid is mixed and blown into reactor with high digestion temperature (55±1 °C). The removal rate of COD could reach more than 87% while gas production is about 3 $m^3 \cdot m^{-3} \cdot d^{-1}$.

Table 4 The situation of AD stations on the farm (Xu J., 2000)

Manure	Working temperature	Process type and Volume /m^3	COD removal rate	Gas generating rate /$m^3 \cdot m^{-3} \cdot d^{-1}$	Province
swine manure	35-38	USR, 300	89	1.7-2.2	Zhejiang
swine manure	18-28	UBF, 500	86	0.94	Zhejiang
swine manure	9-13°C	USAB, 1000×2	80	0.2-0.3	Guangdong
Fowl manure	50°C	UASB, 100	56	3.0-3.6	Beijing
Fowl manure	23-25°C	UBF, 128	85	1.0	Shanghai
Cattle manure	Moderate	Anaerobic tank, 300	---	1.5	Zhejiang
Cattle manure	Near moderate	Advanced tank, 450×6	75-85	1.2	Shanghai

UASB and UAF is often used to treat monosodium glutamate wastewater (MGW). Low concentration and high load feedwater is adapted for UASB to treat MGW with low ratio of carbon and sulfur($COD:SO_4^{2-}=2$) in case that high concentration of NH^3, N and SO_4^{2-} may spoil and restrain the methane bacteria under anaerobic condition.

EGSB specially is suitable for low concentration and low temperature beer lees treatment. AF, AFB and UASB is often used to treat bean product wastewater, the removal rate of COD reaches 95.5% under 35±1℃.

2.1.3. Sludge (Mei X., 1994; Dong Y., 1998)

There is a great content of water and organic matter in sludge, organic compound is only decomposed 45%-55% in the first stage of moderate temperature digestion. After the first stage digestion, a large disposal field is needed for drying of the residues while a large amount of heat produced during the digestion. The followed second stage of anaerobic digestion, without any additional heat, will need 7-10 days and methane production is about 80% of the total gas generated in both phases. If under airtight condition, gas productivity is 2-3$m^3/m^3 \cdot d$.

2.2. Technique for sulfur removal from methane

The pipe and equipment is likely corrupted with high H_2S in methane. High sulfur content will restrain the activity of anaerobic microorganisms. The limit of sulfur concentration in methane is 20 mg/m^3 according to the national standard.

Technologists have experienced ten years to develop a new technique which can keep sulfur concentration as low as 1.0g/m^3. This technique has been patented in 1993 and applied in more then ten molasses wastewater treatment station both in China and abroad. For example, in a alcohol factory in Bengal, 200 tons of wastewater is treated with AD technology and 5000-6000m^3 methane each day is generated. Six hundred tons of wastewater from a sugar refining factory in South China is treated anaerobically and 15000-18000m^3 methane is generated daily. This AD station has kept running stably for four years (Ren A., 1994).

An advanced technique to separate H_2S from methane has been developed to reach an efficiency of 5.8 times of the ordinary technique.

2.3. Methane generating power

To use methane as the fuel of power engine is wildly found, Tianjin Wastewater Treatment Plant is a successful example. The mixed gas fuel for methane-electricity transform is the main problem. The mixed fuel goes against methane combustion; the outlet gas temperature is too high for CO_2 is not separate from methane that leads to heavy smoke. Recently a newly developed generator uses methane itself (instead of fix gas fuel) as fuel. To redesign the portfire system in a diesel engine can get a methane-diesel mixed fuel engine (Zhang B., 1999).

2.4. The equipment relevant to methane engineering

Many types of solid-liquid separators for different AD engineering have been widely used, such as vibrating screen, rotating screen and fixed bed filter. The dehydrating rate is 10%-40% after depositing and extruding. Many types of transporting equipment are advanced in the world, the efficiency of 75YE-10 bump is about 65%, for an example.

Wet and dry gas storage tank are both developed. A new methane press storage system--compressed storage tank with buffer tank is put into use on a pigfarm. The cost of this system is only 60% of that of wet gas storage tank (Fang R., 1999).

3. COMPREHENSIVE AD APPLICATIONS

3.1. Utilization of methane and its byproducts

Besides being used for rural and industry energy, methane can be used in such fields as supplying heat in greenhouses, supplying CO_2 gas fertilizer, fostering silkworm, persevering fruit and food. Experiments on CO_2 enrichment revealed that the leaves photosynthesis intensity is enhanced by 7%-20%; the productivity of cucumber, pepper, tomato and shepherd's-purse is increased by 49.8%, 36%, 21.5% and 5% respectively. Good result could be reached to use methane to keeping fruits fresh. More than two hundred thousand tons of grain was stored with methane in China in 1994 and save 25 thousands tons of grain and 220 thousand Yuan of the running cost (Zhang Y., 1998).

There are many kinds of mineral and organic components in AD effluent that can be used to soak seed, spray onto leaves, foster vegetable, irrigate orchard and feed livestock and fish. The liquid, with appropriate treatment, can be used as pesticides. The plant louse can be killed up to 95%, for an example; the egg productivity is increased by 7%-14% with the effluent used as additives in the feed; the fededing cost will decrease 20-26 yuan for each pig if the effluent is fed. Solid residues from AD process can be used to produce organic fertilizer, cultivate mushroom and earthworm. Table 5 is the component of AD effluent and solid.

Table 5 Component of residue (Wang K., 2000)

Compo net	Effluent	Solid residue
Total N /%	0.03~0.08	0.78~1.61
Total P /%	0.02~0.07	0.39~0.71
Total K /%	0.05~1.40	0.61~1.3
Organic content /%	---	36~49.9
Humic acid content /%	---	10.1~24.6
VB_{12} /mg·L^{-1}	9.3×10^{-6}	---
VB_{11} /mg·L^{-1}	6.42×10^{-6}	---
Protein /vitality unit	1.43	---
Vitamin enzyme /vitality unit	7.65	---
Grothhormone /mg·L^{-1}	8022	---
Gibberellins /mg·L^{-1}	3.510	---

3.2 Ecological AD application patterns (Dong R., 1998)

AD currently acts as a key link to lead the whole ecosystem to a good circulation. More than 80% of users prefer AD in an ecological pattern (Fig.5). The swine-methane-apple tree pattern is welcome in South China while "Four-parts System" pattern that integrates animals, vegetables and ADR inside a greenhouse is popular in North China.

3.3. Benefit analyses

Let us take a moderate pig farm to build a 100m^3 conventional AD plant as an example. Unit cost is 350 yuan/m^3 and the whole investment is 35 thousands Yuan. Its gas productivity is 0.2-0.3m^3/m^3·d and the whole gas production is 20-30m^3 every day, and 7000-10000m^3 for one year which is equal to 7000-10000 Yuan of liquefied petroleum gas every year. So the investment can be covered in 4 or 5 years. In addition, effluent and solid byproducts can be sold as fertilizer with much more profit. Table 6 is finance analysis of some AD stations (Han K., 2000; Zeng B., 2000).

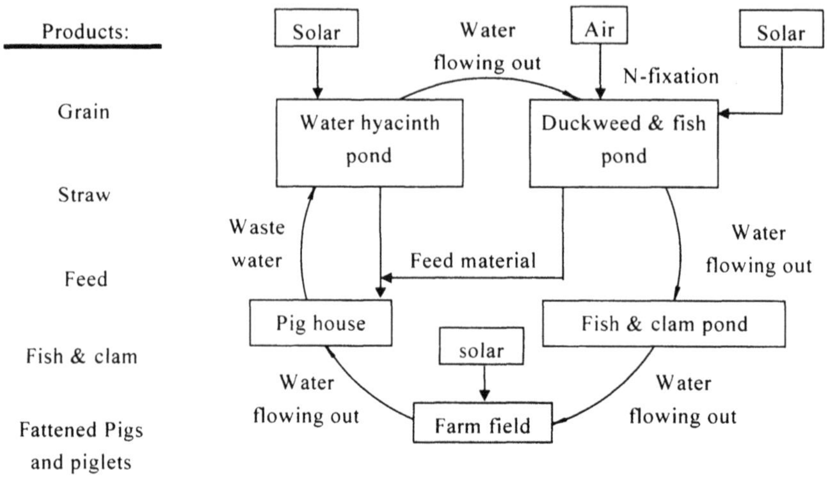

Fig.5 Farm ecological agriculture with AD waste treatment method

Table 6 AD engineering financial analyses (Yin Q., 1999; Zhang Y., 1998; Xu J., 2000)

Parameter	Hangzhou	Hangzhou	FuShang	ShangHai	Shengyang	Beijin	Beijin
Farm scale	200,000 fowl	10,000 fowl	3,000 swine + 130,000 fowl	2,000 cattle	20,000 swine + 25,000 cattle	15,000 swine	30,000 swine
AD station scale /m^3	750	500	700	2700	2000	1000	2000
Total cost /10^4Yuan	95.0	281.45	77.15	912	200	263.2	440.00
Running fee /10^4Yuan	6.0	21.59	5.03	73	21.00	25.0	41.70
Total profit /10^4Yuan	12.75	65.5	11.41	198	61.67	75.60	142.13
Gas profit /10^4Yuan	6.75	20.5	6.66	144.1	10.95	57.60	11.52
Gas price /Yuan·m^{-3}	0.25	0.80	0.25	1.20	0.50	0.80	0.8
Electricity profit /10^4Yuan	---	----	---	---	---	---	94.61
Granule feed /10^4Yuan	6.0	33	4.75	4.7	49.30	18.0	36.0
Feed /10^4Yuan	---	---	26.9	---	---	---	---
Unit price /Yuan	---	---	---	1000	---	---	---
Fine reduction /10^4Yuan	---	12	---	15.4	---	---	---
Usage life /year	21	21	21	---	21	21	21
FNPV /10^4Yuan	-37.6 5	-73.17	-24.38	7.8	62.53	74.12	206.67
FIRR /%	2.67	4.53	4.28	12	15.07	14.54	17.40

4. PROBLEMS

The operation efficiency of some special equipment such as separator, purifying equipment and all types of bump is still low and the performance is not stable. Further work is needed for automatization. Theoretical and basic researches on microorganisms and necessary additives for AD process is not enough. The comprehensive application of AD process in the global sustainable development is sufficient.

Inorganic fertilizer occupies the market and the prevalence of LPG affects the market of methane. There is little profit in farm running. Without finance support, it is difficult to build AD stations and develop the relevant products (Li Q., 2000).

5. DISCUSSION AND CONCLUSION

AD technology has been mature after a hard time experiencing a circuitous way. Although the fundamentals are still active, current ADR has been much different from the primary one for the introduction of new equipment and microorganisms and promoting additives. A single anaerobic digestion technology is usually integrated with others so as to get a high efficiency. Anaerobic technology is not only treated as an energy technique, but more importantly, is thought as one important clasp in the ecological circulation. Through this step, the harmful organic materials are transformed to be the useful one to the environment.

China is one of the leading and experienced countries in AD researches and application. Series of ADR have been developed and proved to be practical and beneficial to both developing and developed countries from the different viewpoints of energy and ecological benefits.

REFERENCES

Bian Y., 2000. Solid Residues Re utilization. Chemical Industry Press. Beijing: pp324

Chen H., 2000. Researches and application on soybean factory waste water treatment. China Methane (3): 13-20

Dong R., et al., 1998. Agricultural waste management and Chinese-ecological agriculture. Biosystem Study 1(1):38-45

Dong Y., 1998. Practical Waste Water Treatment. Chemistry Industry Press, Beijing: pp293

Wang K., 2000. Fermentation Industry Waste Water Treatment. Chemistry Industry Press, Beijing.

Fang R., 1999. Integrated technology for pigfarm manure digestion. 17(3): 17-19

Han K., etal., 2000. Household rural energy comprehensive application. Rural Energy (3): 13

He Y., 1998. Anaerobically Digest Waste Water. China Light Industry Press, Beijing.

Huang X., 2000. Technologies for monosodium glutamate waste water treatment. China Methane (3)

Ji B., et al., 1994. Farm ecological raising and waste purification in Liaoning Province. Agro-environmental Protection. 13(6): 285-286

Li Q., 2000. The current problems and countermeasures for anaerobic digestion. New Energy (8)

Liu C., 2000. Researches on the treatment of VB_{12} and protein mixed wastewater with anaerobic digestion process. China Methane (3): 24-34

Mei X., et al., 1994. Making methane from household residues with AAFEB. China Methane (3): 7-11

Ren A., et al., 1994. Researches on separating H_2S with SW additives. Hebei Chemistry Industry (3): 59-63

Xu J., 2000. Large scale animal farms anaerobic digestion.China Methane (3): 27-30

Yin Q., 1999. The key direction of anaerobic digestion development in Zhejiang. Power Engineering (3): 13-14

Zeng B., 2000. Anaerobic digestion and comprehensive utilization of animal wastes in Shanghai. China Methane (3): 31-34

Zhang B., 1999. Rural Energy Engineering. China Agriculture Press, Beijing.

Zhang G., 1996. Urban solid wastes treatment by anaerobic fermentation. Rural Ecological Environment (121): 46-49

Zhang Y., 1998. The application of methane the its by-products in the ecological agriculture. Agricultural Environment Protection 17(2): 94-95

Zhao S., et al., 2000. Waste water treatment and application II—Types and examples. Urban Environment and Ecology 13(6)

Greenhouse Gases and Animal Agriculture
J. Takahashi & B. A. Young (Editors)

Manure and biowaste digestion in Germany - history, trends, and practical verification -

G. Langhans

Linde-KCA-Dresden GmbH, Environmental Technology D-01277 Dresden, Germany

Anaerobic digestion as natural process enables the re-entering of organic materials into the environmental cycle. As an industrial process it can be helpful to solve problems of environmental pollution and recovery of natural ressources but needs special efforts for stable and efficient operation.

The process design and economic efficiency of biogas plants is primarily related to the possible accuracy in prediction of biogas yields from organic compounds and waste compositions. With the help of experimental results and the analytical description of complex organic compounds it is possible to develop models of energy and mass balances for the digestion of wastes like manures or sewage sludge as well as heterogeneous bulk biowastes.
Process design and plant economy are discussed for the full scale application of a thermophilic anaerobic digestion system applied for the processing of farm manure and industrial organic wastes as well as agricultural crop residues.

This co-fermentation approach results in enhanced biogas generation which is used for electrical and thermal energy production to meet the plant consumption and in addition enables use of the CHP-waste heat for farm and residential heating at the plant location while surplus of electric power is sold to the public grid.

By the references of numerous plants in operation mainly in Europe full scale manure digestion is shown to be a proven and reliable technology.

Keywords: anaerobic digestion, biogas plant, methane, manure, Germany

1. INTRODUCTION

The worldwide revival of the anaerobic process for the treatment of biowastes and organic residues has enhanced the general interest especially to find out the capabilities and limits of this technology.
The anaerobic microbial populations are species of a very archaic type of life without oxygen use in their metabolism. That's why their life-energy generation is rather inefficient and the stepwise incomplete destruction of organic matter causes products of furtheron high energy content which cannot be used by those microorganisms without the ability of aerobic metabolism based on oxygen as electron acceptor.
This has important consequences for the ecology of earth because it is the prime guarantee for the permanent recycling of organic matter within the natural succession and results in the following items:

- mineralization of high amounts of organic mater with only low reproduction of bacterial cells (that means very small generation of new organic matter in opposite to aerobic processes).
- generation of intermediate as well as final products of furtheron high energetic (caloric) content like methane, alcohol's, or organic acids, which will be used by aerobic microorganisms as substrate but are also basic compounds of manifold processes helpful for life of mankind.

That's why anaerobic processes have been used in all times of human history consciously or instinctively, e.g. for
- beer brewery
- wine growery
- leaven for dough
- conservation of food like pickled cabbage or annual ensilage
- fermentation of spices
- methane, generated from human sewage for instance at old Chinese fire places
- last but not least development of mystic ideas by observation of will-o'-the wisps on swamps caused by burning methane.

The use of anaerobic processes for the reduction of human organic wastes in an industrial scale only started at the beginning of the twentieth century.
There was a first culmination in anaerobic treatment of municipal sewage sludges in the twenties and thirties. But after World War II the aerobic stabilisation processes dominated because of the more simple process handling and the cheap energy for the aeration technology at that time.

Meanwhile the raise in energy prices as well as the dangerous destruction of global environment have initiated a renaissance of anaerobic digestion again also in the treatment of organic sludges, biowastes and agricultural residues including animal manures.

In this connection it has to be remarked that the composting process which has long been successfully used in practice is now in critical discussion for its industrial scale application because the ecological balance of large composting plants due to their electric power and fuel consumption accounts for a high demand of energy produced from fossil ressources. Digestion processes or combined fermentation/composting processes offer clear advantages in this respect because they can cover their own electric power and heat demand on basis of the methane in the biogas without CO_2 generation from the use of fossil fuels and normally there can be exported remarkable surplus of energy.

So besides the recycling of organic matter especially the use of methane as energy source of a CO_2-neutral process reduces the consumption of fossil fuels like coal, crude oil or natural gas and is said to damp the greenhouse effect.

Last but not least anaerobic processes offer a reduction in annoyance by bad odour during treatment because of the operation in closed reaction tanks compared with the aerobic composting processes always generating a high amount of odour-loaded exhaust air.

Although there has been a drastic reduction in pollution from municipal and industrial point sources a lot of European as well as worldwide waterways still do not meet the goals set for fishing and/or swimming and also groundwater ressources are furtheron more or less contaminated. One of the leading sources of the remaining pollution is from agricultural operations involving animal confinement facilities and the land application of manure because these wastes from confined animal operations have the potential to contribute pollutants such as organic matter, sediments, nutrients and pathogens. In addition, confined animal operations may produce objectional odors.

To limit those pollution problems the European Community is issuing relevant permits. These permits are intended to control pollutants from both discrete source as well as assure that manure is properly applied for agricultural benefit and does not lead to water quality or public health problems.

If the impact of animal manure on the environment can be minimized, the manure is a valuable ressource as a fertilizer, soil conditioner and as an energy source. However, due to intensive stock farming technologies, animal manure is no longer obtained in the conventional bulk form with high amounts of chaff and straw, but as a more or less pumpable slurry, the so-called liquid manure. As a consequence special technologies must be employed for storage, transportation and utilization.

Manure is continuously produced all the year long, whereas its utilization in grasslands and agricultural fields is regulated and restricted to certain periods of the year. These periods depend on the needs of crop plants, their ability to take up the nutrients, seasonal climate changes, crop rotation and plant ripeness.

To reduce environmental problems linked to manure, like

- the annoying smell
- pathogens, which might be contained in the manure and can be spread over huge areas together with the manure to be distributed finally with wind driven aerosols
- the organic load of manure which can have a negative influence on the soil if applied in too high amounts
- the high amount of nutrients which are contained in manure and can be a risk for the environment if not spread correctly (wrong timing, concentration too high, losses by glueing of high viscouse manure on plants)

in an economical way, during the last decades the anaerobic fermentation process has gained great importance for the treatment of agricultural wastes in Europe. As mentioned before it is the concept to combine the treatment of manure with the production of renewable energy in the form of methane. Existing installations range from smaller plants mainly built by the users themselves to large industrial-scale co-digestion complexes.

Lately some digestion plants are purely designed for the production of renewable energy. For those installations, the major input material is formed by energy-rich plants which are specifically cultivated to be fed to the fermenter.

While inquiries in the middle of the eighties detected in the European Community and Switzerland about 550 biogas plants with 420 of them in the agricultural area including about 60 plants for manure treatment within the German territory (Palz, 1985), similar observations in the year 2000 registered in Germany ca. 1000 biogas plants in agriculture with at least 30 of them in industrial scale (digestion volume over 1000 m^3), the others smaller individual farm types between 50 m^3 to 500 m^3 reactor volume.

At the moment there is a strange situation in Europe; while governments in Germany and other members of the EC support by law energy production from renewable sources (Biological Treatment of Biowaste, 2001) (including biogas) in kind of guaranteed revenues for the power sold to the public grid, on the other hand BSE, MFD as well as scandals because of illegal use of antibiotics etc. bring intensive stock farming into the twilight of low acceptance. By that it is not clear at the moment if animal life stocks will be limited in heads in future.That for instance would hinder the present trend to industrial scale plants. The typical Danish concept to install centralized plants with manure collection and distribution of digested effluent to several smaller farms is not everywhere accepted by European farmers. Indeed there is sorrow maybe to contaminate the own farm yard with unknown matter or germs from a central plant which is not under individual check and control.

2. BASIC PRINCIPLES OF MANURE DIGESTION

The microbiological process which is described as „fermentation" or „anaerobic digestion" in „black box" terms is a sequence of complicated metabolic procedures of a great number of populations of microorganisms living more or less in symbiosis (Fig.1).
There can be distinguished the steps of

- decomposition of solid organic matter into water-soluble compounds by extracellular bacterial enzymes (so-called hydrolysis)
- utilizing of that soluble high molecular substrate by different species of microorganisms with set free of intermediates like alcohols, sugars, and primarily C_3- to C_5- organic acids (acidogenic species)
- production of mainly acetic acid, H_2, and CO_2 (acetogenic species)
- use of these intermediates by the end-of-metabolic-chain methanogenes with the gaseous end products of CH_4 and CO_2.

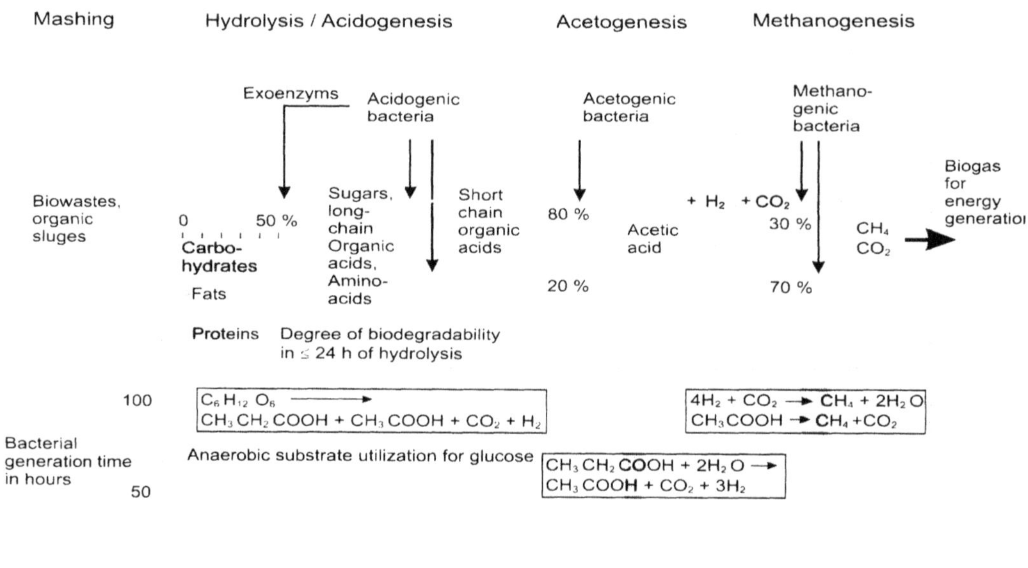

Figure 1: Metabolic chain of anaerobic microorganisms (Langhans, 1995)

Especially acetogenes and methanogenes are living in strong symbiosis because the acetogenic process will be depressed by the enrichment of H_2 which has to be utilized by the methanogenes simultaneously with its production.

It is clearly to understand that the first steps in this metabolic chain would operate strictly in the acidic range of pH if the methanogenes are absent which neutralize the local habitat by utilizing the acetic acid (Langhans, 1997c).

So the whole process runs in a labile pH-equilibrium and disturbances will cause a non-reversible gliding towards acidification with often finally total failure of the system.
This risk is raised by the low growth rates of acetogenes and methanogenes compared with those of the other species which can cause enrichments of acidic intermediats.
Main factors of possible disturbances are instabilities in process temperature, destruction of bacterial flocs by shear stress, overloading in organic input easy to hydrolyse, enrichment of biotoxic intermediate compunds (NH_3, H_2S, heavy metal ions), and intolerable intake of oxygen, as well.

Therefore, this process as an industrial procedure oriented fully on performance makes great demands on strict control.
For manure digestion just ammonia toxicity is a relevant problem because in dependence of pH and temperature a defined fraction of the naturally high ammonium content is present as undissociated ammonia (Fig.2). That results first in hindered bacterial growth beyond NH_3-concentrations about 700 mg/l and finally causes complete failure of the microbiological process at higher values.

As shown in Fig.2 bacterial growth at thermophilic conditions is more sensitive to NH_3-toxicity and at higher concentrations the principally faster growing but hindered thermophilic flora needs the same hydraulic retention time for digestion like the undisturbed mesophiles.
That´s why pig and poultry manure for better process stability should be digested preferably at mesophilic temperatures.

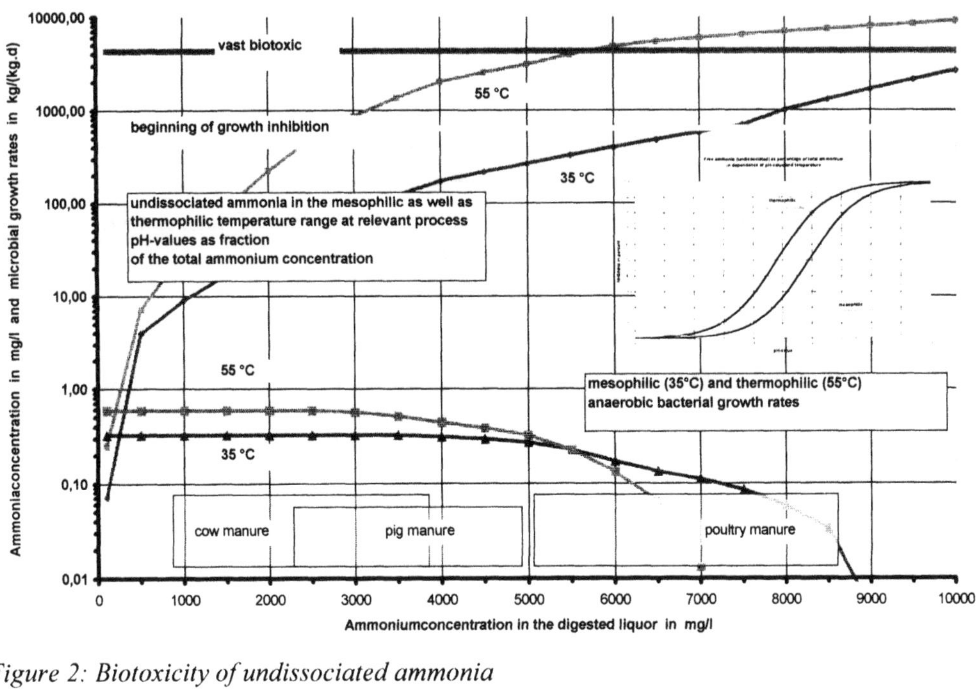

Figure 2: Biotoxicity of undissociated ammonia

Because of the content of intestinal anaerobic bacterials cow manure is a well-operating natural inoculum for digestion. On the other hand the manure just from dairy cattle contains a lot of fibre in the organic fraction which is difficult to utilize by the anaerobic flora.

That means low reduction efficiencies of the organic fraction (Fig.3) and results in low biogas yield.

But those discussed attributes make cow manure to be the ideal basic substrate for co-fermentation with other organics from energy crops or biowaste fractions of the food processing industry.

There the manure guarantees the pH-buffering of the process due to the nitrogen content and also the delivery of other essential nutrients and prevents the process as a result of the continuous inoculation in a wide range from failures by overloading. The other co-substrates are the necessary carbon sources for high biogas production.

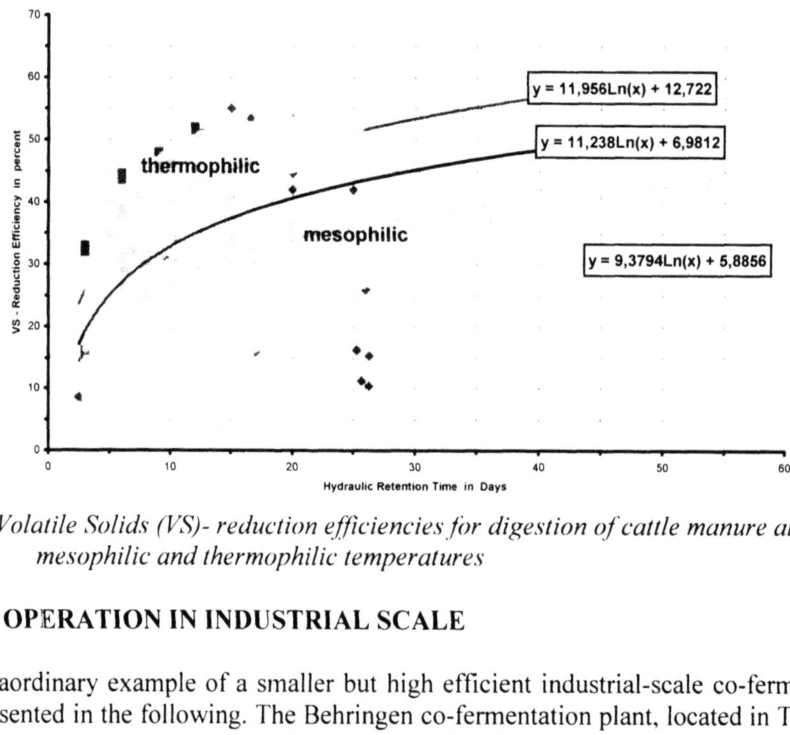

Figure 3: Volatile Solids (VS)- reduction efficiencies for digestion of cattle manure at mesophilic and thermophilic temperatures

3. PLANT OPERATION IN INDUSTRIAL SCALE

The extraordinary example of a smaller but high efficient industrial-scale co-fermentation plant is presented in the following. The Behringen co-fermentation plant, located in Thuringia County, Germany, processes manure from a confined animal operation of about 1,000 dairy cows together with other industrial and agricultural organic residues. This facility was designed, constructed and commissioned by Linde AG of Germany for BEUG Biogas and Energie GmbH, and started operation during December, 1995.

A general view of the Behringen plant is provided in Figure 6 and a process flow diagram in Figure 7. The overall plant process consists of waste storage and feed composition followed by anaerobic digestion, conversion of biogas to electrical and thermal energy and storage of the stabilized liquid/fiber nutrient product.

Typical calculated bio-chemical parameters (Buswell et al., 1933; Boyle, 1976; Langhans, 1997a; Langhans, 1997b) of accepted and often used co-substrates are summarized in Fig. 4 and Fig.5, respectively.

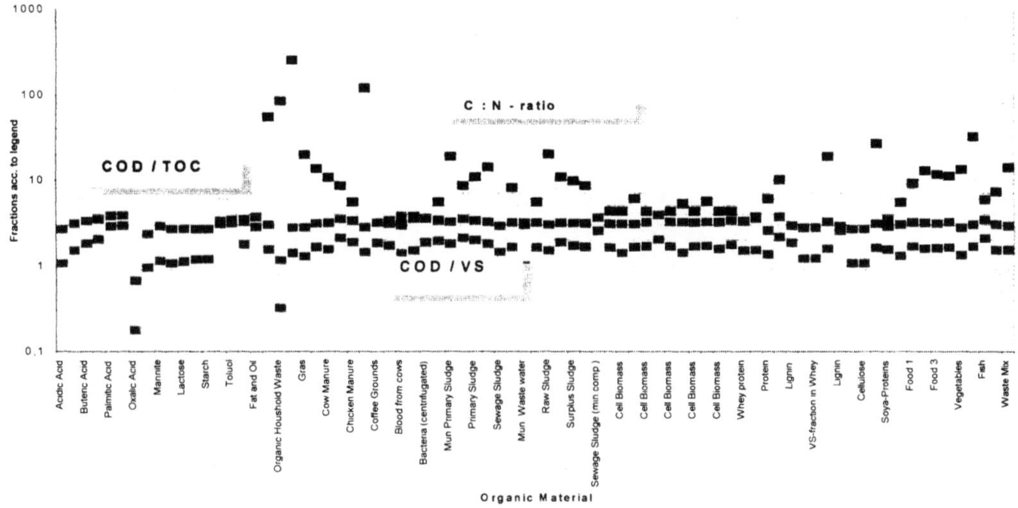

Figure 4: Biochemical parameters of important co-substrates

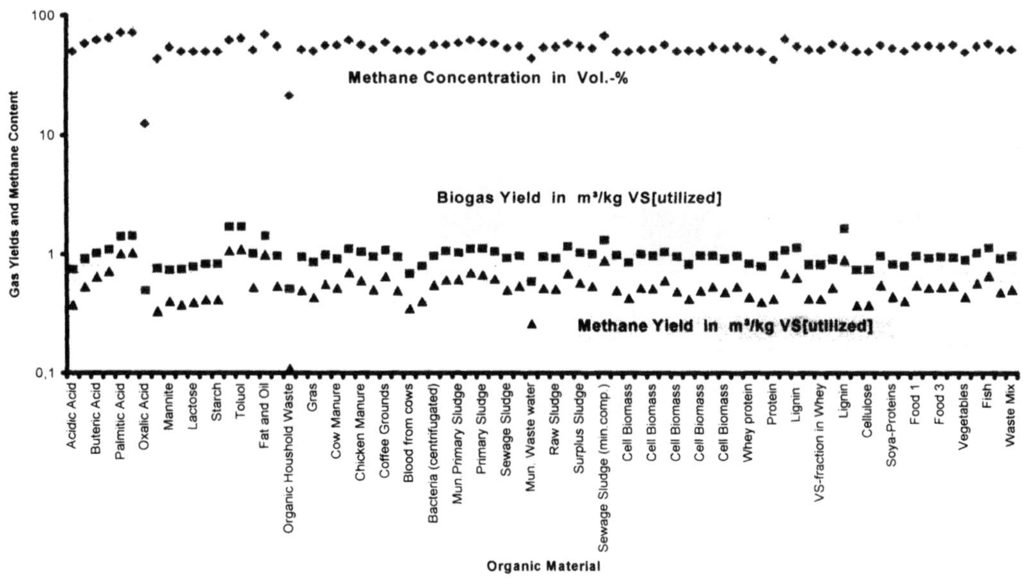

Figure 5: Biogas and methane yield for important co-substrates

Figure 6: General view of the Behringen co-fermentation plant in Germany

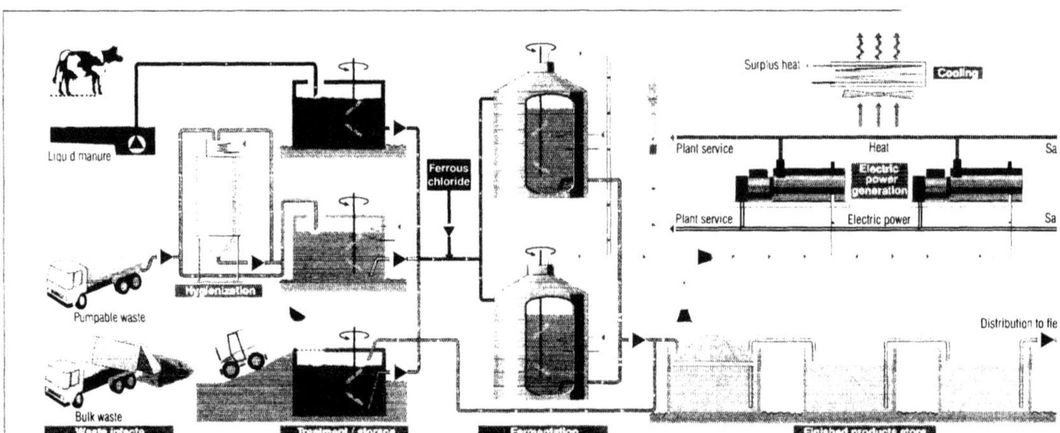

D01/ D04/ D02 **R01A/ R01B**

Figure 7: Process flow diagram of the plant

The Behringen plant design was based on an anaerobic digestion process that was applied to (a) reduce the odour of the liquid manure, (b) improve the storage stability of the manure, (c) alter the manure nitrogen components to be more easily assimilated by the plants and (d) recover useable energy. Recognizing that dairy cow manure has already undergone a digestive phase in the cow's stomach, and thus has a limited amount of degradable organic available which results in low biogas production, the Behringen plant was designed as a co-fermentation plant. In this case the manure was supplemented with organic wastes from the several industrial sources for processing in one step thermophilic anaerobic digesters with an additional final "anaerobic maturing" of the effluent in a covered storage tank with connection of the gas space to the biogas system. The key waste sources are:

- Liquid manure from a contained animal operation of about 1,000 dairy cows that are housed in closed stables containing slotted floors. The manure is pumped to a storage tank (DO1).
- Fat trap residues and spent cooking oils are delivered to the plant in tank trucks as an emulsion and pumped to a heated storage tank (DO4).
- Bentonite clay that is loaded with fats and oils, derived from the bentonite being used as a filter media in plant-oil refinery, is delivered to the plant in container trucks, stored as bulk and loaded into an underfloor storage tank (DO2).
- Solid Fraction of piggery manure that has been dewatered with a screw press at the piggery is delivered to the plant with container trucks, stored as bulk and loaded into a storage tank (DO2). The piggery residue is added as a substitute for the bentonite feed material.
- Brewery sludges, residues of maize silage, low-grade seed potatoes, and rape press cake as chargewise delivery, stored as bulk and loaded into a storage tank (DO2) to be mixed there with the above mentioned wastes and manure or digester effluent to realize a pumpable slurry.Before mixing the bulk materials, such as the fat- saturated bentonite clay and piggery residue, are stored on a concrete base. In addition the bulk products are converted into a pumpable state by using a mascerator if necessary and adding other waste components or digester effluent, and then stored (DO2).

Characteristics of feed components together with the typical feed composite resulting from combining of the feeds is provided in Figures 4, 10, and 11, respectively.The Behringen plant was designed to process 13,000 t/a of dairy cow manure and up to 20,000 t/a of other industrial organic wastes and agricultural residues from the own farm production (Langhans, 1995; Poggi-Varaldo & Estrada-Vazques, 1997)

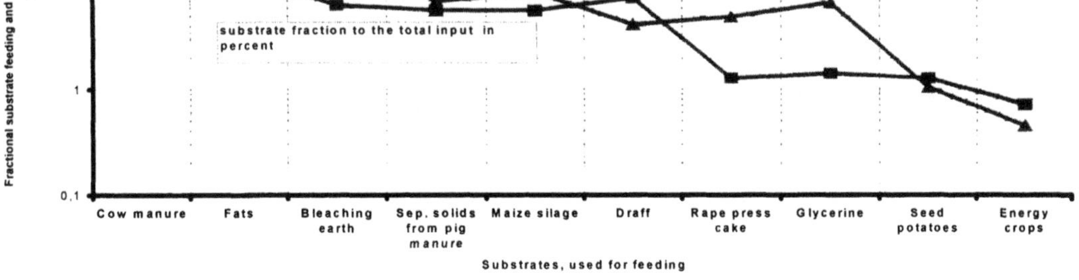

Figure 8: Mass fractions and energy yield of the digested co-substrates

Taking advantage of the co-fermentation synergistic effects allows the biogas yield to be substantially increased relative to the digestion of only liquid manure (Fig.8).

The anaerobic digesters are charged with the various component feeds in accordance with a metering control concept that was developed specifically for the Behringen co-fermentation plant to ensure maximum biogas production. Prior to the combined feed entering the digesters, ferrous chloride is added to control the level of hydrogen sulfide maintained in the digesters. The two digesters (RO1A and B) are fabricated of welded steel plate and insulated to minimize heat leak at the 57 °C operating temperature. Each digester has a liquid volume of 800 m³ and is equipped with a slow speed mixer mounted on the digester tank top to slowly mix the digester contents. The digester temperature is maintained by using coil-pipe heat exchangers located within each digester. Hot water from the plant hot water systems (CHP cooling water) is utilized in these heat exchangers on an as needed basis. At the design feed capacity the digesters provide an 18 to 22 day retention time (in dependence of the capable industrial waste input and consistency) and have a volumetric organic loading of about 5 to 7 kilograms of volatile solids per day per cubic meter of reactor volume.

The biogas generated by the fermentation process is supplied directly from the digesters to two engine-generator sets (CHP) to produce electrical power and thermal energy. Additional biogas for supply to the engine-generators is collected by a membrane gasholder that is installed floating on top of the initial storage tank for the digester effluent. The quantity of biogas collected by the membrane represents about 10 percent of the total biogas produced.

The co-fermentation of the dairy cow manure and the industrial organic residues provides for a biogas yield of about 900 to 950 m³ NTP per ton of converted volatile solids. The methane content of the biogas is about 68 percent. The biogas yield depends upon the composition of the feed material and its biodegradability and can vary over a ten-fold range (Fig.8). Since the Behringen fats/oils feed stream contains a substantially larger quantity of convertible carbon matter then the manure, the co-fermentation approach to waste conversion results in substantially higher biogas yields.

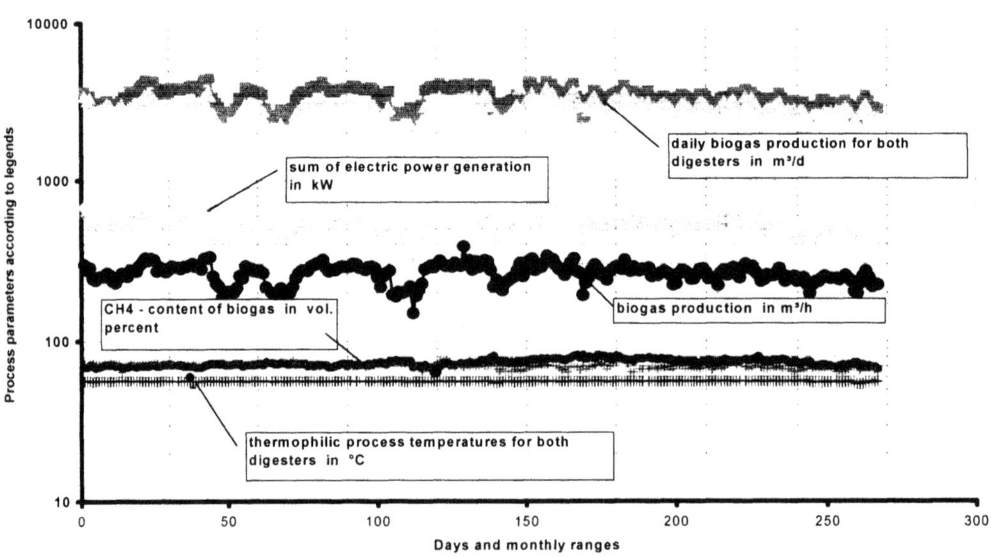

Figure 9: Process parameters and energy production

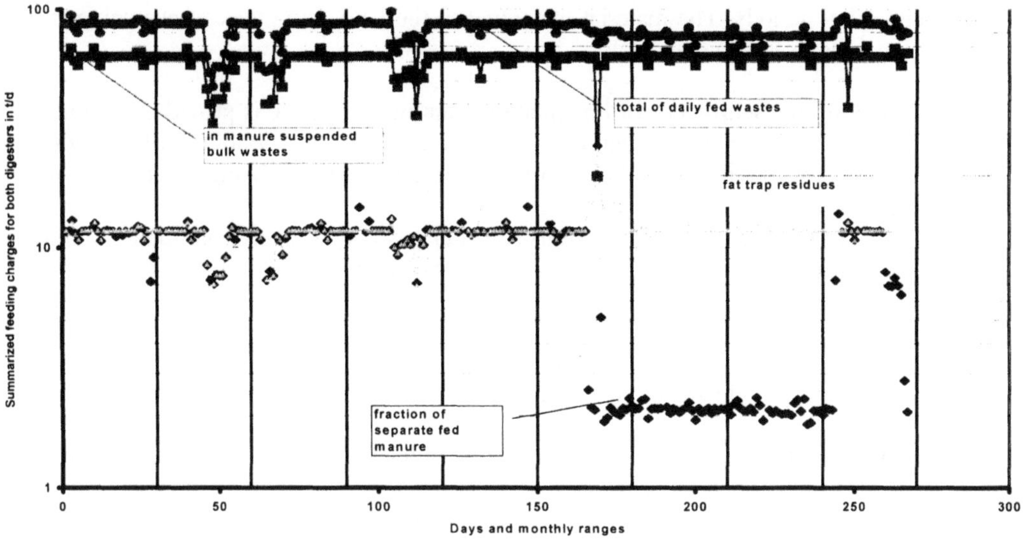

Figure 10: Daily input feeding of the stored substrates

The reciprocating engine-generator sets were supplied by Jenbacher Energiesystem GmbH, Austria, and each has a rated electrical energy of 450 kilowatts and a the thermal energy of 750 kW. Based on the biogas production rate the Behringen power station typically produced about 600 to 700 kilowatts of electrical energy (Fig.9) with peaks up to 800 kilowatts by the present feeding of glycerine. The heat rejected by the engine-generators is contained by a hot water system and used by both the co-fermentation plant as well as the confined animal facility.

About 30 percent of the electrical energy produced is used to satisfy the co-fermentation plant and confined animal facilities. The remaining electrical energy is supplied into the local utility energy grid for which Behringen receives about 0.2 DM per kilowatt-hour. This electrical energy purchase price is fixed by the German government by law as the electrical energy price for "green" power from renewable sources.

The energy production is typically limited by the variability of the industrial organic wastes. As previously discussed, the thermal energy produced from the engine-generators is used for maintaining digester temperature, heat treatment of specific industrial wastes and heating of facilities relating to both the co-fermentation plant and the confined animal operation as well as some residences in the neighbourhood of the plant.

The effluent liquid nutrient stream has a total solids concentration of about 9 percent with about 3.5 percent of the total solids available as total nitrogen. This total nitrogen has a higher ammonia content than the feed, allowing the nitrogen to be assimilated easier by the crops.

The fermentation effluent, rich in nutrients, is pumped to storage tanks and held until applied directly to the local agricultural land. There are four effluent tanks, each having a volume of about 2,000 m^3. Each storage tank contains a mixer which is normally operated prior to withdrawing the liquid nutrient for land application. The storage tanks have sufficient volume to hold the liquid mutrient for over 100 days, which avoids land application during winter and rainy periods. The material is applied to the land using tank trucks eqipped with drag-piping systems to avoid aerosols.

A summary of the Behringen co-fermentation plant input balance is provided in Figure 10. The feed composite can vary substantially due to availability and variabilitiy of the industrial organic wastes. This results from the majority of the carbon matter being associated with the industrial organic wastes.

The feed composite of the bulk materials (Fig.11) guarantees always the same potential of digestibility independent from the daily or weekly waste delivery.

The time period that was required to complete the Behringen plant, from start of engineering to mechanical completion, was ten months. In addition, about three months was required to achieve pseudo-steady state operation of the co-fermentation process.

Operation staff is two people mainly organizing unloading and material specific storaging of the incoming industrial waste charges.
Night and weekend operation is full automatically with included telephone alarm service for significant disturbances or malfunction in main systems of the plant.

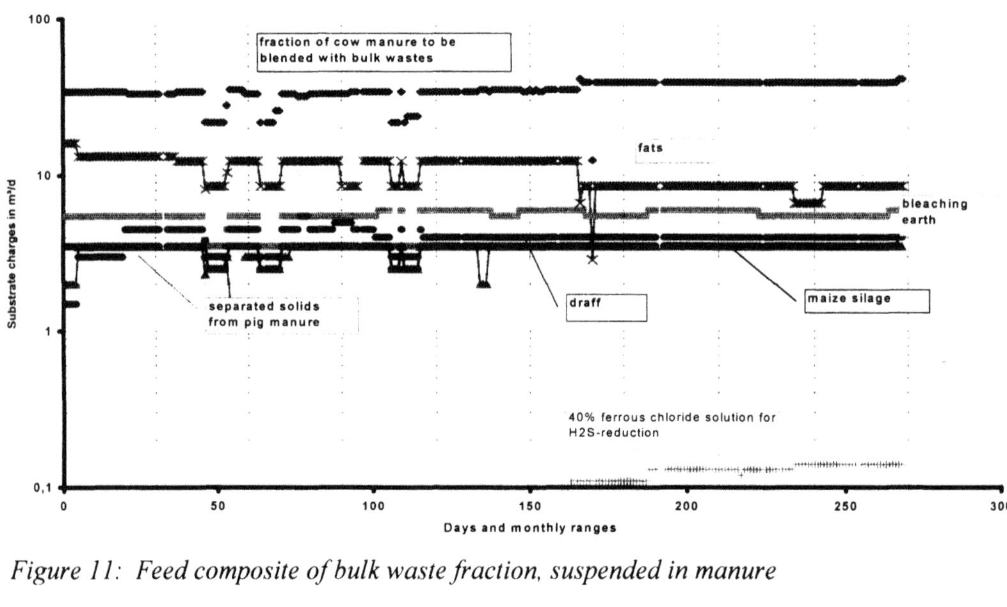

Figure 11: Feed composite of bulk waste fraction, suspended in manure

4. CONCLUSIONS

The full-scale experience from that farm-owned biogas plant utilizing a complex manure/biowaste substrate demonstrates the possibility of high and stable biogas production with technological efforts acceptable also for an agricultural company.

Cost efficiency of such plant depends on the governmental support for the idea of „green" energy production and the peoples´ acceptance to use the digested products in agriculture as fertilizer. To offset the plant annual costs, income is received from the sale of electrical energy and payment for processing of industrial organic wastes. However, whereas the value for electrical energy has been guaranteed now for a longer period by governmental decision, the value for the industrial organic wastes has been steadily decreasing because former industrial organic wastes become more and more the status of a wanted "fuel" for energy production on the European market.

It is evident that a true mass and energy balance based on detailed input specification with due account of the long-term development of the waste situation in the territory should be taken into account at the stage of such a plant design already in order to achieve a high plant efficiency.

The possibility of theoretical calculation of biogas yields also for complex waste mixtures enables the modelling of process dependencies on input changes in quantity and quality.

It is an instrument to reduce design failures in the planning phase as well as enables the plant operator to foresee fluctuations in biogas production and methane content. Manure treatment on industrial scale and technology reduces harmful and hazardous impacts on the environment. On the other hand this process safety has its price and makes necessary consequent process optimisation with respect to the individual local conditions.

Integrated treatment/production plants will help to compensate treatment expenses by the receipts for sold products. In addition savings for energy, fertilizer, and so on will economize the overall efficiency of such a project.

That's why all possible efforts have to be made not only to realize a „manure treatment" but to enable directly or indirectly the generation of useful products to be sold on the market.

REFERENCES

Palz, W. (Editor), 1985. Biogasanlagen in Europa. Verlag TÜV Rheinland GmbH, Köln, Germany

Biological Treatment of Biowaste. Febr. 2001. 2nd draft, European Commision, Directorate A – Sustainable Development and Policy Support, Brussels, Belgium

Buswell, A. M., Symons, G. E., 1933. J. Amer. Chem. Soc. (55) pp.2028

Boyle, W. C.,1976. Microbial Energy Conversion, pp. 125
Erich Götze Verlag, Göttingen, Germany

Langhans, G., 1995. Know- how for plant stability and operational efficiency. Linde Technical Report, Germany

Langhans, G., 1997a. Bemessung und Bilanzierung der Biogasausbeuten in der Abfallvergärung. AbfallwirtschaftsJournal, 1/ 2

Langhans, G., 1997b. Dimensioning and Balancing the Biogas Yields in Waste Fermentation. Linde Technical Report, Germany

Langhans, G., 1997c.Bio-chemical reduction of heavy metals in organic wastes by advanced digestion process. Anaerobic Conversions for Environmental Protection, REUR Technological Series 51, FAO Regional Office for Europe, Gent, Belgium

Poggi-Varaldo, H., Estrada-Vazques, C., 1997. Agricultural Wastes.
Water Environment Research (69) 4: 575-603

Greenhouse Gases and Animal Agriculture
J. Takahashi & B. A. Young (Editors)

Large scale manure based biogas plants in Denmark Configuration and operational experience

L. Ellegaard

Environmental Design & Development Dept., Burmeister & Wain Scandinavian Contractor A/S, Gydevang 35, P.O. Box 235, DK 3450 Allerød, Denmark

In Denmark a number of large scale manure based biogas plants has been established in recent years as part of a government sponsored demonstration program. This activity has naturally led to valuable experience being accumulated and incorporated into new designs. Although each biogas plant is configured in a unique way, in response to local project conditions and to improve previous designs, it is possible to identify a number of characteristic process and equipment elements, which have emerged or matured into solutions, which can be found in most new biogas plants now under planning or construction.

In this paper the most important aspects governing the overall plant configuration and process selection are presented. Also the most common technical solutions applied in key functions of a manure based biogas plant are presented along with guiding dimensioning figures and principles. Finally, typical operational production results for some of the newest biogas plants in Denmark are presented.

Keywords: biogas plant, manure, methane, heat exchanging, Denmark

1. INTRODUCTION

In 1985 the Danish ministries of Energy, Environment and Agriculture launched a development and demonstration programme in an effort to demonstrate the potential of large scale manure based biogas plant as suppliers of renewable energy, a method for better utilisation of manure as fertiliser and for a more flexible manure redistribution among farmers. The programme has included various reporting and follow up activities, which have ensured a fruitful exchange of experience to the benefit of development.

The programme has since then led to the construction of approx. 20 plants, at present treating an annual amount of manure and other biomass of approx. 1.2 mio. tonnes per year and producing biogas equivalent to approx. 25 mio. m^3 of methane per year. The potential amount of manure in Denmark could support a large number of new plants, and the continued growth of this sector is foreseen in the governmental planning as an important contribution to achieve a higher degree of sustainability in the energy and environmental sector.

2. PROCESS & PLANT CONFIGURATION

2.1. Process temperature

The first biogas plants were of the mesophilic type with a process temperature of 35-37°C and a hydraulic retention time of 20-25 days. Thermophilic process temperatures (50-60°C) were known to be beneficial for the conversion rate, but were originally feared to be more difficult to control. After one of the early plants (Vegger) had been converted to thermophilic operation, demonstrating the viability of thermophilic process temperatures, this process has gained a dominating position today. With a thermophilic process temperature the hydraulic retention time is typically reduced to approx. 15 days, and the necessary veterinarian sanitation at temperatures above 50°C can be integrated into one single thermophilic methanization and sanitation process step.

2.2. Ammonia inhibition

Early operational experience from thermophilic plants, indicated that the optimal process temperature varied among plants. Some plants could operate at 60°C with good results while others had difficulties reaching 55°C. This was later identified to be a result of ammonia inhibition. Free ammonia (NH_3) is an important inhibitor which controls the rate and efficiency of the biogas process (ref.2). Free ammonia exists in equilibrium with ammonium (NH_4^+) and the concentration depends on the total NH_3/NH_4^+ concentration, pH and temperature (fig. 1). Pig manure and some types of organic waste contribute to an increased ammonia level and high temperatures promote a higher fraction of free ammonia. At levels of free ammonia above 0.6-0.8 g/l NH_3 serious signs of stress occur (ref.3). If the total ammonia concentration (after digestion in the fermenter) exceeds 3 g/l this limit is reached at thermophilic conditions, where pH is typically 8.0, and the useful process temperature range is reduced to below 55°C. If a stable high rate digestion cannot take place at temperatures above 50-52°C, veterinarian benefits fade away and mesophilic operation, with a separate thermophilic sanitation step, become a more attractive alternative. At present conditions in Denmark, above limit is approached when pig manure in a mixture with cow manure reaches a fraction above ¾, in which case it is worth considering the mesophilic process, despite added process complexity.

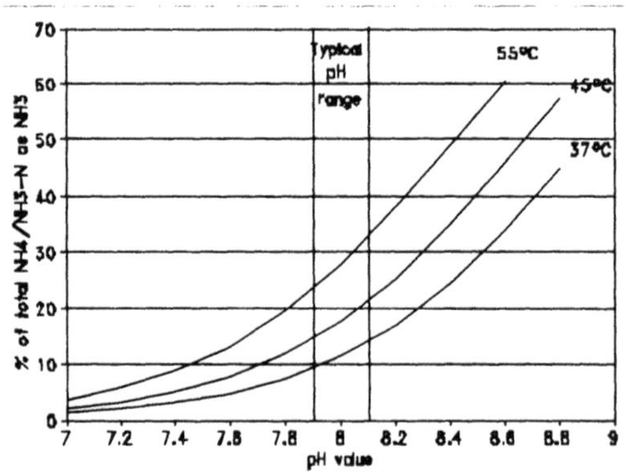

Figure 1 : Ammonia dissociation as a function of pH and temperature.

2.3. Veterinarian aspects

When mixing and redistributing manure from several farms, it becomes important to adopt a sufficient level of sanitation in order to prevent the biogas plants from spreading pathogens.

The veterinarian requirements concerning manure are to ensure that the manure treated is kept at a thermophilic temperature (>50°C) for a minimum of 4 hours. The required sanitation can be obtained directly in a thermophilic process by observing special pumping routines, whereas a mesophilic process requires a passive pre- or after sanitation stage (fig. 2). A veterinarian follow up programme has documented that above requirements ensure an effective pathogen reduction, reaching a level equal to or better than 4 log-units when using Faecal Streptococcus as an indicator organism. Certain types of supplementary waste suitable for co-digestion in manure based biogas plants, such as sewage sludge, household waste or certain types of abattoir waste require more strict sanitation. Originally such types of waste required heating to 70°C for a minimum of 1 hour. A special veterinarian research effort, utilising high temperature resistant bacteria and vira as indicators, has resulted in the definition of alternative temperature/time combinations which in conjunction with a biogas process are considered equal to a 70°C/1h heating (ref. 4). For a thermophilic biogas process (minimum 52°C), a guaranteed temperature/retention time of 52°C/10 hours, 53.5°C/8 hours or 55°C/6 hours is considered equal to 70°C/1 hour. The retention can take place in separate sanitation tanks before or after the fermentor stage or directly in the fermenter if pumping sequence allow the necessary pauses. For mesophilic biogas plants slightly higher retention times are required and must take place in a separate tank. While inclusion of "problematic" waste in the past required a separate pretreatment step, all the biomass can now be sufficiently pasteurised in a single thermophilic process, provided the longer guaranteed retention times is built into the design.

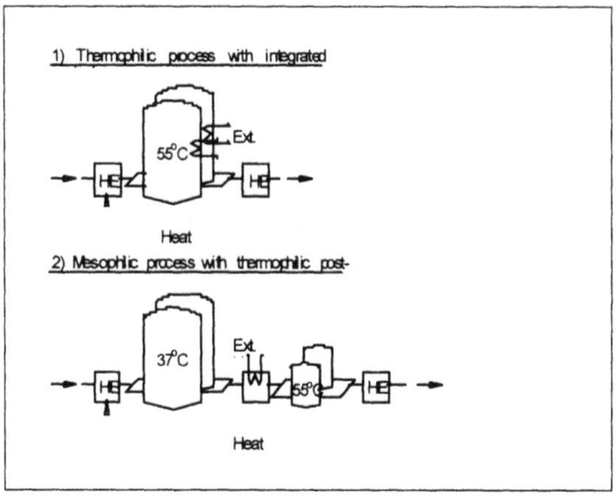

Figure 2 : Thermophilic and Mesophilic configuration.

2.4. Process tank configuration

When applying heat exchanging continuous feeding/pumping is often preferred in order to optimise the use of heat exchangers. In order also to obtain the necessary guaranteed retention time for sanitation purposes, configurations with 3 fermenters (or sanitation tanks) operated in parallel is often seen. At a given time one is being fed, one emptied and one resting, i.e.

ensuring the sanitation retention time.
As fewer larger digester tanks are often preferred for economical reasons, various schemes with discontinuous pumping or combinations with smaller buffer tanks are also seen.

2.5. Co-digestion

Manure alone usually yields an amount of biogas of approx. 25 m^3 per m^3 of manure treated. A slightly higher production is considered necessary under prevailing Danish conditions in order to secure a reasonable plant economy. All biogas plants take in different types of organic waste, primarily from the food processing industry in order to boost the biogas production and plant economy. Co-digestion offers the opportunity to deal with such types of waste in an environmentally sound way and a more economic scale (ref. 1), and ensures the recycling of nutrients back to the agricultural sector. Some types of waste can be treated as manure and require no additional installations. Other types containing high concentrations of oil/grease or proteins must be treated with care and fed to the process at a controlled rate. Most plant therefore have one or more special reception and pre-storage tanks for industrial waste with a separate feed line to the process.

3. EQUIPMENT CHARACTERISTICS

3.1. Mixing Technique

Initially submerged electrically driven medium speed mixers, mounted on guiding rails were utilised for most mixing tasks. These mixers soon proved rather costly to operate, especially where access and routine service inspection is difficult as is the case for digesters or other closed process tanks. Slow moving, top mounted central mixers with a freely suspended shaft with two propellers have become the preferred solution for fermenter mixing. The mixers usually work continuously with a mixing power input of 3-4 W/m^3. Service requirements are very limited, but a correct fermenter level relative to the upper impeller is very critical in order to avoid annoying floating layer formation. In storage tanks several types of mixers are in use, including improved submerged models. The mixing energy input is often discontinuous (high power for a short period) and average input typically varies from 10 W/m^3 in prestorage/mixing tanks to 1 W/m^3 in after storage tanks. The position and type of mixers in combination with tank geometry have proven to be very critical. Hydrodynamic favourable solutions allowing the material to flow to the mixers generally work best in combination with mixing at different depths. Fig. 3 schematically shows the most common mixer configurations.

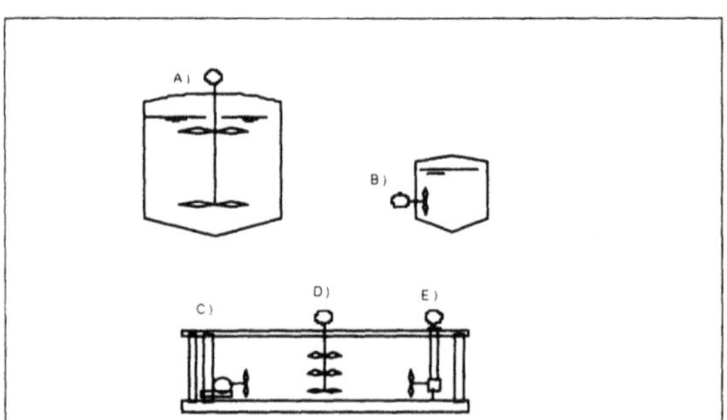

Figure 3 : Various mixer configurations: A) Digester, B) Buffer tank, C)-E) Storage tanks.

3.2. Pumping

Eccentric worm pumps have been used for continuous feeding/extraction of biomass (low flow, high pressure) while centrifugal pumps with open impellers and cutters have been used for biomass transfer between storage tanks (high flow, low pressure). Especially feeding pumps handling fresh biomass have proven to be rather costly to operate due to high wear rates, but despite several efforts no viable alternative has been found so far. Low pump speed (<300 rpm) and a pressure drop as low as possible can somewhat ease the problems.

3.3. Heat exchanging

Most biogas plants in Denmark utilise heat exchanging between in-going fresh biomass and outgoing digested biomass, although a few plants have been built with no or limited heat exchanging. The choice depends on the value of heat, i.e. the possibility of utilising heat for other purposes.

Due to the high viscosity of manure and other biomass, especially on the fresh side, it is seldom possible to obtain turbulent flow without generating an excessive pressure drop. Laminar or transition flow is the result with rather poor heat transfer coefficients. The most common types of exchangers utilise curved flow channels to break up the thermal boundary layers by secondary flow or multi pass design to break up the flow at regular intervals. Most designs attain heat transmission values of approx. 300 $W/m^2/^oC$ with flowing velocities up to approx. 1 m/s, when exchanging fresh biomass against digested biomass, or approx. the double if exchanging against water. Dimensioning, especially for pressure drop, relies very much on experience, as prediction of biomass viscosity is very uncertain. A good design margin and spare pump capacity is necessary to avoid problems. In designs with rather narrow flow passages (< approx. 30 mm) it is considered wise to macerate the fresh biomass, which is also likely to improve the biogas yield slightly.

Typical dimensioning practise (in Denmark) leads to heat exchanger installations with a resulting temperature difference of 10-15°C. The remaining heating up to process temperature and for heat loss compensation is by hot water coils in the fermenters or with a final biomass/water heat exchanger plus fewer coils for heat loss compensation. In this connection it is important to recognise and include the heat loss due to evaporation of the water contained in the humid biogas leaving the fermenter.

On the fresh side of heat exchangers (and pumping systems) the risk of clocking is manly due to the risk of foreign objects. On the digestate side of heat exchangers one must reckon with the formation of struvite ($MgNH_4PO_4$) which forms due to over-saturation when the biomass is cooled. The scaling mainly occurs in the cold sections and will gradually close flow channels. The struvite need to be dissolved by flushing with a circulating weak acid at regular intervals. Residual scaling also occurs in cold pipes down stream of heat exchangers, although at a much slower rate.

3.4. Biogas cleaning

Dry biogas from manure generally consist of approx. 60-70%v/v methane (CH_4), 30-40%v/v carbon dioxide (CO_2), 1-2%v/v nitrogen (N_2), 1000 - 3000 ppm hydrogen-sulphide (H_2S) and 10-30 ppm NH_3. The concentration of hydrogen-sulphide stated is typical when operating on manure alone. Other substrate, even in relatively small amounts, can result in quite different levels, both lower and higher.

Raw biogas is a wet gas, which needs to be cooled and carefully drained before utilisation. Early attempts to dry biogas using regenerative absorption equipment designed for pressurised air proved difficult due to accelerated absorbent decomposition, and the preferred solution has instead become only to cool the biogas and to design the gas system with care to avoid undrained low points. In case the biogas is to be utilised in a boiler, and emission requirements on sulphur are not to strict, no further cleaning is necessary. In most Danish cases, however, the biogas is utilised in gas engines for Combined Heat and Power (CHP) generation. In this case the hydrogen-sulphide content needs to be kept below approx. 700 ppm on conventional gas engines in order to avoid excessive corrosion and too rapid (and costly) lubrication oil deterioration. One method to reduce the content of hydrogen-sulphide is to add a commercial ferrous solution to the fermenter feed. Ferrous compounds bind sulphur as insoluble products in the liquid phase, which reduces the evolution of gaseous hydrogen-sulphide. The method has been utilised on a number of plants, but is rather costly if applied continuously, as the consumption of ferrous material on a stoichiometric basis has proven to be 2-3 times the desired reduction in gaseous hydrogen-sulphide. In some cases, however, ferrous containing waste products can be obtained for co-digestion most of the year, with commercial ferrous addition acting as a back up possibility only.

Often H_2S biogas cleaning is necessary, and in this case a biological oxidation process has become the dominating solution. By injection of a small amount of air (2-8 %v/v) into the raw biogas, the hydrogen-sulphide content can be biologically oxidised either to free (solid) sulphur or (aqueous) sulphurous or sulphuric acid according to following reactions :

(1) $2H_2S + O_2 \rightarrow 2H_2O + 2S$
(2) $2H_2S + 3O_2 \rightarrow 2H_2SO_3$
(3) $2H_2S + 4O_2 \rightarrow 2H_2SO_4$

The reaction will occur spontaneously and can take place in the fermenter head-space on the floating layer (if any) and fermenter walls, if air is injected directly in the head space. Products are a mixture of above, and the degree of oxidation/ratio of products probably depend on amount of excess air/oxygen. Due to the acidic nature of the products with a risk of corrosion and the dependency of a stable floating layer, it is often preferred to isolate the process in a separate reactor as shown schematically in fig. 4. The reactor is somewhat similar to a scrubber, consisting of a porous filling (randomly packed plastic elements) where micro-organisms can grow and a sump, pump and nozzle arrangement allowing regular showering of the filling. Showering has the function to wash out acidic products and to supply nutrients to the micro-organisms. The sump must therefore contain a liquid with a high alkalinity and a content of essential nutrients. Screened digested manure has been chosen for this purpose, but also recently biogas condensate with addition of commercial nutrient mixtures has proven suitable and with less risk of clocking the reactor filling. A reactor loading of approx. 10-12 m^3/h of biogas per m^3 of reactor filling and a process temperature around 35°C have been the normal choice, and the process has proven very efficient, provided sufficient air is injected (slightly more than stoichiometrically needed) and sump pH is maintained at 6 or higher. A washing procedure, where the filling elements are gurgled through with an air/water mixture, has to be carried out at regular intervals in order to prevent free sulphur deposits from closing the reactor filling.

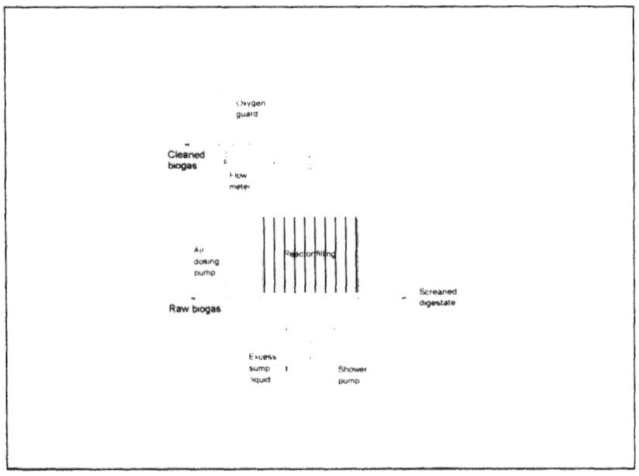

Figure 4 : Typical biological H2S reduction reactor.

In some cases where biogas is stored or passing a digestate after storage, the H_2S reactor is omitted and only air injected. Cleaning is then relying on the formation of a floating layer in the after storage, on which the micro-organisms can grow and perform the oxidation. A floating layer can usually be maintained with the choice of a low mixing intensity, without too many problems in operating the tank as a buffer storage. This solution is certainly more cost effective, but can also be more unreliable as floating layers can be rather unstable, i.e. sinking overnight without notice perhaps to resurface some days later. At least some periods with reduced cleaning efficiency must be reckoned with.

3.5. Biogas transmission

Biogas Plants are usually situated at some distance from the nearest town, while biogas utilisation in an engine CHP installation often takes place inside or in the vicinity of the town in order to connect to a district heating system. As a consequence biogas often has to be transported up to several km between the plants. The first plants utilised high pressure (2-4 baro) dry transmission with piston compressors and absorption drying equipment. As mentioned earlier, absorption drying presented problems, and as also piston compressors proved vulnerable and operational expensive, solutions based on low pressure wet transmission have gradually taken over. This solution usually involves a transmission pressure in the range 400-700 mbaro in order to deliver at 100-200 mbaro, utilising Roots blowers and a carefully planned underground transmission line executed with a minimum 3-5 o/oo fall (preferably in the flow direction) to a limited number of condensate wells. The bulk of condensation occurs in the first few wells within 100 - 500 meters from the plant. Wells are equipped with valves for regular manual emptying, and the wells are usually designed to hold up to a week's amount of condensate, meaning that the first wells must be quite large (typically 2 m^3), while the last only need to be small (typically 100 litres).

At distances up to a few hundred metres centrifugal blowers and a transmission pressure of approx. 100 mbar can be employed with advantage.

3.6. Fibre separation

Manure and other types of biomass contain fibres (from straw) and other organic structures which are difficult for anaerobic organisms to access. A typical degradation "efficiency" of

50-60% is often seen in manure based systems, meaning that the digestate still contain 40-50% of the original organic content primarily as fibres. Some of the early plants included separators to take out a part of the fibres for production of commercial compost. Marketing this compost in relatively small scale proved quite difficult, and compost production has ceased on most plants. It has also been attempted to include separation equipment with the intention to produce a fibre fraction with a dry matter content in excess of 45%, which can be utilised as a supplementary fuel in wood chip boilers. This way the overall energy efficiency can be raised by approx. 15% through additional heat production. However wood chip boiler operators has so far been sceptical and it has been difficult to pursuit this option. A side benefit, which in the near future might add to the feasibility of this scheme, is the removal of surplus phosphorous which predominantly is a attached to the fibres.

Screw type separators are dominating, sometimes in combination with a pre-separation in order to increase the capacity.

3.7. Odour control systems

One of the benefits from digestion of manure is a considerable reduction in odour nuisance when spreading manure on fields. The biogas plant itself can, however, give rise to local odour nuisance if care is not taken to limit odour emissions.

As a first preventive measure it is of course desirable to situate the plant (if possible) so as not to have immediate neighbours. In addition all storage tanks should be covered, but due to service openings and as unloading/loading and internal pumping generates tank breathing, it is necessary to arrange a weak suction from storage tanks and other contaminated sections of the plant. This air must be treated in order to avoid emission.

Air treatment is usually chosen either as biological cleaning in a compost filter or as combustion. Compost filters are relatively cheap but operational experience is ambiguous. Careful attention and maintenance are required in order to ensure satisfactory cleaning as filters some times fail due to acidification (oxidation of sulphide), might collapse and become too dense or might run dry. Combustion in a boiler in conjunction with generating process heat for the biogas plant has been tried, but the need for combustion air is limited, enforcing a compromise on venting flow. Combustion in engines (together with the biogas) has so far not been attempted, due to the fear of affecting the engine negatively. In one plant (Blaabjerg) a novel solution have been attempted, employing a system involving catalytic combustion with regenerative heat exchanging, a technique commonly utilised for the combustion of gaseous solvents in industry ventilation air. Although a rather expensive solution, results are so far positive.

3.8. Road transport equipment

After some early disappointing results with pump based tanker trucks, which proved too sensitive to foreign elements in the manure, most plants now utilise vacuum tanker trucks with a 10" suction crane. Since transport equipment and driver wages are dominant cost elements, focus has been on arranging facilities both at the plant and at the individual farmers as to avoid any waste of time. With proper installations a typical round trip takes between 30-45 minutes, allowing two 20 m3 trucks to service a plant of capacity approx. 100,000 tone/year. It has however proven essential with some organisational flexibility of drivers to allow for optimisation, often having 2½ - 3 drivers to operate two trucks.

4. OPERATIONAL EXPERIENCE AND RESULTS

4.1. Start up procedure

Start up of mesophilic plants has proven to be quite straight forward as manure has a natural content of the anaerobic micro-organisms needed. It is possible to start up without any seeding material, although some initial seeding amount can speed up the start up procedure. Often near full load conditions can be reached within 3 months.

For thermophilic plants it is more necessary and efficient to obtain a suitable seed material in a quantity as large as practical. Once this has been supplied to the plant and the temperature is re-established, the initial filling can be loaded with approx. 5% fresh material every day. The fresh material fed can gradually (i.e. in the course of 15 days) be regarded as activated and included in the loading calculation, thus gradually building up the plant fermenter(s) filling. The level of loading can be adjusted in accordance with the concentration of organic acids in the fermenter, which should not exceed approx. 5 - 8 g/l during start up. Start up time depends on the initial amount of seed material. With a 1/6 initial seed filling, the time until full production is reached will typically last 3-5 months.

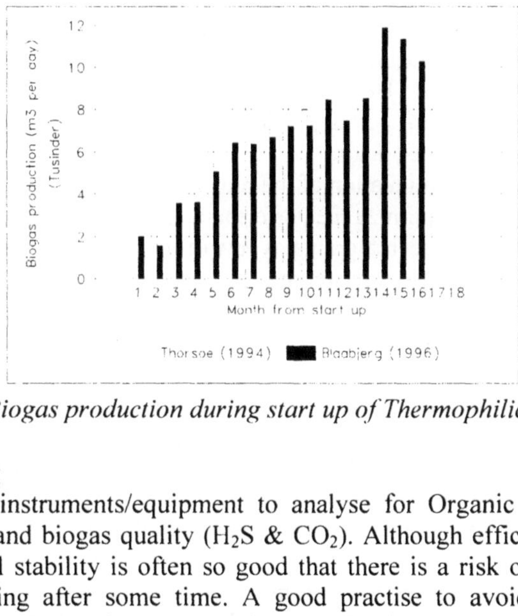

Figure 5 : Biogas production during start up of Thermophilic plants.

4.2. Process monitoring

Most plants possess instruments/equipment to analyse for Organic Acids conc. in the fermenter (by titration) and biogas quality (H_2S & CO_2). Although efficient to "catch" most disturbances, operational stability is often so good that there is a risk of the operating staff neglecting this monitoring after some time. A good practise to avoid disturbances on a preventive basis is to seek (as far as possible) stability in input mixture and operating conditions, and to watch that the biogas production correspond to expectance. In addition all new types of feed (in particular industrial waste) should be handled with care and only be introduced gradually.

4.3. Production results

Fig. 5 shows the average daily biogas production during the first operational period of the Thorsoe and Blaabjerg Biogas Plant, two thermophilic plants put into operation in 1994 and 1996 respectively, both with a biomass treatment capacity of 100,000 tonnes biomass per year. The figure shows typical running in periods of 4-6 months and illustrates the relative stability of the production, once normal operation is obtained. The sudden rise in biogas production of the Blaabjerg plant after month 13 were due to the introduction of residual fish oil as a co-substrate.

Commissioning tests of these relatively new plants showed a process electricity consumption of approx. 4-5 kWh per m^3 biomass treated for mixing, pumping, control etc., and a heat consumption of 15-25 kWh per m^3 of biomass treated, including consumption for hot water, building heating etc. This should be compared to a biogas yield of approx. 30 m3 per m3 biomass, corresponding to a LCV yield of approx. 200 kWh, i.e. 10-15 % of the energy produced is consumed in the process . These figures represent the state of the art level for large scale manure based biogas plants in Denmark, achieved as a result of the technical development described shortly in this paper.

5. DEVELOPMENT & TRENDS

5.1. General

While nearly 20 plants were established in the period 1988-1997, progress has been slow since then, despite the fact that the biogas plant sector is foreseen in the national Energy Plan 21 to be expanded considerably. An important reason has been the uncertainty of the future electricity price structure in Denmark due to privatisation of the electricity sector, to be fully implemented until year 2003. However, now the future structure of pricing of sustainable electricity production is more certain, which together with higher international fossil fuel prices has increased the interest again.

In order to support the continuation of the biogas sector expansion, research and tests have been carried out to further optimize the performance of Biogas Plants in Denmark, often financed through national research grants. A few of the activities foreseen to play a role for future biogas plants in Denmark are summarized below.

5.2. Post treatment of digestate

Pilot tests has been conducted with various methods of reducing the content of specific nutrients in the digested manure, specifically N and P. Separation of the solids fraction after digestion has been demonstrated to take out the major part of P. By centrifugation 60-70% of the P content can be extracted with the solid fraction or somewhat less with other types of separation such as brush sieve and screw presses.

Some intensive pig raising regions have a significant surplus of P compared to the need of crops. So far only N application (approx. 200 kg-N/ha/year as natural fertiliser) has been limited by manure handling regulations, but soon also P limits (approx. 30 kg-P/ha/year) are expected to be enforced by new regulations. This will make after-treatment attractive in some cases to avoid excessive transport distances to distribute manure.

5.3. Optimisation of reactor configurations

Since the middle of the 1990's it has become "standard" too arrange biogas collection from the digestate storage tanks on biogas plants, where residual biogas formation may occur until the digestate is completely cooled to ambient level. In some cases relatively high biogas yields from such installations have been reported (on some plants >10 %). However, recent tests and studies seem to indicate that this will only be the case if the main process is not working properly. If the main process is healthy, one can not expect any significant contribution of residual biogas from after-storages, once the digestate is cooled (by heat exchanging) significantly below the process temperature.

However the same investigations reveal that there are still typically 5% residual biogas potential left in the digestate extracted from conventional reactor systems, partly in the form of VFA, but also due to un-degraded particulate matter, which leave the fully mixed CSTR reactors with a relatively short retention time.

It is now foreseen that a two stage digestion process will be advantageous, with a relatively large main process step (12-15 days average retention) and an additional approx. 3 days post-digestion in the form of a series CSTR or batch step. With such a configuration the residual loss can be minimised, especially if the main process for some reason is out of balance. Biogas collection from the after-storage tanks will still be arranged as a final "insurance" and also for improved odour management.

A side benefit of such series reactor concepts will be a higher required degree of sanitation.

5.4. Improved process monitoring

Several research projects have tried to develop improved process monitoring methods, ranging from Fuzzy logic concept to automated on-line sampling for continuous VFA measurements. An on-line VFA monitoring system involving a special filter technique and GC based analyser has already been developed in a test version, and is currently being developed further for commercial marketing. However, the cost an reliability of such systems still need to documented, and it remains to be seen if biogas plants want to invest in such advanced instrumentation, since most plants work stable and satisfactory for long periods without worrying much about process health.

5.5. Road transport

Until recently 20 m^3 vacuum tanks, mounted on regular 32 ton trucks have been the standard solution. The service life of the truck itself is typically for 5-7 years, after which it is necessary to exchange the truck and shift the vacuum tank and suction equipment to the new truck. Recently also trailer trucks have been employed with a loading capacity of approx. 30 m^3, with the benefit that the pulling truck can be renewed without needing to exchange vacuum tank. It is expected that larger biogas plants can benefit from having at least one such truck to handle the suppliers in the outer region of the supply area, although also manoeuvring conditions at the farms have to be considered.

REFERENCES

Ahring BK, Angelidaki I, Johansen K (1992). Anaerobic treatment of manure together with industrial waste. Water Sci Technol 25:311318

Angelidaki I, Ahring BK (1993). Thermophilic anaerobic digestion of livestock waste: the effect of ammonia. Appl Microbiol Biotechnol 38:560564

Angelidaki I, Ahring BK (1994). Anaerobic digestion of manure at different ammonia loads: effect of temperature.Water Res 28:727-731

Bendixen HJ (1996). Hygiene and sanitation requirements in danish biogas plants. Chartier, P., Ferrero, G. L., Henius, U. M., Hultberg, S., Sachau, J., and Wiinblad, M. Biomass for Energy & Environment, 9th European Bioenergy Conference. 1, 296-301. Copenhagen, Pergamon

Greenhouse Gases and Animal Agriculture
J. Takahashi & B. A. Young (Editors)

Reuse of fibrous agricultural biomass for energy resources by methane fermentation

A. Oda, Y. Sakata & A. Oida

Graduate School of Agriculture, Kyoto University, Kitashirakawa-Oiwake-cho, Sakyo-ku, Kyoto-shi, 606-8502, Japan

In order to clear characteristics of wheat bran for a substrate of methane fermentation, some laboratory scale experiments were conducted under a continuous fermentation. The experiments made progress in a steady fermentation under all conditions of variations of organic solids concentration and hydraulic retention time. The volumetric ratio of methane and carbon dioxide in the generated gas was nearly one to one under all the conditions. But the state of the fermentation became worse and the volume of methane became less, as the organic solids concentration became higher. Application of Chen-Hashimoto model for analysis of the fermentation almost cleared the optimum organic solid concentration and hydraulic retention time.

Keywords: methane fermentation, agricultural biomass, alternative energy

1. INTRODUCTION

Methane fermentation, originally developed as one of the methods of wastewater treatment, is recently regarded as an energy production technology, because the generated gas contains a lot of methane for gaseous fuel. Therefore the methane fermentation must be suitable for a local energy system to produce alternative energy from agricultural biomass and to decrease the organic wastes.

Wheat bran, which was much valuable for feeds of livestock before, is recently getting to be regarded as wastes from flour milling industry in Japan. Therefore the wheat bran was adopted as a substrate for the methane fermentation as a typical agricultural biomass. We studied to make clear the characteristics and performances of solubilized wheat bran for the substrate of the methane fermentation. Chen-Hashimoto model (Chen & Hashimoto, 1978), which was widely appreciated as a model of methane fermentation, was applied for analysis of the characteristics.

2. MATERIALS AND METHODS

The schematic of methane fermentation system is shown in Figure 1. Each fermenter was made of polyethylene with a liquid volume of 1 liter and was placed in a water bath so that the fermented temperature was maintained at 36 ± 1°C in a mesophilic range by thermostat and heater. Gas holder was a polyethylene bottle filled with saturated salt water in order to prevent a generated gas from solving. Reservoir was a graduated cylinder to measure a volume of displaced salt water by the generated gas in gas holder. So the volume of generated gas production was determined by a volume of displaced water at a feeding time of substrates everyday. The generated gas composition was measured using a gas chromatograph (TCD) with activated charcoal column once per a week.

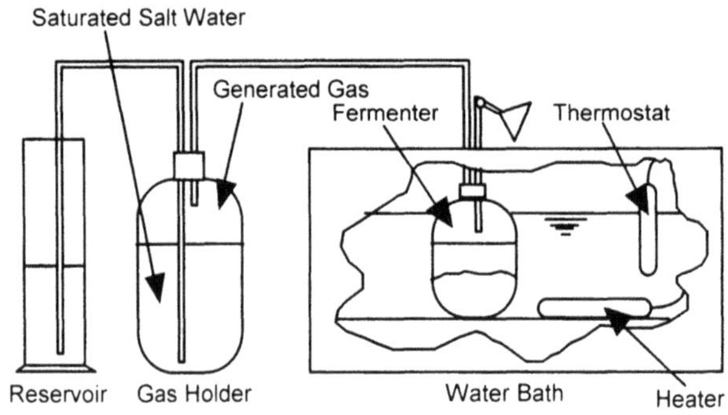

Figure 1. The experimental apparatus

Figure 2. A picture of wheat bran

A picture of wheat bran as a substrate for the system is shown in Figure 2. It contains a lot of low biodegradable fiber and is not suitable for the methane fermentation. To easily ferment, boiling with HCl solution is done for solubilization as a chemical pretreatment (Esaka *et al.*, 1997). The boiling wheat bran was neutralized by NaOH solution and finally diluted with water to an intended organic solids (wheat bran) concentration.

The organic solids concentration was set to five levels of 10 [g/L], 20 [g/L], 30 [g/L], 50 [g/L] and 70 [g/L]. At each level, hydraulic retention time was set to three conditions by varying feeding volume. Table 1 shows these experimental conditions mentioned above. At each condition, four fermenters in entirely same situations were placed in water bath and the mean value of four results was adopted as a result of each condition.

Table 1. The experimental conditions

Organic Solids Concentration [g/L]	Feeding Volume [ml]	Hydraulic Retention Time [day]
	40	25.0
10	50	20.0
	60	16.7
	30	33.3
20	40	25.0
	50	20.0
	30	33.3
30	40	25.0
	50	20.0
	25	40.0
50	30	33.3
	40	25.0
	25	40.0
70	30	33.3
	40	25.0

To analyze the characteristics of the fermentation, we applied Chen-Hashimoto model appreciated as a model of methane fermentation widely. First, This model found out maximum methane production per added organic solids by plotting methane production per organic solids with specific growth rate of microorganism. Next, minimum hydraulic retention time and kinetic parameter were decided by plotting hydraulic retention time with maximum methane production per organic solids. Lastly, from using these parameters, methane production per organic solids and volumetric methane production rate with hydraulic retention time were simulated by the following two equations.

$$B = B_0\left\{1 - \frac{K}{\theta/\theta_m - 1 + K}\right\} \tag{1}$$

$$\gamma_v = \frac{BS_0}{\theta} = \frac{B_0S_0}{\theta}\left\{1 - \frac{K}{\theta/\theta_m - 1 + K}\right\} \tag{2}$$

θ hydraulic retention time [day]
θ_m minimum hydraulic retention time [day]
B methane production per added organic solids [L/g]
B_0 maximum methane production per added organic solids [L/g]
γ_v volumetric methane production rate [L/L/day]
S_0 organic solids concentration [g/L]
K kinetic parameter

3. RESULTS

The experimental results about gas production per added organic solids, methane contents and methane production per added organic solids at all conditions are shown in Table 2. By applying methane production per added organic solids in Table 2 to Chen-Hashimoto model, relationships with methane production and hydraulic retention time can be simulated.

Table 2. Experimental results of methane fermentation

Organic Solids Concentration [g/L]	Hydraulic Retention Time [day]	Gas Production per Added Organic Solids [L/g-TS]	Methane Contents [%]	Methane Production per Added Organic Solids [L/g-TS]
	25.0	0.40657	58.842	0.23924
10	20.0	0.41208	55.439	0.22845
	16.7	0.38213	53.970	0.20624
	33.3	0.41165	58.639	0.24139
20	25.0	0.41688	56.867	0.23707
	20.0	0.39643	55.717	0.22088
	33.3	0.39213	54.568	0.21398
30	25.0	0.39365	53.272	0.20971
	20.0	0.38767	52.765	0.20455
	40.0	0.41081	52.877	0.21723
50	33.3	0.41042	52.425	0.21516
	25.0	0.40096	52.616	0.21097
	40.0	0.44003	51.328	0.22586
70	33.3	0.43693	50.768	0.22182
	25.0	0.42413	51.821	0.21979

Figure 3 illustrates the relationship with methane production per added organic solids and hydraulic retention time at expression (1) and Figure 4 illustrates the relationship with volumetric methane production rate and hydraulic retention time at expression (2). These figures indicate that methane production per added organic solids slightly decreases and volumetric methane production rate gradually increases, as the organic solids concentration becomes higher. They also suggest that minimum hydraulic retention time is about 10 days in order to maintain a steady and continuous fermentation. Furthermore, Figure 4 also shows optimum organic solids concentration is 50 [g/L] when maximum volumetric methane production rate can be obtained in this fermentation system. But this maximum volumetric methane production rate should be considered to be low as an energy production system through the methane fermentation. To obtain a higher volumetric methane production rate, we try to apply a dry methane fermentation method to this system for a further practical study.

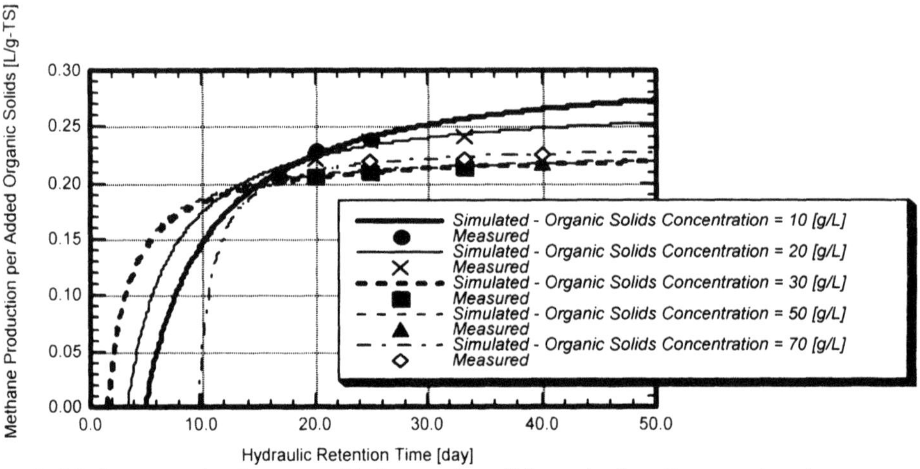

Figure 3. Methane production per added organic solids vs. hydraulic retention time

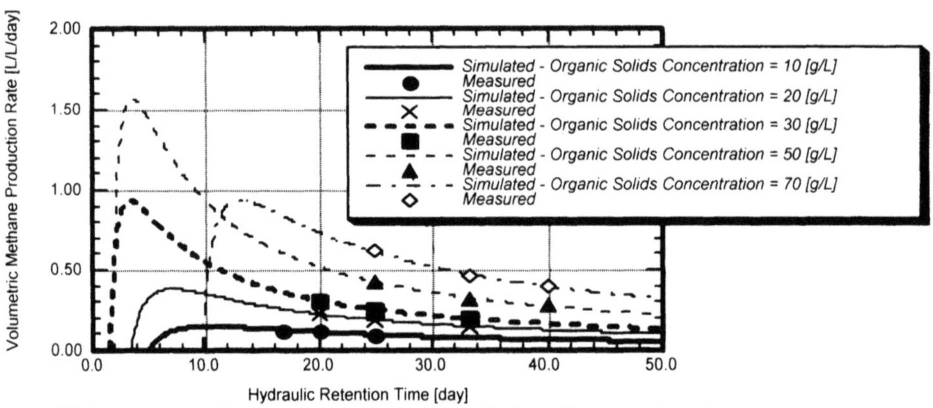

Figure 4. Volumetric methane production rate vs. hydraulic retention time

REFERENCES

Chen, Y. R. & A. G. Hashimoto, 1978. Kinetics of Methane Fermentation. Biotechnology and Bioengineering Symp. 8: 269-282.

Esaka, A., A. Oda & M. Yamazaki, 1997. Application of Agricultural Wastes as Methane Fermentation Substrate - Application of Wheat Bran -. Proceedings of International Symposium on Agricultural Mechanization and Automation. 2: 319-324.

Greenhouse Gases and Animal Agriculture
J. Takahashi & B. A. Young (Editors)

Biogas production from unconsumed milk -Reduction of CO_2 emission by utilization of unconsumed foodstuff-

Y. Kitamura[1], M. Takimoto[1,2] & Y. Sasaki[1]

[1]Dept. of Regional development, Shimane University, Nishikawatsu-cho, Matsue, Shimane, 690-8504 Japan, [2](Present) Dept. of Environmental Machinery, Satake Corporation, Nishijyonishihon-machi, Higashihiroshima, Hiroshima, 739-8602 Japan

Biogas production from food wastes is believed to offer the solutions to the problems such as energy saving, environmental protection and CO_2 emission. Unconsumed milk was treated by the mesophilic methane fermentation system with sludge return operation to find its characteristics of organic removal and biogas production. For the 3 to 8 fold diluted milk with water, the substrate removal ratios were about 98 % and methane concentration of the biogas were from 65 to 68 % at 15 days of HRT. The methane yields were ranged from 0.84 to 0.81 L/g-COD while the ammonium and volatile acid inhibition did not occur. The kinetic parameters also showed the feasibility of the unconsumed milk to produce clean energy effectively by the biogas production system.

Keywords: biogas production, CO_2 emission, methane fermentation, milk, organic removal, gas productivity

1. INTRODUCTION

The PL law and food safety programs as HACCP leads to the safe and high quality food supply while they increase the returned or discarded foodstuff. When they are left in the yard, not only the waste but also the greenhouse effect gas, CO_2 are produced during their aerobic degradation. Although these unconsumed foods are not appropriate for meal, they could be useful as bio-energy resource by the biogasification or methane fermentation. The CO_2 involved in the biogas is easy to control because the system is usually closed to the ambience. As one of these foodstuffs, unconsumed milk discharged from dairy facilities or processing factories was chosen for this study. The characteristics of organic removal and biogas production of the feeding were experimentally determined to find its feasibility for bio-energy resource by methane fermentation.

2. MATERIALS AND METHODS

2.1. Experimental apparatus and operational condition

Figure 1 shows the schematic design of the bench scale methane fermentation system that is composed of four methane fermentors (Sibata, 3L-glass flasks), gas circulation equipments with suction pumps (Nihon Rikagaku-kiki, UP-5), a hot water bath with a temperature controller (Iuchi, TR1) and gas storage acrylic holders controlled by water replacement method. The gas circulation equipments stirred each fermentor for 15 minutes at 3 times a day by blowing the produced biogas from the holders into the ferementors. Mesophilic sludge from the sewage digester (Lake Shinji East Treatment Center, Shimane) was used as the seeding sludge for the fermentors. The fermentors were put in the water bath to maintain the temperature at 36±1℃. Feeding and withdrawing was conducted once in a day by the draw and fill method after the stirring of the fermentors. Feeding material or influent was the expired milk (Dairy Matsue, Shimane) diluted with tap water at 3, 4, 6 and 8 fold. Each influent was fed until the steady states were obtained. The solid or sludge in the fermentor effluent was separated by the centrifuge (Hitachi, C4) and returned to the fermentor with the influent to keep the sludge or microbes high concentration in the fermentors. HRT (Hydraulic Retention Time) was set at 15 days.

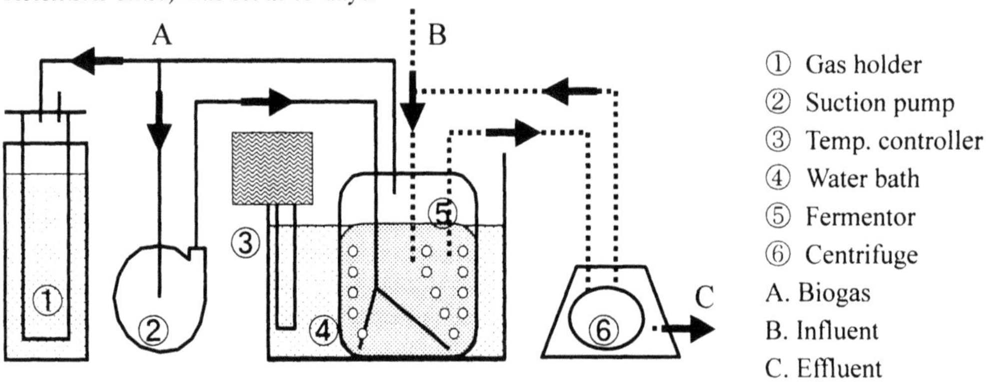

Figure 1. Schematic design of methane fermentation system with sludge return

2.2. Measurement parameters

pH, TS (Total solid), VS (Volatile solid), TVA (Total volatile acid), $NH_4^+$and COD (Chemical Oxygen Demand) as water quality parameters for the fermentor influent and effluent, and CH_4, CO_2 and H_2S (Sulfide hydrogen) as gas quality parameters for the biogas were determined. The pH was measured with a glass-probe pH meter (Horiba, D-21). The TS and VS were determined by the weighing after drying method. The VA was measured by the water vaporization and titration method. The COD measurement was based on the colorimetric analysis by using the colorimeter (HACH, DR/700). All the water parameters except COD were measured according to the municipal wastewater test method. The biogas composition as CH_4 and CO_2 were measured by using the gas chromatograph (Shimadzu, GC-14) analysis mentioned by Maekawa et.al (1984). The H_2S was determined by use of Kitagawa-gas detector (Gas tech).

3. RESULTS AND DISCUSSION

3.1. Steady state parameters for influent and effluent

The steady state parameters of water qualities are summarized in Table 1. They are the arithmetic mean values of 3 to 5 data points during the steady state for each influent. All the pH values were higher than 7.0 so that the single-phase methane fermentation was confirmed to occur in the system. The sludge density ranged from 12.6 to 18.1 g-VS/L and they seem to be higher than the values for the conventional fermentors without sludge return. The organic or substrate concentration based on COD for the influent were between 10.52 and 28.05 g/L while the effluent substrate concentration varied from 0.21 to 0.48 g/L. Accordingly, the substrate removal ratios were determined to be almost constant as 98.0 %. The effluent substrate concentrations were still high so that the 2nd stage treatment should be necessary for the discharge of the effluent. Although the ammonium ion concentrations from 261 to 847 mg/L were not so high, they seem to gain with the increase in the loading rate of COD. It suggests to have incorporated an ammonium reduction process into the system at higher loading rate of COD.

Table 1. Steady state parameters for water

	Fermentor1	Fermentor2	Fermentor3	Fermentor4
COD loading rate(mg-COD/L・d)	1870	1403	935	701
pH_{in} (-)	6.92	6.97	7.04	7.09
pH_{out} (-)	7.47	7.27	7.1	7.02
Sludge density (g-VS/L)	18.1	16.8	13.6	12.6
$Substrate_{in}$ (g-COD/L)	28.05	21.04	14.03	10.52
$Substrate_{out}$ (g-COD/L)	0.48	0.37	0.30	0.21
Substrate removal ratio (%)	98.3	98.2	97.9	98.0
NH_4^+concentration (mg/L)	847	594	416	261

(Subscript, in : influent out : effluent=fermentor)

3.2. Steady state parameters for biogas

Table 2 shows the gas parameter values in the steady state. They are also the arithmetic mean values of 3 to 5 data points during the steady state. The gas production rates increased with the increase in the loading rate of COD because the gas yields were maintained at 1.2 or 1.3 L/g-COD. The methane concentrations were ranged from 65 to 68 % so that the methane yield from 0.81 to 0.85 L/g-COD was obtained. By using the methane yield of 0.84 L/g-COD for the fermentor1 and the substrate concentration of 5.61 g-COD/L, the methane production from 1 L unconsumed milk was calculated to be about 4.7 L It has heat amount of 187.5 kJ and corresponds to the energy to heat up 3.0 L of milk (C=4.2kJ/kg·K) from 20 to 35 ℃ that is the temperature for the mesophilic methane fermentation. H_2S concentration increased with the gain in the loading rate of COD, and which suggests that the H_2S scrubber should be equipped with the system.

Table 2. Steady state parameters for biogas

	Fermentor1	Fermentor2	Fermentor3	Fermentor4
COD loading rate (mg-COD/L·d)	1870	1403	935	701
Gas production rate(L/L·d)	2.36	1.77	1.12	0.83
Biogas yield (L/g-COD)	1.28	1.28	1.23	1.21
CH_4concentration (%)	65.5	66.8	66.6	67.3
CH_4yield (L/mg-COD)	0.84	0.85	0.81	0.81
H_2S concentration (ppm)	822	735	620	588

3.3. Kinetic evaluation of the system

Being based on the reaction model bellow, the relationship between the effluent substrate concentration and the specific rate of substrate consumption were determined by using the steady state data in Table 1. The obtained rate parameters as the maximum specific rate of substrate consumption, ν_{max} and the saturation constant K_s are shown in Table 3 with the rate parameters for the washed rice wastewater. The ν_{max} for the milk was larger than the one for the washed rice wastewater so that the milk was found to be better substrate for the microbial growth activity. The K_s of the milk was also higher than that of the washed rice wastewater, and which means the affinity of the substrate to the microbes is lower. It seems to be modified by frequent stirring of the fermentor in order to increase the contact efficiency between the substrate and microbes.

$$(S_{in}-S_{out})/(HRT*X)=\nu_{max}*S_{out}/(K_s+S_{out})$$

Where, HRT: Hydraulic retention time (d), X: Sludge density (g-VS/L), S: Substrate concentration, ν max: Maximum specific rate of substrate consumption (mg-COD/mg-VS・d), Ks: Saturation constant (mg-COD/L). The subscripts in and out represent the influent and effluent, respectively.

Table 3. Apparent rate parameters

	ν_{max} (mg/mg・d)	Ks (mg/L)
Washed rice wastewater	0.27	177
Unconsumed milk	0.68	1147

REFERENCES

Chang, J.E., Noike, T., Matsumoto, J., 1982. Effect of retention time and feed substrate concentration on Methanogenesis in anaerobic digestion. Proceedings of the Japan Society of Civil Engineers. 320(4): 67-76. (In Japanese)

Kitamura Y., Suzuki K., Tagawa A., Hayashi H., 1996. Anaerobic contact treatment of dairy processing wastewater with low organic matter. Journal of Agricultural Science, Tokyo Nogyo Daigaku. 41:103-107

Kitamura Y, Tagawa, A., Hayashi H., 1996. Treatment Characteristics of the Washed Rice Wastewater by Methane Fermentation, Journal of the Japanese society of agricultural machinery. 58: 87-93

Maekawa, T., 1982. Methane fermentation. In: Biomass energy, Yamazawa, S., Asakura Syoten, Tokyo, Japan. 115-161. (In Japanese)

Greenhouse Gases and Animal Agriculture
J. Takahashi & B. A. Young (Editors)

How the biogas plants based on the manure in E.U. work in practice

H. Soma

Kaname Engineering Co., Ltd.,Ichizawa-Danchi2-225,Asahi-ku,Yokohama 241-0014 Japan

In general, the animal manure in E.U. is sprinkled on the lands as the live slurry or the compost, however several environment pollutions have been noted. The biogas plants are applied as the manure treatment system with some economical profits, which may depend on the conditions of the farm holding lands, the located condition between the farms each other, treatment fee of the organic waste and the political subsidy, such as the investment and the selling price of the electric power in Japan.

The specifications and the operation manuals of the farm-scale biogas plant that works in practice are introduced.

Keywords: biogas plant, manure, slurry, methane, E.U.

1. INTRODUCTION

In Netherlands, each of the stockbreeding farms and the cultivating farms is specialized in stockbreeding or planting, so that the amount of manure and the lands to be sprinkled are unbalance. In general, the live slurry is transported by the vacuum cars from the stockbreeding farms to the cultivating farms and sprinkled on the lands. This is the most economical control in Netherlands, based on the special cheap transportation fee. However, some farms, which have enough lands to be sprinkled, applied the farm-scale biogas plant.

In Denmark, the some farms having enough lands for the manure to be sprinkled have applied the farm-scale biogas plants before about 15 years ago. However these biogas plants had been difficult to start again by the poor engineering of the farms at the trouble.

Now the 20 centralized biogas plants for avoiding the environmental pollution are treating 1 million tons of the manure in totals a year, which is 2.5% of the manure in Denmark. The treated slurry is distributed to the farms, depending on the holding land areas. However, the most farms still sprinkle the live slurry or the compost on the lands.

In Germany, almost farms are sprinkling the manure on the owned lands. The manure is the live slurry condition or the treated compost condition. About 1000 farms that have enough lands for the manure to be sprinkled, apply the farm-scale biogas plants from a viewpoint of the environmental pollution, with the economical profits.

2. EXISTING MANURE CONTROL SYSTEM AND BIOGAS PLANT MANURE CONTROL SYSTEM

The summery of the background of the manure treatment system including the biogas plants system shows as *Figure 1*.

In general in E.U., the manure stored in the slurry tanks at the slurry condition for long times, and sprinkled on the lands in the suitable season using the large sprinkling machines. On the other hand, the manure is also separated to the slurry and the solid part, which is treated as the compost and stored on the ground without any cover in some cases.

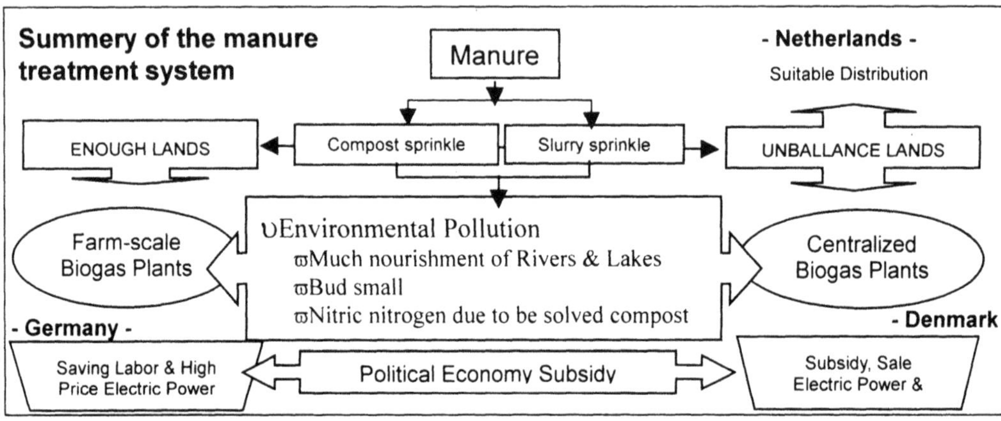

Figure 1. Summery of the manure treatment system

The compost take the long times for fermenting and then sprinkled on the lands. The slurry stored in the tanks for long times and also sprinkled on the lands.

These compost and slurry is decomposed to HH_4, PO_4, KO_3, etc. on the ground, taking the long times. Therefore there are risks of the environmental pollution that un-decomposed manure may spill out to the river, lake and underground water due to rain, or sinking By the way on the manure control system by the biogas plant, the manure stored in the slurry tanks for a few days and put in the fermenting tanks with pre-heating for the centralized type. The manure is decomposed by anaerobic fermentation to HH_4, PO_4, KO_3, etc and the biogas, which consists of methane (60%) and carbon oxide (40%).

The digested slurry is stored and sprinkled on the lands, but there is not any risk from the spillage of un-decomposed manure. In E.U., this digested slurry is evaluated as the higher efficient fertilizer.

2.1. Centralized Biogas Plants

In the typical centralized biogas plants, the manure is transported from the manure slurry tanks of the 2 – 80 farms to the manure storage tank, using the vacuum cars. The manure is homogenized with the other industrial organic wastes and then put into the anaerobic fermenting tanks through the heat exchangers.

The separated biogas is soled out to the power station as the fuel that generates the electric power and heat energy for the local consumption. The digested slurry is transported to the slurry storage tank of the farms and then is sprinkled on the lands at the suitable season. Some times, the digested slurry is stored in the centralized slurry storage tanks.

In any case, the manure is gathered from the farms and mixed. After fermenting, the digested slurry is distributed to the farms again, and then sprinkled on the ground, so that the sanitation control of the manure is necessary essentially for the centralized type.

The main reasons applied with the centralized biogas plants in Denmark, are as follows:

- To avoid the environmental pollutions and the unbalanced lands problem can be solved, by redistributing the digested slurry, depending on the land area of each farm.
- To ensure the economical profits by the scale merits, taking and selling the energy of the electric power and heat with higher efficiency and then getting the higher incomes.
- To got the incomes by treating the industrial organic wastes.

The operating days of the centralized biogas plants is improved recently, adopting the intelligent design and operation, so that the operating costs included the maintaining works of the plants are reduced and then the modern profits is improved as shown on Table 1.

Table 1. Profits of each centralized biogas plants in Denmark unit: 1,000DKK

Plants	Farms	Capacity t/d	1990/91	1991/92	1992/93	1993/94	1994/95	1995/96	1996/97	1997/98	Turning point
Hjermitslev	5	41+13	488	330	591	904	837	583	1,516	879	950
Vegger	7	42+17	-131	999	891	1,270	1,225	1,037	1,574	1,482	700
Davinde	6	25+ 3	173	300	223	283	255	150	77	17	270
Fangel	26	124+19	1,225	1,458	2,422	2,247	2,842	2,221	2,279	2,317	2,800
Revninge	2		35	544	262	445	309	217	289	254	250
Ribe	79	352+68	472	2,256	2,657	2,895	3,230	3,161	3,298	3,855	2,600
Lintrup	62	410+137	2,762	2,350	2,822	3,622	3,605	2,639	3,865	3,457	3,600
Lemving	80	362+75		318	2,393	3,927	4,642	3,848	4,444	4,353	4,200
Hodsager	6	42+ 6				786	1,106	1,132	1,195	1,204	1,200
Hashoj	17	100+38				490	1,502	1,341	1,882	1,982	1,500
Thorso	75	230+31				303	1,482	1,756	2,023	2,495	1,800
Arhus	45	346+46						-3,994	-2,207	-1,860	3,500
Fiskov	7	61+18					742	1,394	1,526	1,674	1,500
Studsgaard	50	230+36						1,762	2,974	3,327	6,400
Blabjerg	58	222+87						1,676	3,090	3,809	3,000
Snertinge	14	66+42							1,655	2,180	2,850
Blahoj	15	70+17							574	1,845	1,900

2.2. Farm-scale Biogas Plant

The manure from the livestock is collected in a manure storage tank, and put an anaerobic fermenting tank by the feeding pump automatically with the certain intervals.

Simultaneously the digested manure is discharged to the slurry storage tank, which will be sprinkled on his land of the farm at the necessary times, and the biogas is supplied to the power unit that generated the electric power and heat energy. The electric and heart energy is used for the biogas plant and his farm, and the excess electric power is sold to the power company with the certain price (about 11yen / kWh in Germany)

The economical profits of the farm-scale biogas plants are as follows:

- Free from the manure handling works, such as gathering and treating works of the manure, and then these working times are spent usefully for the other works.
- Increase the products by sprinkling the digested slurry on the lands as a fertilizer.
- Sale the excess electric power to the electric power company with higher price, and then gets the much profits.

In Germany, the economical profits of the farms are ensured by the higher selling price of the electric power, even if the excess heart energy is not used economically.

3. BEST TYPE OF THE BIOGAS PLANT FOR EACH FARM CONDITION

3.1. Type of the biogas plant

The best type of the biogas plants between the centralized type and the farm-scale type depends on the land area condition held by the farms, the located condition of the farms each other and the transporting cost of the manure & digested slurry.

In general, the best type of the biogas plants is selected from the condition of the farm shown on the Table2.

However, the farm will select the type of biogas plants after each economical profit in total of the both is confirmed, depending on the condition of each farm.

Table 2. The Suitable type of the biogas plants, depended on the conditions of the farms

<table>
<tr><td colspan="2" rowspan="2">Location Condition of Farms</td><td colspan="3">The holing lands of the farms are unballanced</td><td colspan="2">Each farm holds the enough land for sprinkling</td></tr>
<tr><td>Many Farms located in the wide area</td><td>Some Farms located in the limited area*1</td><td>Each farm with long distance</td><td>Many Farms located in the wide area</td><td rowspan="1">Other Farms</td></tr>
<tr><td rowspan="2">Transport Cost</td><td>Expensive</td><td rowspan="2">Large Scale, Centralized</td><td>Farm-scale</td><td rowspan="2">Farm-scale</td><td>Farm-scale</td><td rowspan="2">Farm-scale</td></tr>
<tr><td>Cheap</td><td>Middle Scale*2, Centralized</td><td>Centralized</td></tr>
</table>

*1 Farms are located in the limited area, but in so the wider area

*2 Middle scale centralized type, if any economical merit

3.2. Temperature of the fermenting process

There are the mesophil and the themophil processes of the anaerobic fermenting on the biogas plants, each of which has their merits and demerits. In case of the centralized biogas plants, the themophil process is recommended because the plant may be constructed with the less capital cost by the smaller size and the simple process for the sanitation.

In case the farm-scale biogas plants, the mesophil process operated within the wider temperatures is recommended with no sanitation process. However, from a viewpoint of the sanitation, the themophil process is recommended when the digested slurry is sprinkled on the other farmlands.

4. THE TYPICAL OF THE FARM-SCALE BIOGAS PLANT WORKED IN PRACTICE

4.1. Specification of the existing farm-scale biogas plant

The farm-scale biogas plant is applied instead of the live slurry sprinkling, increasing the incomes by the electric power sale, saving the manure handling work, and increased production of the plants (*Figure 2.*). The design concept of this plant is the simple system that consisted of a primary manure storage tank, a fermenting tank combined with the gasholder, the power-generating unit, and a digested slurry storage tank. The fermenting tank of the vertical type is applied with the mesophil process of a single step, mounted on the ground for saving the construction cost. The power unit module is operated 24 hrs automatically.

The particulars of the farm-scale biogas plant, located at Boddin, Germany

Manure: 16ton/day (200milk cows)	Electric power: about. 910kwh
Fermenting tank capacity: 400m3	Heart energy: about. 1,370kwh
Generated biogas volume: 480m3/day	Digested slurry
Ferment temp & days: 35℃ 25days	Storage tank: 4,000m3
Electric generator: 50kw/hr	Sprinkled rate: 3ton/ha

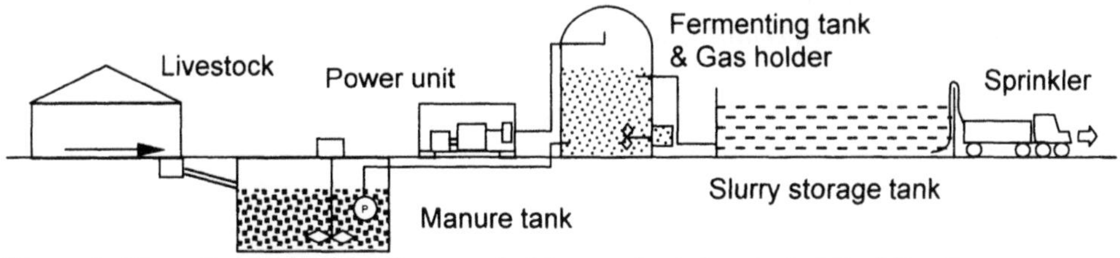

Figure 2. Flow diagram of the farm-scale biogas plant, located at Boddin, Germany

4.2. Recording and checking works of the plant

All operations of the plant, such as supplying the manure to the primary manure tank, agitating it homogeneously, feeding the manure to the fermenting tank by 3 times a day, discharging the digested slurry to the slurry storage tank is controlled automatically. The manure in the fermenting tank is agitated in the certain interval and is kept at the temperature between 30 – 38℃ automatically.

However, the several data following the maintenance manual supplied from the engineering company should be recorded or checked

The examples are follows:

- The temperatures at the several levels located in the fermenting tank
- The slurry level in the fermenting tank
- Gas meter level in the gas holder
- The rate of CH_4 in the biogas
- The rate of H_2S in the biogas
- The level of the Electric power gage
- Check the seal water level of the gas holder and feeling the water, if necessary
- Check the level of the lubricating oil of the generator and the agitating drivers

4.3. Maintain works of the plant

For operating continuously with no stop, the maintain works for the power unit module are required, which are the similar maintenance works to the ones for the motorcars.

Example maintenance works are as follows, picked up the maintenance manual.

- Every 300 hours
  - Exchange the filter of the line filter
- Every 600 hours
  - Exchange the oil and oil filter
  - Check the battery
- Every 1800 hours
  - Exchange the material of the air filter
  - Exchange the fuel oil filter
  - Cleaning the oil supply pump and the filter

These maintain works are depended on the type of the machineries supplied the power unit, and therefore there are not the special maintenance works for the biogas plant.

REFERENCES

Kurt Hjort-Gregersen, M.Se. Danish Institute of Agricultural and Fisheries Economics, 1999, centralized Biogas Plants -Integrated Energy Production, Waste Treatment and Nutrient redistribution Facilities-, Danish, table8.1

Greenhouse Gases and Animal Agriculture
J. Takahashi & B. A. Young (Editors)

Bioenergy supply potential and bioenergy utilization costs

H. Yamamoto[1], J. Fujino[2] & K. Yamaji[3]

[1]Socioeconomic Research Center, Central Research Institute of Electric Power Industry, 1-6-1 Otemachi, Chiyoda-ku, Tokyo 100-8216 Japan, [2]Global Warming Response Team, Global Environment Division, National Institute for Environmental Studies, 16-2 Onogawa, Tsukuba, Ibaraki 305-0053 Japan, [3]Department of Advanced Energy, School of Frontier Sciences, The University of Tokyo, 7-3-1 Hongo, Bunkyo-ku, Tokyo 113-0033 Japan

In this study, we evaluate the global bioenergy supply potential in the future using a multi-regional global-land-use-and-energy model (GLUE) and investigate costs of bioenergy resources and utilization technologies.

Concerning the bioenergy supply potential, the following results are obtained. (1) Supply potential of energy crops produced from surplus arable land will be available in North America, Western Europe, Oceania, Latin America, and Former USSR & Eastern Europe. However, the potential of energy crops will be strongly affected by variation of parameters of food supply and demand such as animal food demand. (2) Bioenergy supply potential of biomass residues will be stable against a change of a food demand parameter. The ultimate bioenergy supply potential of biomass residues will be 265 EJ/year in the world in 2100.

Concerning the bioenergy-related costs, we obtain the following results. It is expected that the gasifier combined-cycle (BGCC) will take an advantage compared with the steam power plant when BGCC technologies are matured. The biogas power plant may have economic benefit when it uses biomass resources with large disposal costs.

Keywords: bioenergy supply potential, land use competitions, biomass flow, biomass costs, biomass utilization technology.

1. INTRODUCTION

Bioenergy is expected to become one of the key energy resources in the future because bioenergy maintained adequately is renewable and free from net CO_2 emissions.

In the study we evaluate global bioenergy supply potential and bioenergy-related costs. We evaluate global bioenergy supply potential using a global-land-use-and-energy model (GLUE) that considers land use competition and overall biomass flow including those of biomass residues (Yamamoto et al., 1999, Yamamoto et al., 2000, Yamamoto, 2000, Yamamoto et al., 2001). We explain the outline of GLUE, and simulation results.

In addition, we investigate bioenergy-related costs concerning biomass resources, biomass utilization technologies, and biomass-derived secondary energy.

2. MULTI-REGIONAL GLOBAL-LAND-USE-AND-ENERGY MODEL (GLUE)

Outline of GLUE

In this section, we explain the outline of the multi-regional global-land-use-and-energy model (GLUE).

We divide the world into eleven regions (Table 1) in order to analyze land use competitions and bioenergy supply potential regionally.

GLUE consists of two sectors (a food sector and a forest sector) and describes land use competition among various uses for biomass applications such as paper, timber, food, feed, and energy. The model covers a wide range of land uses and biomass flow including food chains from feed to meat, paper recycling, and discharge of biomass residues (Figure 1).

The model calculates bioenergy potential from 1961 to 1990 based on the past records and that from 1990 to 2100 based on the data explained in the next section with a one-year time step.

Table 1 Regions in the model (GLUE)

No.	Regions
1	North America
2	Western Europe
3	Japan
4	Oceania
5	Centrally Planned Asia
6	Middle East & North Africa
7	Sub-Sahara Africa
8	Latin America
9	Former USSR & Eastern Europ
10	Southeast Asia
11	South Asia

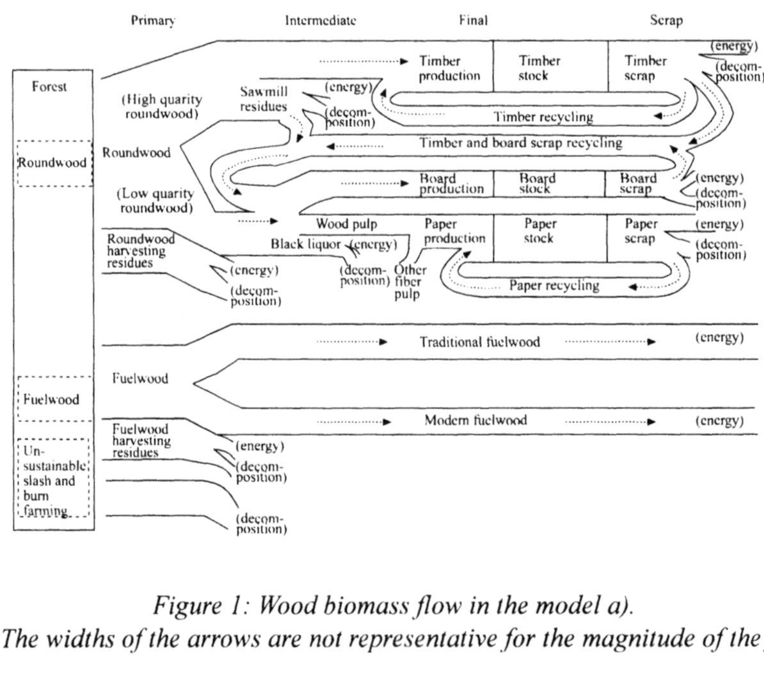

Figure 1: Wood biomass flow in the model a).
a) The widths of the arrows are not representative for the magnitude of the flow.

3. SIMULATION RESULTS

We conduct simulations using GLUE and analyze the simulation results such as bioenergy supply potential and land uses.

3.1. Supply potential of energy crops

In the reference case there will be supply potential of energy crops produced from surplus arable land in North America, Western Europe, Oceania, Latin America, and Former USSR & Eastern Europe. The bioenergy supply potential in the world will be 110 EJ/year in 2050 and 22 EJ/year in 2100 (Figure 2).

The reasons of the decrease of the potential between 2050 and 2100 are as follows.
We assume that crop productivity will mature in the world but animal food demand per capita would grow continuously in the developing regions after 2050. Therefore, the increase of the food demand will exceed the increase of the food supply in the world, and the supply potential of energy crops will decrease between 2050 and 2100.

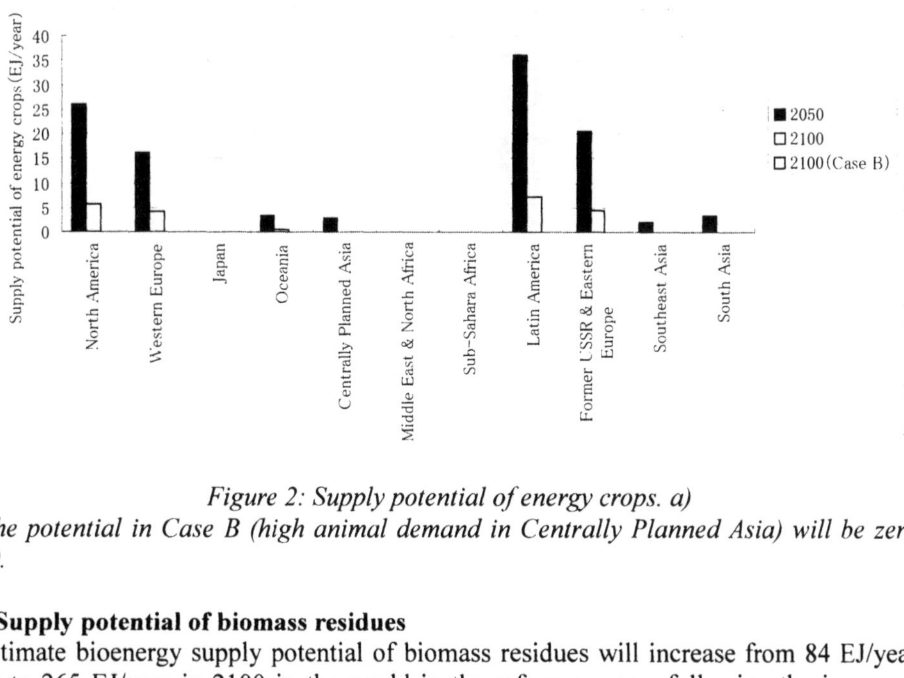

Figure 2: Supply potential of energy crops. a)
a) The potential in Case B (high animal demand in Centrally Planned Asia) will be zero in 2100.

3.2. Supply potential of biomass residues

Ultimate bioenergy supply potential of biomass residues will increase from 84 EJ/year in 1990 to 265 EJ/year in 2100 in the world in the reference case, following the in-crease of biomass consumption in the future.

The values of the potential will be large as shown in Figure 3 in North America, Centrally Planned Asia, Latin America, and South Asia where will be major consumers or exporters of biomass. The values of the potential in those regions will be larger than 30 EJ/year in 2100. Cereal-harvesting residues will take the highest share at 42 % of the total residue potential in the world in 2100. Besides, industrial roundwood felling residues, timber scrap, and animal dung will take large shares above 10 %, respectively (Figure 4).

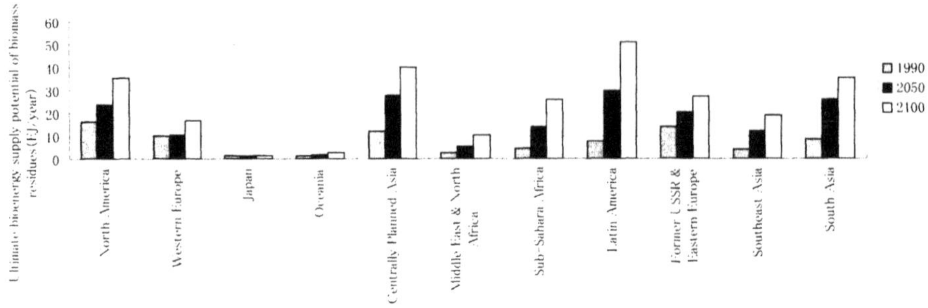

Figure 3 Ultimate bioenergy supply potential of biomass residues.

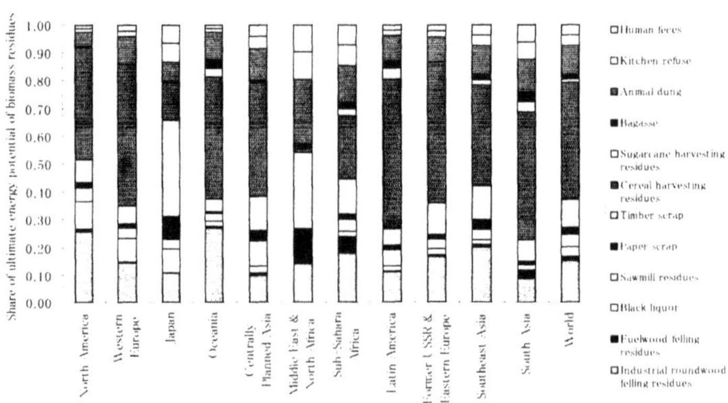

Figure 4 Share of ultimate bioenergy supply potential of biomass residues (in 2100).

4. COSTS OF BIOENERGY

We explain economics of bioenergy concerning bioenergy resources, bioenergy utilization technologies, and bioenergy-derived secondary energy.

4.1. Costs of biomass resources

Kinds of biomass resources that can be used for energy can be used also for material or food. Therefore, we set bioenergy costs considering opportunity costs of biomass for material or food. If the production cost of a kind of biomass is less than the price of the biomass for material or food, the biomass cost is priced the opportunity cost for material or food. On the other hand, we consider disposal costs concerning some kinds of biomass residues that have no use except for energy.

As an example we explain prices of roundwood, timber, wood chips (for wood pulp), and saw dust. An average price of roundwood is about $84/m^3$ (about $7/GJ). Market prices of timber and wood chips for wood pulp are about 45,000 yen/m^3 (about $31/GJ) and about 8,000 yen/m^3 (about $5/GJ), respectively. A market price of sawdust for fuel or manure is about 1,000 yen /m^3 (about $0.7/GJ).

Generally, the wood prices for non-energy are higher the wood price for energy.

Based on the concept of the opportunity cost, we set representative costs of bioenergy resources (Figure 5).

Modern fuelwood (wood chips) is the most expensive among the bioenergy resources we considered, since the opportunity cost as well as the quality of the modern.fuelwood is high.

Next, cereal harvesting residues are the second expensive because of the low density (t/m^3) and the high cost about transportation and storage.

Next, the energy crops is the third expensive because they needs production cost.

Some kinds of biomass residues that are discharged at factory or are collected using public garbage-collection system can be used for energy without additional expense.

Furthermore, some kinds of biomass residues that have no use except for energy can be used for energy at negative cost.

The costs of the biomass residues with manure-grade are smaller than the cost of coal.

However, since the number of the cost data we collected is limited, we need to be collecting the data about the biomass resources continuously.

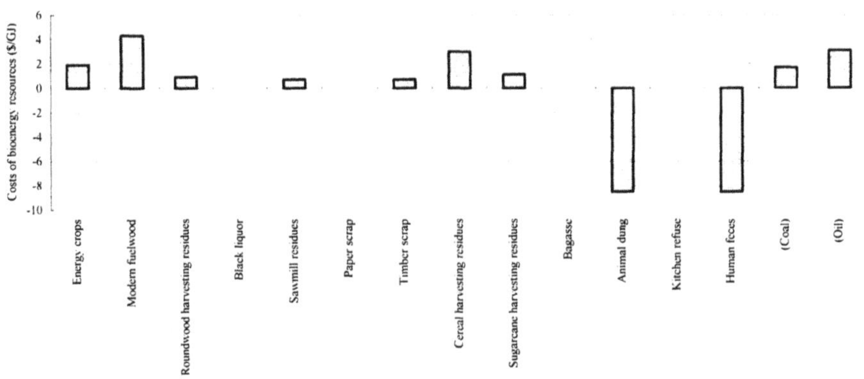

Figure 5: Costs of bioenergy resources.

4.2. Costs of bioenergy utilization technologies

Modern bioenergy resources are finally consumed after they are converted to the forms of liquid fuel, gaseous fuel, and electricity using bioenergy utilization technologies, while traditional bioenergy resources are finally consumed in the forms of raw wood or charcoal.

We consider costs and efficiencies of bioenergy utilization technologies such as power generation (steam power, gasified combined-cycle, and co-firing), and gasified methanol synthesis, biogas (anaerobic digestion) power generation, and ethanol fermentation using cellulosic biomass.

We compare the costs of the bioenergy utilization technologies considering the costs per unit outputs ($/kW) (Figure 6).

Consequently, biogas power generation is the most expensive because of the small power output compared with the large facility size. However, the biogas power generation improves the cost condition when it uses biomass resources with disposal costs such as animal feces.

Concerning steam power, gasified combined-cycle, and gasified methanol synthesis, these technologies are developing and the expectations of the cost reduction contain uncertainty.

Finally, co-firing is the cheapest among the power generation technologies, but it is assumed that the co-firing system is to retrofit the existing coal power plant without increase in the plant capacity and the cost of the co-firing cannot be compared with that of the other power plant technologies in a simple way.

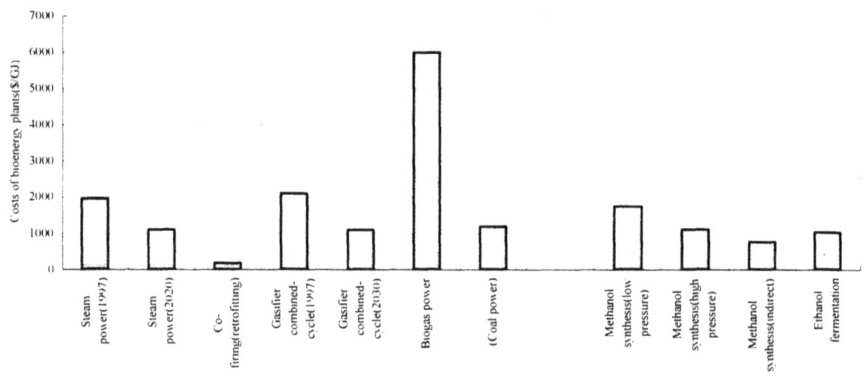

Figure 6: Costs of bioenergy utilization technologies.

4.3. Costs of biomass-derived secondary energy

We evaluate secondary energy costs derived from biomass resources considering costs of primary biomass resources and costs and efficiencies of biomass utilization technologies (Figure 7).

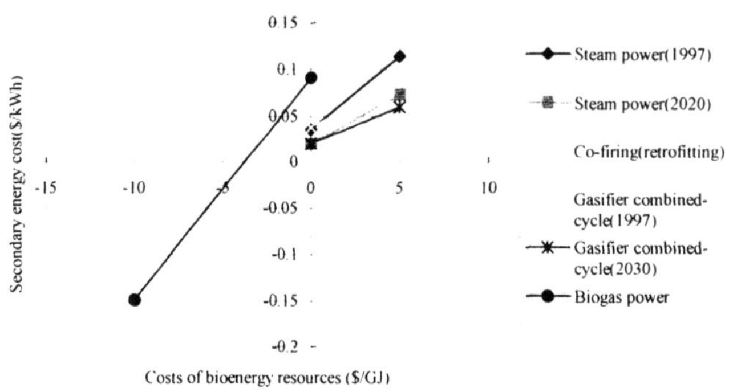

Figure 7: Costs of biomass-derived secondary energy (about power generation).

We covert the equipment costs into the constant costs per year during the life of the plants. It is assumed that the life of the power plants except the biogas power plants is 30 years, the life of the synthesis fuel production plants and the biogas power plant is 20 years, and the interest rate is 5 %/year. It is assumed that the ratios of the annual costs (annual plant costs plus annual operation costs) to the plant costs are at 12 % for all the power plants except the biogas power plant, at 10 % for the biogas power plant, at 17 % for gasified methanol synthesis, and at 27 % for ethanol fermentation.

Consequently, concerning the power generation, it is expected that the gasifier combined-cycle (BGCC) will take an advantage compared with the steam power plant when BGCC technology is matured. Especially, BGCC will take an advantage on the condition of high fuel cost since BGCC has high conversion efficiency.

The cost of the co-firing is the cost of retrofitting without increase in capacity of the power plant, and the cost cannot be compared with the cost of the other plants in a simply way.
The biogas power plant may have economic benefit when it uses biomass resources with large disposal costs (see Figure 7).

However, these secondary energy costs contain large uncertainty since they are based on the cost expectations about the developing technologies.

5. CONCLUSION

In this study, we evaluate the global bioenergy supply potential in the future using a multi-regional global-land-use-and-energy model (GLUE) and investigate costs of bioenergy resources and utilization technologies.

Concerning the bioenergy supply potential, the following results are obtained. (1) Supply potential of energy crops produced from surplus arable land will be available in North America, Western Europe, Oceania, Latin America, and Former USSR & Eastern Europe. However, the potential of energy crops will be strongly affected by variation of parameters of food supply and demand such as animal food demand. (2) Bioenergy supply potential of biomass residues will be stable against a change of a food demand parameter. The ultimate bioenergy supply potential of biomass residues will be 265 EJ/year in the world in 2100.

Concerning the bioenergy-related costs, we obtain the following results. It is expected that the gasifier combined-cycle (BGCC) will take an advantage compared with the steam power plant when BGCC technologies are matured. The biogas power plant may have economic benefit when it uses biomass resources with large disposal costs.

REFERENCES

Food and of the United Nations (FAO), FAO yearbook Production. FAO Statistics Series No.104.FAO, Rome.

Yamamoto, H., 2000. Global bioenergy systems including energy derived from animal wastes under constraints on greenhouse gas emissions. University of Queensland.

Yamamoto, H., Fujino, J., and Yamaji, K., 2001. Evaluation of bioenergy potential with a multi-regional global-land-use-and-energy model. Biomass and bioenergy.

Yamamoto, H., Yamaji, K., and Fujino, J., 1999. Evaluation of bioenergy resources with a global land use and energy model formulated with SD technique. Applied Energy, 63(2):101-113.

Yamamoto, H., Yamaji, K., and Fujino, J., 2000. Scenario analysis of bioenergy resources and CO_2 emissions with a global land use and energy model. Applied Energy, 66:325-337.

Greenhouse Gases and Animal Agriculture
J. Takahashi & B. A. Young (Editors)

The biogas plant using animal effluent as fuel

Y. Nagamori[1], K. Iyasu[1] & S. Sano[2]

[1]Power and Industrial Systems Research and Development Center, Toshiba Corporation, 1, Toshiba-cho, Fuchu-shi, Tokyo 183-8511, Japan, [2]Environmental System Engineering Dept. Toshiba Corporation, 1-1, Shibaura 1-chome, Minato-ku, Tokyo 105-8001, Japan.

We develop the biogas plants as waste management systems to prevent pollution. Thees systems integrated the technologies of anaerobic digestion, fuel cell power generation, wastewater treatment and compost making. The biogas plants generate more than enough electric power to satisfy operational energy needs for the biogas plant from the anaerobic digestion and fuel cell power generation. These plants reduce pollutants in wastes to levels less than required by effluent standard for wastewater treatment.

In this paper presents plans and designs of biogas plant facilities for biogas generated from piggery wastewater. in Japan. More impurities exist in the biogas from piggeries than in the standard gases of fuel cells and the impurities cause the fuel cells performance to decrease. The removal of the biogas impurities is important. We have found a suitable impurity removal method.

Keywords: piggery waste, anaerobic digestion, biogas, fuel cell, removal of impurities

1. INTRODUCTION

Pollutants from unmanaged animal and bio-based industrial wastes can degrade the environment, and methane emitted from decomposting wastes may influence global climate change. The desired waste management system should prevent pollution and converts wastes into energy resources. We have developed a biogas plant that has integrated the technologies of anaerobic digestion, fuel cell power generation, wastewater treatment and compost making. This integrated system recovers biogas from wastes discharged from animal industries by anaerobic digestion first, and generates clean energy source utilizing methane gas as fuel for fuel cells, then purifies digested wastewater to satisfy requirements of effluent standards and then composts excess sludge coming out of the system for reuse. The biogas plant provides electricity utilizing unused resources like methane gas and animal wastes and minimizes the impact on the environment thus reducing the amount of pollution discharged.

This system that utilizes biogas recovered from piggery waste water, has been progressed in Japan recently. We are also proceeding a joint project between Japan and China. The project is called "Practical Use of Methane Gas Fermentation Fuel Cell Power Generation Systems". IEEJ has been entrusted with the implementation of this project by NEDO for the study period from fiscal year 1999 to 2004. IEEJ is responsible for the system engineering, providing wares in the plant and plant construction. Our Chinese counterparts are responsible for providing the site, making of the documents for submittion to the regulatory authorities and plant operation. Toshiba accepted an order for the basic engineering of the system from IEEJ in fiscal year 1999. At this time Toshiba has the consent of IEEJ and herein reports content of the international joint project..

2. OUTLINE OF THE BIOGAS PLANTS

The process flow sheet of the biogas plant is illustrated in Figure 1. It consists of four components; digester facilities, fuel cell facilities, wastewater treatment facilities and compost making facilities. Biogas is recovered from piggery wastewater in the digester. The biogas is reformed and then used for fuel producing electricity and heat in the fuel cells power plant. The heat is used partially to keep the temperature of the digester. The digested waste sludge is treated in the wastewater treatment facilities. The excess sludge is converted to manure then and returned to field. Each facility has been obtained commercially. The main subjects of the project are the development of an optimized system.

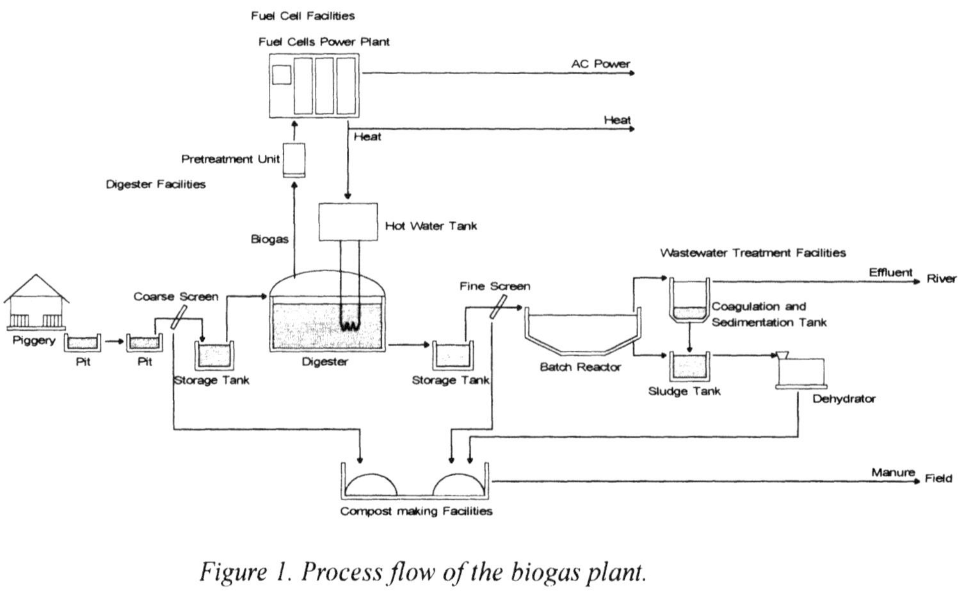

Figure 1. Process flow of the biogas plant.

2.1. Digester facilities

Anaerobic digestion of animal and bio-based industrial wastes has been well know as a method of energy recovery from waste matters, and has been already put into practical use in many regions. Digesters are commonly designed so as to operate in either the mesophilic temperature range (30°C to 45°C) or the thermophilic temperature range (45°C to 60°C). Thermophilic systems offer several advantages, including higher methane production, faster throughput, as better kill pathogen and virus and the prospect of compost production to a consistent standard. However, thermophilic systems are more expensive and require high level of control management. We select the mesophilic system because of its simple structures, low energy consumption and satisfactory results in treating animal wastes. Temperature of the digester is kept by the hot water at 60°C from fuel cell facilities

Table 1 shows expected biogas composition recovered from the digester. The biogas contains not only principal components of methane and carbon dioxide, but also nitrogen, oxygen and hydrogen sulfide et al as the impurities. Even at a small amount, these impurities cause fuel cell performance decrease. In order to reuse biogas, hydrogen sulfide must be removed to prevent that is corrosive and forms solid of sulfur oxide.

Table 1. Expected composition of biogas.

CH_4	66%
CO_2	31%
N_2	2%
O_2	0.5%
CO	0.25%
H_2	0.25%
H_2S	0.2%

2.2. Fuel cell facilities

2.2.1. What is a fuel cell ?

A fuel cell is an electrochemical device that combines fuel hydrogen and oxygen from air to produce electricity, heat and water. Fuel cells operate without combustion, so they are virtually pollution free. Since the fuel is converted directly to electricity, a fuel cell can operate at much higher efficiencies than internal combustion engines, extracting more electricity. The fuel cell itself has no moving parts that results making the fuel cell a quiet and reliable source of power.

2.2.2. Fuel cell benefits

The fuel cells power plants have much advantage to internal combustion engines, because a fuel cell converts fuel directly to electricity and heat. The fuel cell has high efficiency. Fuel cells extract more energy than combustion engines from the same amount of fuel. The fuel cells power plant converts 40 percent of available fuel to electricity and 40 percent to heat recovery and then overall energy recovery is 80 percent. The fuel cell performs clean emissions and quiet operations. Compared with traditional combustion generator such as diesel engine, gas turbine and gas engine, the fuel cells power plant reduces such any emissions as carbon dioxide, NOx (Nitrogen Oxides) and SOx (Sulfur Oxides), and then the fuel cells power plant is very quiet and makes no noise so that the plant is installed even indoors. The fuel cell can use various fuels. The owners are allowed to make their own fuel selections. The owner can choice not only such standard fuels as NG (Natural Gas) and LPG (Liquefied Petroleum Gas) but also a biogas, an anaerobic digester gas and methane that is reformed in a pretreatment unit.

2.2.3. Performance and limitation of the fuel cell

The fuel cell facilities consist of the fuel cells power plant and the pretreatment unit. It uses the 200kW phosphoric acid fuel cell (PAFC). The biogas plant is planed to spend itself 120kW of 200kW generated from the fuel cells power plant and is able to provide rest 80kW as the surplus.

The pretreatment unit reduces such biogas impurities as hydrogen sulfide, ammonia et al. These impurities cause the fuel cell performance decrease. The fuel standard for the fuel cells power plant is shown in Table 2.

Table 2. Specification requirement in the fuel standard.

Methane (%)	>55
Carbon dioxide (%)	<45
Sulfide (ppm)	average <6
Ammonia (ppm)	<1

2.2.4. Reforming processes of the biogas

Because hydrogen sulfide concentration of the expected gas composition, 0.25% (2500ppm), in Table 1 is higher than that of the fuel standard, 6ppm, in Table 2. It is needed to reduce hydrogen sulfide to meet the fuel standard. There are some applicable processes for removal of hydrogen sulfide from biogas. A dry process is applied to the biogas plant in Japan generally, because of easier obtaining its catalysts, no emission of waste fluid, low cost of constructions, and suitable for a small volume of gas that should be reformed in the system. Equation 1 shows chemical reaction to remove hydrogen sulfide in the dry process. Figure 2 shows the process flow sheet of reforming biogas in the pretreatment unit. Biogas is transferred to the desulfurizer and hydrogen sulfide reacts with Fe_2O_3 catalyst to form Fe_2S_3 and immobilize it onto the catalyst bed. Additionally, activated carbon adsorber may remove impurities in the desulfurizer gas. The pretreatment unit of desulfurizer and activated carbon adsorber is able to reduce not only hydrogen sulfide but also other impurities in the biogas. It provides purified gas to the fuel cells power plant to prevent performance decline.

$$Fe_2O_3 + 3H_2S \rightarrow Fe_2S_3 + 3H_2O \quad (1)$$

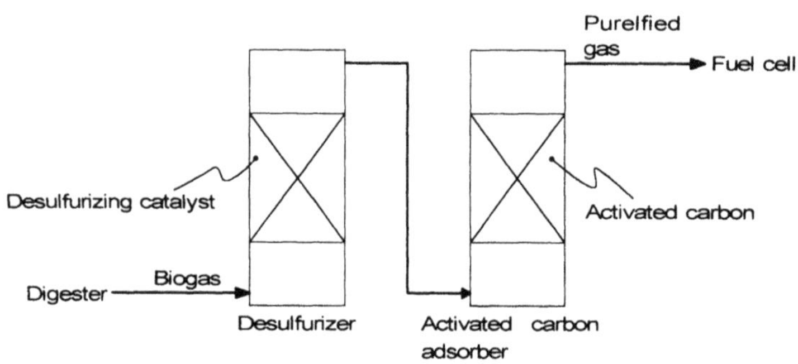

Figure 2. Process flow of reforming biogas.

2.3. Wastewater treatment facilities

The wastewater treatment facilities are able to purify digested waste sludge to discharge into the rivers environment. It consists of the sequencing aerobic batch reactor (SBR) and the following coagulation and sedimentation tank.

The SBR system is an activated sludge process that operates in the batch mode with waste sludge feed, aeration, sludge settlement and effluent discharge, occurring in the same tank. The SBR process is applied to the biogas plant because of many system advantages. The SBR system allow for better control of waste sludge results, may be designed for BOD and nitrogen removal, carbon oxidation, nitrification and denitrification.

During coagulation-sedimentation chemical coagulants, iron salts, are used to improve the rates of waste constituents removal. Flocculation of colloidal particles followed by sedimentation has been indicated to be effective in reducing levels of suspended solids, BOD, phosphorus. The sedimented sludge is collected and the part of them is returned to the reactor by the sludge pump and the rest of the sludge are pumped up to the sludge tank as excess sludge. The excess sludge is converted into manure in the compost making facilities.

The wastewater treatment facilities have been designed to satisfy the requirement of the effluent standard in Japan. It is showed in Table 3.

Table 3. Effluent standard[1] for water treatment facilities in Japan.

pH	5.8 - 8.6
BOD (mg/L)	160 (day average < 120)
COD-Mn (mg/L)	160 (day average < 120)
SS (mg/L)	200 (day average < 150)

[1]Water pollution prevention law

3. PLANING ON THE BIOGAS PLANT

Material balance

Figure 3. shows the material balance in the biogas plant that has been estimated from the actual experiences in Japan. The volume of 150 m^3/day piggery wastewater that has BOD value of 25,000 mg/L is handled and 2,200 m^3/day biogas is recovered from digester. Although expected biogas composition has been shown in Table 1, it is supposed that digestion gas consists of 60 % methane and 40 % carbon oxide in this estimation in consideration of the fluctuation of gas composition. The fuel cells power plant generates 200 kW electric power utilizing 2,200 m^3/day biogas as fuel. Anaerobic digestion reduces BOD of waste sludge down to 4,500 mg/L. Furthermore the waste sludge is purified to the level of less than the effluent standard in Table 3 in the wastewater treatment facilities. The compost making facilities convert suspended solid collected by screens and excess sludge into manure with a volume of 3 m^3/day. Consequently, only treated water is discharged to the environment, that has been purified to satisfy the requirement of the effluent standard.

4. CONCLUSIONS

The biogas plant makes good use of piggery residues. The plant generates all of its energy needs from the fuel cells power plant and the rest can be provided to the other usage. The less pollutant in the effluent of the system, results in the minimum impact on the environment. The biogas plant converts old disposal problem into new profits such as electricity, hot water and manure.

It is important that the planning and designing of plant systems should be done after analyzing sufficiently the nature of raw wastewater and contents of biogas as well. And then the reforming processes are very important for the removal of impurities such as hydrogen sulfide exists in the biogas and they causes fuel cells performance decrease.

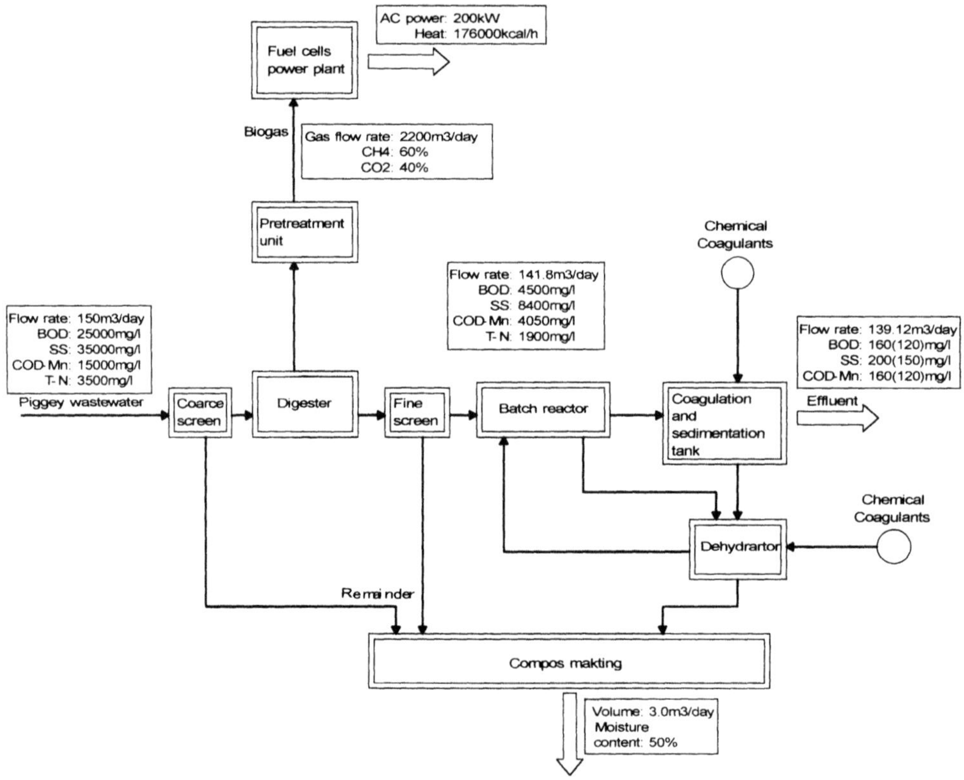

Figure 3. Material balance of the biogas plant.

REFERENCES

Institute of Energy Economics, Japan, 2000. Practical Use of Methane Gas Fermentation Fuel Cell Power Generation Systems.

Institute of Energy Economics, Japan, 2001. Practical Use of Methane Gas Fermentation Fuel Cell Power Generation Systems.

Greenhouse Gases and Animal Agriculture
J. Takahashi & B. A. Young (Editors)

Environmental load gas emissions from swine waste treatment

T. Osada & K. Haga

Lab. Animal Waste Management, National Institute of Livestock and Grassland Science, Japan
Ikenodai 2, Kukizaki-machi, Inashiki-gun, Ibaraki-ken 305, Japan

Recently, great concern has gathered in generating from the livestock farm of environmental load gases, such as ammonia , methane and nitrous oxide. In order to calculate the total amount of each gas per capita which occurs from swine waste, three processes considered to be the main generation sources of each gas was examined. According to the results of those experimental studies, 1) The NH_3-N, N_2O-N and CH_4 emissions from the swine keeping unit and its manure contributes (a full fattening period of 8 weeks) were estimated at 271~438 , 16.5~49.4 and 270~438 g/head respectively. And, 2) Those emission could be reduced by adequate manure contributes, such as; weekly discharge of slurry in swine unit, employment of intermittent aeration process for wastewater treatment and keeping of aerobic condition during swine manure composting.

Keywords: ammonia, nitrous oxide, methane, livestock waste, greenhouse gas

1. INTRODUCTION

Unsuitable handling and/or treatment of livestock manure have caused an emission of harmful gases from livestock production systems. A number of studies have evaluated ammonia (NH_3) emission from livestock production (Rom, 1995, Ryden *et al.*, 1987, Hartung and Phillips, 1994, Sømmer *et al.* 1996), and it has proved necessary to find solutions to the problems. In recent studies it was suspected that livestock manure contributes significantly to the emission of methane (CH_4) and nitrous oxide (N_2O), which are effective greenhouse gases. Some efforts have been made to quantify the emissions from livestock manure stores (Safley and Westerman, 1988, Husted, 1994) and from treatment systems (Yamamoto, 1995, Osada *et al.*, 1995). The data collected have been restricted to a few hours or days, and therefore, it has not been possible to define the emission rates from manure.

The purpose of this study is to evaluate NH_3, N_2O, and CH_4 emission rates from the swine waste treatment during all of its fattening periods. In order to calculate the total amount of each gas per capita that occurs from swine waste, we had studied three waste treatment processes below;

1) Emission from the waste in the swine barn (unit), 2) Emission from the wastewater purifications for liquid part of swine waste, and 3) Emission from the composting for solid part (feces) of swine waste.

2. MATERIALS AND METHODS

2.1. Calculation method for emission rate of each substances

The rate of emission (E) of each substance (NH_3, CH_4 and N_2O) from each treatment process was computed from the amount of ventilation of the experimental room or apparatus and the concentration differences of each substance between the inlet (fresh) and outlet air (exhaust) samples.

E (mg/5min.) = (Conc. of outlet air (mg/m^3) – Conc. of inlet air (mg/m^3)) x 5/60 x Ventilation rate (m^3/hour)

2.2. Emission from swine barn

The experiment was performed in autumn 1996 in the experimental swine units at Research Centre Bygholm, Danish Institute of Agricultural Sciences. The piglets were raised from 28 to 90 kg (8 weeks). The Gas Emission and Climate Laboratory was equipped with two almost equal experimental units. During the experiment, the slurry in the experimental unit was removed weekly and the slurry in the reference unit was left in the pit during the whole fattening period. Each unit was 45.8 m^2 and a volume of 165 m^3. Each unit had equal pressure ventilation systems with mechanically forced air inlet and outlet ducts. The animal occupation area was 35 m^2 including 11.4 m^2 slatted floor. Each unit had accommodation for 40 fattening swine divided into 4 pens. The ventilation rate in each unit was evaluated with a full size free-running impeller unit placed in the exhaust ducts. Concentration of NH_3, CH_4 and N_2O were recorded with IPD (Bruel &Kjaer, Multi-gas Monitor Type 1302, Multipoint Sampler and Doser Type 1303) three times per hour on outside (inlet air) and in the exhaust ducts of each experimental unit. At least 100 hours measurement of each gas was carried out every week during the experiment.

2.3. Emission from the wastewater purification for liquid part of swine waste

Two bench-scale activated sludge units (3 L operational volumes) were operated at 5 and 20°C. A mixture of swine feces and urine was used as influent wastewaters. The contents are as follows: 1500 mg L^{-1} BOD, 207 mg L^{-1} Total Nitrogen. The similar operational conditions of each run are as follows: 3.3 days of hydraulic retention time (HRT); 15~20 days of sludge retention time (SRT); 6000~9000 mg L^{-1} of MLSS; 0.5 Kg m^{-3} d^{-1} of BOD loading; 1.2 L min^{-1} 3 L^{-1} aeration rate. Exhaust gases were collected from each head space (3 L) of the units opened (aeration periods) or closed (non-aeration periods) for 5 minutes intervals with IPD.

The units were set for a 24-hour cycle. Two types of aeration programs were conducted. For the conventional one, a continuous aeration for 21 hours was adopted, while for the intermittent aeration, aeration was done at one-hour intervals. The aeration periods started just after daily charging. After the end of the aeration periods, the sludge in the mixed liquor was allowed to settle for two hours, and then the supernatant was discharged. Each unit was operated for a period longer than two SRTs (about 50 days) to obtain a steady-state condition. The water qualities of the influent and the effluent were analyzed two or three times a week for pH, TOC, TN, TP, NH_4-N, and NOx-N, and weekly for BOD.

2.4. Emission from the composting for solid part of swine waste

NH_3, CH_4 and N_2O emissions from swine waste composting were determined by an experimental composting apparatus with continuous forced aeration (19.2 ~ 96.1 L m^{-3} min^{-1}).

The composting apparatus consisted of a small reactor (52L) and a data logger connected with thermocouples. The reactor was covered with Styrofoam (30mm) and supported with a plywood board to prevent heat loss.

Swine waste or a mixture of swine wastes and matured compost were conditioned at moisture content of 65% by mixing with the corrugated paper. The compost mixture was removed from the reactor, re-mixed (by turning) and replaced weekly until the temperature of the materials in the reactor had not risen to 40°C. It took 3 weeks at more than about 40 L m^{-3} min^{-1} aeration or 5 weeks under less than about 20 L m^{-3} min^{-1} aeration. The exhaust gases from the reactor were collected every 6 hours over 3 days from the start of the composting and every 24 hours until the end of the experiment. N_2O and CH_4 were determined by gas chromatography by use of an electron capture detector (ECD) and flame ionized detector (FID), respectively. NH_3 was monitored with indicator tubes.

3. RESULTS AND DISCUSSION

3.1. Emission from the swine barn

Emission of NH_3, N_2O and CH_4 from swine units have been investigated continuously by use of IPD. The emission of all gases varied during the growth of the swine. For example, the highest value of CH_4 emission rate was four times as high as the lowest one during the experimental period. In addition to this, all gas emissions have a typical diurnal fluctuation. The gas emission value observed at the peak hours (1:00 - 2:00 P.M.) was twice as high as that observed around 6:00 A.M., although the room temperature was kept at around at 17 °C. The NH_3 (NH_3-N), N_2O (N_2O-N) and CH_4 emissions from pig units during a full fattening period of 8 weeks were estimated at 149, 5.8 and 302 g/swine (reference units) and those were reduced slightly at 138, 5.4 and 268 g/swine (experimental units), respectively with weekly discharge of slurry (Osada *et al.* 1998).

3.2. Emission from the wastewater purification for liquid part of swine waste

About 5% of the influent nitrogen was emitted as N_2O gas using the conventional process under 0.5kg m^{-3} d^{-1} BOD loading and 20°C condition. Most of the N_2O was emitted during the first hours of aeration started just after daily charging . To reduce nitrous oxide (N_2O) emissions from fill-and-draw type activated sludge units treating swine wastewater, the N_2O emission was compared between the continuous aeration process (conventional process) and the intermittent aeration process. Only less than 0.05% of the influent nitrogen was emitted as N_2O gas during the intermittent aeration process both of temperature condition (5°C and 20 °C). The total emission of other gases (NH_3, NO, NO_2 and CH_4)
was negligible (Osada *et al.* 1995).

3.3. Emission from the composting for solid part of swine waste

The amount of gas emissions from the composting process of swine waste was determined by use of an experimental composting apparatus. Forced aeration (19.2 ~ 96.1 L m^{-3} min^{-1}) was carried out continuously, and exhaust gases were collected and analyzed periodically. The temperature of most f the contents rose to 70°C and composting was complete within three to five weeks. NH_3, CH_4 and N_2O emissions were high in the early stage of composting. About 10 to 25% of the nitrogen in the raw material was lost as NH_3 gas during swine waste composting. The emission rate of NH_3 mainly depended on the aeration rate, so that as the aeration rae rose, the level of NH_3 emissions increased. The CH_4 and N_2O emissions could be kept lower with adequate treatment at more than 40 L m^{-3} min^{-1} aeration. N_2O may be mainly the result of the denitrification of NOx-N in the additional matured compost used as a composting accelerator (Osada *et al.* 2000).

According to those results from this experimental studies,

1) The NH_3-N, N_2O-N and CH_4 emissions from the swine keeping unit and its manure handling (a full fattening period of 8 weeks) were estimated at 271~438, 16.5~49.4 and 270~438 g/swine respectively.

2) Those emission could be reduced by adequate manure handling, such as; weekly discharge of slurry in pig unit, employment of intermittent aeration process for wastewater treatment and keeping of aerobic condition during swine manure composting with adequate aeration.

REFERENCE

Hartung, J., and V.R. Phillips. 1994. Control of gaseous emissions from livestock buildings and waste stores. J. Agric. Eng. Res. 57, 173-189.

Husted, S.. 1994. Seasonal variation in methane emission from stored slurry and solid wastes. J. of Environ. Quality, 23, 585-592.

Osada, T., K.Kuroda, and M.Yonaga. 1995. Reducing nitrous oxide gas emissions from fill-and-draw type activated sludge process . Water Research 29: 1607-1608.

Osada T., H. B. Rom and P. Dahl. 1998. Continuous Measurement of Nitrous Oxide and Methane Emission in PigUnits by Infrared Photoacoustic Detection. Trans. of the ASAE. Vol 41, pp1109-1114.

Osada T., Kuroda K., and Yonaga M. 2000. Nitrous oxide, Methane and ammonia emissions from composting process of swine waste. The Japanese Society of Waste Management Experts 2 (1) 51-56

Osada T., 2001 Environmental Load Gas Emissions from Swine Waste Treatment . Ph.D. Thesis. Hokkaido University, Sapporo, Japan and National Institute of Livestock and Grassland Science, Tsukuba,Japan.

Rom.H.B.. 1995. Ammonia emission from pig confinement buildings. System analysis and measuring methods. Ph.D. Thesis. The Royal Veterinary and Agricultural University, Copenhagen, Denmark and Danish Institute of Animal Science, Research Centre Bygholm, Horsens.

Ryden, J.C., D.C.,Whitehead, D.R., Locker, R.B.,Thompson, J.H., Skinner, and E.A. Garwood. 1987. Ammonia emission from grassland and livestock production system in U.K. Environ. Pollut. 48, 173-184.

Safley, L.M. Jr. and P.W.Westerman. 1988. Biogas production from anaerobic lagoons. Biol. Wastes 23,181-193.

Sømmer, S.G., E., Sibbesen, T., Nielsen, J.K, Schjoerring, and J.E. Olsen. 1996. A passive flux sampler for measuring ammonia volatilization from manure storages facilities. J. Environ. Qual., 25, 241-247.

Yamamoto, K.. 1995. Emission of trace gases contributing greenhouse effect from grassland. (Global environment research of Japan, Final report for project completed in 1994) 55-63, Environment Agency Government of Japan.

Greenhouse Gases and Animal Agriculture
J. Takahashi & B. A. Young (Editors)

Hydrogen production from biogas through steam reforming followed by water gas shift reaction and purification with Selective Oxidation reaction

Z. G. Zhang, A. Effendi, X. Chen, K. Honda & T. Yoshida

Hokkaido Collaboration Center, National Institute of Advanced Industrial Science and Technology (AIST), 2-17 Tsukisamu-Higashi, Toyohira-ku, Sapporo 062-8517, Japan

To explore the possibility of producing hydrogen with CO impurity < 10 ppm for polymer electrolyte fuel cells (PEFCs) from CH_4-based fuels like biogas, a series of experiments including steam reforming, both medium and low temperature water gas shift reactions (WGSRs) and the selective oxidation of CO were conducted in the present study over different commercial catalysts with proper simulated mixtures as feeds. The steam reforming with the different ratios of steam to simulated biogas was carried out over Ni/Al_2O_3 catalyst to produce hydrogen-rich syngas at 1 atm, 650-800°C and a GHSV of 18000 hr^{-1} in a micro fluidized- bed reactor. On the other hand, both medium and low temperature water gas shift reactions and the selective oxidation of CO were performed over Fe/Cr-, Cu/Zn- based and Ru/Al_2O_3 catalysts at temperatures of 380°C, 150-260°C and 90-150°C and GHSVs of 3710, 4000 and 9900 h^{-1}, respectively, in a fixed-bed mode. CH_4 conversion in the reforming increased with temperature and steam ratio, and reached as high as 97% under a condition of 750°C and CH_4:CO_2:steam = 1.5:1.0:2.0. A 45 hrs long-term catalytic activity test conducted under the same condition also showed that the activity of used catalyst was very stable and provided very constant H_2 (60%) and CO (27%) production. Further, the results of medium temperature water gas shift reaction showed that H_2 concentration was increased from 60% in the feed to 68% in the product stream through the reaction, and CO, in reverse, reduced greatly from 29 to 1.7% (dry gas basis). With a low temperature WGSR over Cu/Zn- based catalyst at 210°C, this CO was decreased further deeply to a level of < 0.2%. Finally, the selective oxidation reaction used for purification removed the residual CO and provided a product hydrogen mixture with CO ≤ 0.5ppm, a highly quality fuel to PEFCs.

Keywords: hydrogen production, steam reforming, shift reaction, biogas, fuel cells

1. INTRODUCTION

Due to the increasing concern about global climate problems, more attention has been paid to greener sources of energy and new technologies with high energy-conversion efficiencies. Biogas, for example, produced from anaerobic digestion processes with animal agriculture wastes as feed, is usually composed of CH_4 up to -60vol%, and it is a high quality, renewable fuel. Its effective utilization will not only improve the overall economy of the processes and bring animal farmers more benefit, but also, more importantly, make a contribution to reducing CO_2 emission. On the other hand, fuel cells, particularly low temperature (around

80°C) PEFCs are believed to be one of the most promising energy conversion technologies presently under development (Remons, R.A., 1990). PEFCs produce both electricity and hot water with high total efficiencies of 60-80% and are ideal electricity and heat cogeneration systems for individual family homes or small-scale residential applications. Today's PEFCs essentially require CO-free hydrogen as fuel, and therefore, a suitable conversion/CO-removal system is necessary when small scale, dispersed energy sources such as biogas are being considered as on site PEFCs' primary fuels.

It is the final goal of the present work to develop such a small-scale conversion system to produce hydrogen with CO impurity < 10 ppm for PEFCs from CH_4-based fuels like biogas. With the present technologies, chemically four serial reaction steps are essentially needed to reach the goal, feed purification to remove impurities in biogas such as H_2S and NH_3, steam reforming to convert biogas to a hydrogen-rich gas mixture ($CH_4 + H_2O \rightarrow 3H_2 + CO$), WGSRs to react CO in the mixture to produce further hydrogen ($CO + H_2O \rightarrow H_2 + CO_2$) and then hydrogen purification through selective CO oxidation reaction ($CO + 0.5O_2 \rightarrow CO_2$) to reduce CO which did not react in the shift reactor (about 0.5 vol.%) down to a level of <10 ppm and thus to guarantee the quality of hydrogen for the use in PFFCs. In this paper, we report the experimental results on the late three reactions that were separately performed with proper simulated feeds. The main purpose of these investigations is to explore the possibility to produce highly quality hydrogen for use in PEFCs by using commercially available catalysts, and if possible, then to investigate the optimal operation conditions of these catalysts.

2. EXPERIMENTAL

All the catalysts used in the present work are commercially available. Ni/Al_2O_3, Cu/Fe/Cr- and Cu/Zn-based catalysts respectively for the steam reforming, medium and low temperature WGSRs were supplied by Toyo Catalysts and Chemicals Inc., while Ru/Al_2O_3 catalyst for the selective oxidation of CO provided by Sumitomo Metal Mining Co. In addition, high purity CH_4, CO_2, CO, H_2. Ar and compressed O_2 gases were presently used to make proper stimulated feed mixtures for the different reactions or as purge gas.

The reactor system employed in this work was self-designed and set up. It mainly consists of six mass flow meters for the controlling of feed gases, a micro pump and tape heater unit for the supplying of steam when needed, a fixed-bed micro quartz reactor (i.d. of 0.8 or 2.0 cm, depended on the reaction investigated) and a TCD type GC (Yanaco, G3800 GC) equipped with an activated carbon column for analysis of exit gases. The catalyst sample (0.3-2.0 g) in the reactor was held by a quartz disk holder and heated by a tubular gold-coated glass furnace. The bed temperature was measured by a thermocouple inserted inside an in-bed thermowell, and controlled by another thermocouple fixed at a right place of the outside wall of the reactor.

With this reactor system, the reactions mentioned above were experimentally investigated. For the steam reforming conducted at temperatures of above 700°C, a fluidized-bed operation mode was applied to depress the formation of coke in the reaction and thus to have higher conversions (Effendi et al., 2001, Wurzel et al., 2000). In this case, an up-flow feeding of reactant mixture was employed to fluidize catalyst particles (70-140 μm). On the other hand, a down-flow fixed-bed mode was employed for the WGSRs and selective oxidation reaction was performed at lower temperatures. In addition, the pre-reduction of catalyst was also performed in the late cases. The detailed experimental procedure and conditions for each reaction were described individually in the related parts in the [**RESULT AND DISCUSSION**] section.

3. RESULT AND DISCUSSION

3.1. Steam reforming reaction

All experiments were performed over 11.5 wt % Ni/Al_2O_3 (0.3g) at 650-850°C and 1 atm for 5 or 45 hrs. The feed employed was a mixture of steam, CH_4 and CO_2 with a constant CH_4/CO_2 molar ratio of 1.5. The fixed CH_4/CO_2 ratio used here was to simulate a real composition of biogas. In addition, to keep the fluidization state of catalyst particles nearly the same as possible in each run, the flow rate of the mixture was also fixed to be 100 ml/min (STP), equal to a space velocity (GHSV) of 18000 h^{-1}. Since the ratio of steam to CH_4 is the most important factor in the reforming, its effect on CH_4 conversion as well as product selectivity was first investigated at a furnace temperature of 750°C in 5 hrs runs. Figure 1 presents the results obtained. Clearly, the catalyst activity with respect to CH_4 conversion as well as H_2 selectivity increased very quickly with increasing the steam ratio (CH_4:CO_2:steam = 1.5:1.0:x) from 0.5 to 1.0, and then gradually to 2.5. An almost complete CH_4 conversion with the H_2 concentration of $\geq$ 60% was achieved with the ratios $\geq$ 2.0. At a steam ratio of 0.5, the catalyst activity was low, and simultaneously, it was also accompanied by a very poor catalyst fluidization and a massive coke formation. It seems for the catalyst used presently that the steam ratio $\geq$ 1.0 is needed to reach a reasonably high CH_4 conversion as well H_2 selectivity in the reforming of the simulated biogas. Further, a contribution made to CH_4 conversion by CO_2 in the feed through the dry reforming ($CH_4 + CO_2 \rightarrow 2H_2 + 2CO$) also reached a maximum at the steam ratio of 1.0, suggesting that the feed composition at this point may be the best one to give relatively high CH_4 conversion with less steam consumption.

In addition, the temperature effect as well as the life of the catalyst was also investigated in this study. It was confirmed that 100% CH_4 conversion with H_2 selectivity $\geq$ 60% could be reached at temperatures $\geq$ 800°C and steam ratios $\geq$ 2, and that the catalyst activity remained constant in a 45 hrs run under a condition of 750°C and a steam ratio of 2.0.

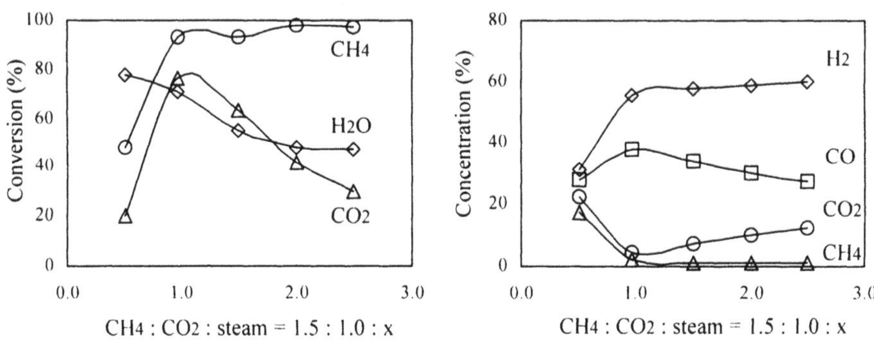

Figure 1. The steam reforming of simulated biogas with different steam ratios.

3.2. Water-Gas Shift Reaction (WGSR)

WGSR, as stated above, is to react CO in the product of the reforming with steam to produce further hydrogen. This is an exothermic equilibrium reaction, and thermodynamically, low temperatures are favorable to having higher CO conversions (Criscuoli et al., 2000). On the other hand, the kinetics of the reaction requires high temperatures to shorten the time toward equilibriums. Therefore, a two-step WGSR was designed presently to provide a product mixture containing a minimum proportion of CO for the subsequent purification operation. A medium temperature WGSR over Cu/Fe/Cr-based catalyst at 380°C was first applied to convert most of the CO to hydrogen, and then followed by a low temperature WGSR over Cu/Zn-based catalyst at around 210°C to reduce CO further to a desired level. Presently, the two reactions were performed separately with proper simulated mixtures plus steam as feeds. Both mixtures were composed of H_2, CO, and CO_2, but with different ratios, 60:29:11 and 67:3:30, respectively. The ratio of steam to the mixture and total feed rate in the two cases were 2.6, 137 ml/min and 1.5, 162ml/min, respectively. In addition, a reactor of 2 cm i.d. was used in these two cases to allow more catalyst charged and thus to have lower GHSVs.

As shown in Table 1, the medium temperature WGSR at 380°C was strongly influenced by GHSV. Lowering GHSV by increasing the amount of catalyst used caused an increase in H_2 and CO_2 concentrations and a decrease in CO concentration in the effluent gas. At GHSV = 3710 h^{-1}, CO concentration decreased to 1.7%, close to the equilibrium value. To make a further CO reduction, some low temperature WGSR experiments were performed over Cu/Zn-based catalyst at 150-260°C and GHSV = 4000 h^{-1} for 5 hrs. As shown in Figure 2, CO conversion in this case was very sensitive to temperature, and the lowest CO concentration (0.18%) in the exit effluent was obtained at 210°C. Furthermore, It should also be noted that the GHSVs were employed in both medium and low temperature WGSRs are smaller in comparison to that in the reforming, suggesting larger WGSR reactors are needed in the presently suggested system to deal with the stream from the reforming reactor.

Table 1. The GHSV dependence of WGSR over Cu/Fe/Cr-based catalyst at 380°C.

Catalyst (mg)	GHSV (h^{-1})	CO conversion (%)	Effluent distribution (%) H_2	CO	CO_2
168	44,100	-	65.0	10.5	24.5
300	24,700	82	66.0	5.0	29.0
1,000	7,360	88	67.4	3.0	29.6
2,000	3,710	93	67.5	1.7	30.8

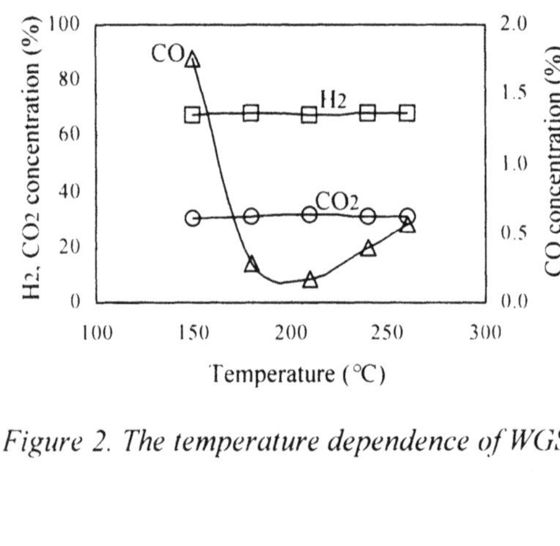

Figure 2. The temperature dependence of WGSR over Cu/Zn-based catalyst.

3.3. Selective oxidation of CO

The stream out of the shift reactor still contains about 0.2% CO and can not be fed to PEFCs directly. Following the suggested system, this stream will be further purified and CO in it will be removed through selective oxidation reaction. To make sure this reaction is indeed effective and then to optimize the condition, some selective oxidation experiments were conducted over Ru/Al_2O_3 catalyst with a simulated feed mixture of 67.1%H_2, 30.9%CO_2, 1.0% CO and 1.1%O_2. After reduction in a 5% H_2/Ar flow at 400°C for 4 hrs, the catalyst (1.0 g) was cooled in Ar and then exposed to a 194 ml/min flow of the mixture (equal to GHSV= 9900 h^{-1}) at temperatures of 90-150°C for 5hrs reaction. In almost all the cases, the temperature of the catalyst bed was observed to be unstable at the initial stage due to the heat of reaction. Stable CO conversion was only observed with a stable bed temperature reached after about 1.5 hrs. Here, all the data were collected at the steady state. It was found that the selective oxidation of CO was very much sensitive to the temperature of the bed. For example, lowering the bed temperature slightly from 148 to 120°C resulted in a dramatic reduction in CO concentration from 1890 to 30 ppm. At 100-110°C, the CO concentration in the product effluent was about 0.5 ppm, much lower than the upper limit of 10 ppm allowed by PEFCs. It clearly demonstrates the effectiveness of the selective oxidation reaction over Ru/Al_2O_3 in removing CO impurity contained in the hydrogen-rich gas mixture.

REFERENCES

Remons, R.A., 1990. Fuel cell for tansportation. J. Power Sources. 29: 251-270

Effendi, A., Z.-G. Zhang & T. Yoshida, 2001. A comparative study on CO_2 reforming of methane using fixed- and fluidized-bed reactors. ACS Symposium Series. in press.

Wurzel, T., S. Malcus & L. Mleczko, 2000. Reaction engineering investigation of CO_2 reforming in a fluidized-bed reactor. Chem. Eng. Sci. 55:3955-3966

Criscuoli, A., A. Basile & E. Drioli, 2000. An analysis of the performance of membrane rectors for the water gas shift reaction using gas feed mixtures. Catalysis Today. 56: 53-64.

Part 4

Processing and use of bio-mass for soil rehabilitation

Greenhouse Gases and Animal Agriculture
J. Takahashi & B. A. Young (Editors)

Ammonia emissions from composting of livestock manure

J. Matsuda[1], T. Maeda[2] and K. Ohmiya[1]

[1]Dept. of Agricultural Engineering, Hokkaido University, Sapporo 060-8589 Japan
[2]Dept. of Agricultural Engineering, Iwate University, Morioka 020-8550 Japan

In separate trials, composting substrates consisting of swine, poultry or dairy cow manure were mixed with sawdust. In each trial the composting period lasted 7 days. Based on these experiments, it was determined that successful composting results in increased ammonia emissions for all manure types. The ammonia emissions obviously decreased as the C/N ratio increased for all substrates. However, in terms of relative ammonia emissions, the differences between the high and low C/N ratios were no longer statistically significant although still varying by up to a factor of two. Ammonia emission was affected by differences in the NH_4^+-N concentration of the initial substrate mixtures and the convertibility of organic nitrogen into inorganic nitrogen during composting.

Keywords: ammonia emissions, livestock manure, high-rate composting, C/N ratio

1. INTRODUCTION

All livestock facilities generate some ammonia gas, but usually the gases released from composting facilities are highly concentrated. These ammonia emissions reduce the fertilizer value of the finished compost, and contribute to both odor problems and acid rain. Ammonia volatilization tends to increase with increased aeration of the composting material. Furthermore, ammonia emissions depend on the C/N ratio of the substrate (Kirchmann *et al.*, 1989; Martins, 1992; Käck *et al.*, 1994; Beck *et al.*, 1997; Osada *et al.*, 2000). This study was carried out to investigate the influence of different types of livestock manure (including swine, poultry and dairy cow manure), C/N ratios and aeration rates on ammonia emissions during high-rate composting.

2. MATERIALS AND METHODS

2.1 Materials

Manures from swine (feces only), poultry (complete with urine) and dairy cows (feces only) were mixed with varying proportions of white birch sawdust. The contents of moisture, total carbon and total nitrogen of the manures and the sawdust were measured prior to mixing to achieve the desired C/N ratios comprising a range between 13 and 56. White birch sawdust was used as a carbon source and a bulking agent because sawdust has uniform physical and chemical characteristics.

2.2 Experimental Apparatus

The composting process was carried out in a closed 1000-mL compact reactor which was submerged in a water bath. The temperature of the water bath was controlled to mimic the temperature of the composting material in the reactor. Humidified air was supplied to the bottom of the reactor at a controlled rate of 0.2 to 1.4 L/min/kg DM. The ammonia emissions from the composting mixture were collected in two stages consisting of a condensation trap and a boric acid solution trap. Initial and final samples were analyzed for moisture content, pH, total nitrogen (T-N) and ammoniacal nitrogen (NH_4^+-N). Each trial ran for 7 days.

3. RESULTS AND DISCUSSION

3.1. Temperature

The temperature pattern as measured over the course of the 7-day trials varied with composting conditions. In general, the maximum and average temperatures decreased as the C/N ratio and aeration rate increased for all livestock manure substrates. Those composts that achieved a thermophilic temperature pattern, indicating a successful composting process, had lower C/N ratios and aeration rates than did composts with less successful temperature patterns. Comparison of the composting success based on the different manure substrates shows that those based on swine manure were able to tolerate the broadest range of C/N ratio and aeration conditions, while those based on poultry manure performed well only in a very limited range of conditions. This may be due to differences in nitrogen concentration and volatile matter in the livestock manure.

3.2. Ammonia emissions

Figure 1 shows the ammonia emissions resulting from the trials based on a mixture of all three types of livestock manure and sawdust. The emissions reached a maximum when the C/N ratio was 13 and the aeration rate was 1.0 L/min/kg DM in composting of swine manure. This maximum value of 2960 mg/kg initial mass represents 32.4% of the total nitrogen in the initial compost mass as shown in figure 2. Figure 1 clearly demonstrates that the emissions decreased as the C/N ratio increased in all three types of substrates. Furthermore, the relationship between the emissions and the aeration rate changed with the C/N ratio. In general the ammonia emissions during successful composting processes were greater than the emissions during unsuccessful composting processes.

As shown in figure 2, the relative emissions refer to the ammonia emissions as a percentage of initial total nitrogen in the substrate (i.e., mg of ammonia-N emitted per mg of initial total nitrogen). As shown in figure 1, comparison of the successful composts using a swine manure substrate at an aeration rate of 1.0 L/min kg.DM shows that the ammonia emission at a C/N ratio of 13 was highest and more than 5.8 times the emissions of that found at a C/N ratio of 43. However, the relative emissions were only 2.0 times greater when the C/N ratio was 13, in comparison to when the C/N ratio was 43, during the same composting process. This fact shows that the relative ammonia emissions do not significantly decrease even when bulking agents are added to increase the C/N ratio.

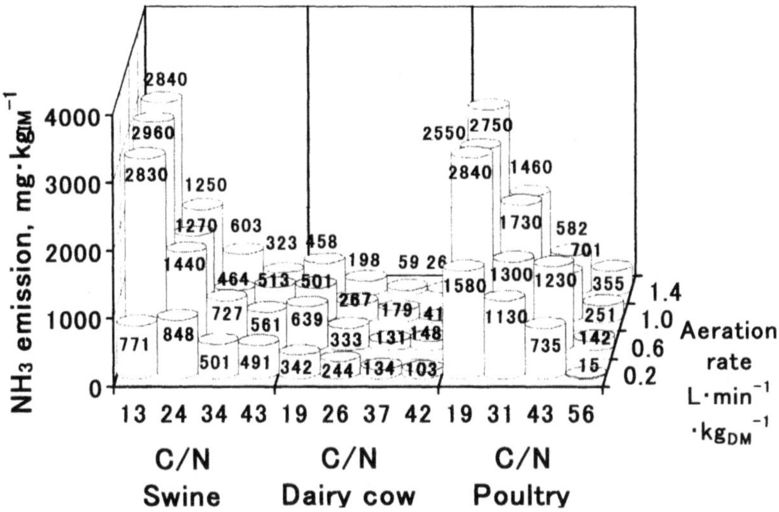

Figure 1. Ammonia emissions from composs

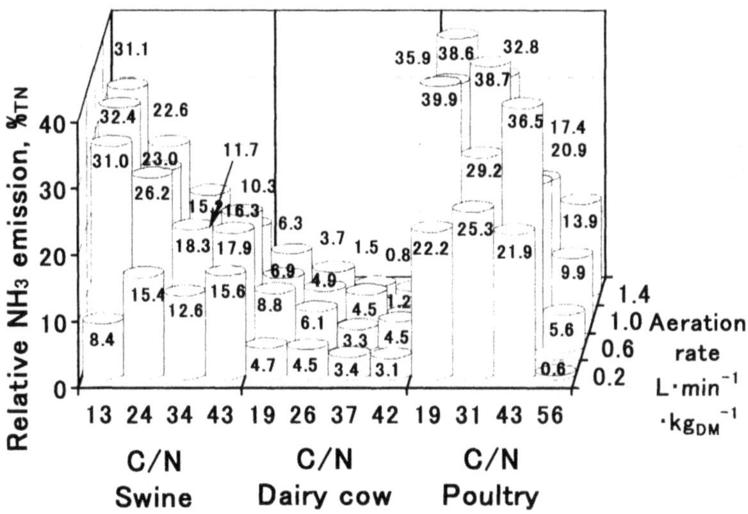

Figure 2. Relative ammonia emissions from composts

3.3. Effect of nitrogen content

The initial ammoniacal nitrogen concentration (NH_4^+-N) of the poultry manure substrate at a C/N ratio of 19 was 3 times that of the swine manure substrate, probably due to the lack of urine in the swine manure. The NH_4^+-N concentration of the dairy cow substrate was very low. The organic nitrogen (Org-N) reduction during swine manure composting was about 2 times that during dairy cow manure composting. Part of the organic nitrogen found in the initial substrate of swine and dairy cow manure was obviously converted to inorganic nitrogen during the composting process and was emitted as ammonia gas. In contrast, the organic nitrogen found in the poultry manure substrate actually increased during composting. It is assumed that part of the NH_4^+-N present in poultry manure is used for growth of the microorganisms during the composting process, and that the remainder of these microorganisms was recovered as Org-N (Käck *et al.*, 1994). The differences in ammonia emissions during composting of the three different livestock manure mixtures can mainly be attributed to differences in the NH_4^+-N concentration of the initial substrate mixtures. This was also affected by the convertibility of organic nitrogen into inorganic nitrogen during the composting process.

4. CONCLUSIONS

This study was carried out to investigate the influence of the type of livestock manure, C/N ratios and aeration rates on ammonia emissions during high-rate composting. The results of the study are as follows:

1. In general, the ammonia emissions during successful composting, as characterized by a thermophilic temperature pattern, were greater than the emissions during unsuccessful composting for all manure types. This suggests that successful composting unfortunately results in higher ammonia emissions.
2. The ammonia emissions decreased as the C/N ratio increased for substrates based on all manure types. However, when measured as relative ammonia emissions, the difference in the emissions generated by substrates with high and low C/N ratios were not so pronounced.
3. The relative emissions from poultry manure composting were the highest, while the relative emissions generated during dairy cow manure composting were the lowest.
4. The differences in the emissions during the composting of substrates based on different types of livestock manure at similar C/N ratio can be attributed to differences in the initial ammonium nitrogen concentrations of the initial compost mixtures. The emissions were also affected by the convertibility of organic nitrogen into inorganic nitrogen during the composting process.

REFERENCES

Beck, J., M. Käck, A. Hentshel, K. Csehi, & T. Jungbluth, 1997. Ammonia emissions from composting animal wastes in reactors and windrows. In: Proceedings of the International Symposium on Ammonia and Odour Control from Animal Production Facilities,. Voermans, J.A.M. and Monteny, G. (editors), CIGR, Vinkelood, the Netherlands, 381-388.

Käck, M., J. Beck, & T. Jungbluth, 1994. Low emission dairy waste management by separating and composting. In: Proceedings of the Third International Dairy Housing Conference, Bucklin., R. (editor), Orland, USA, 443-453.

Kirchmann, H. & E. Witter, 1989; Ammonia volatilization during aerobic and anaerobic manure decomposition. Plant and Soil, 115: 35-41.

Martins, O., 1992. Loss of nitrogenous compounds during composting of animal wastes. Bioresource Technology, 42: 103-111.

Osada, T., K. Kuroda, & M. Yonaga, 2000. Nitrous oxide, methane and ammonia emissions from composting of swine manure. The Japanese Society of Waste Management Experts, 32: 51-56.

Greenhouse Gases and Animal Agriculture
J. Takahashi & B. A. Young (Editors)

Compost processing organic waste for recycling

P. Pittaway

Recycled Organics Consortium, The University of Queensland Gatton, Qld 4343, Australia, and National Centre for Engineering in Agriculture, University of Southern Queensland Toowoomba, 4350, Australia

Composting is most commonly perceived as a waste management process. As a result the biological stability of composts can vary substantially, and their agronomic value is rarely specified. However it is possible to manage the composting process to maximise the fertiliser and soil health properties of the product. Managing the bulk density, the water content and the temperature profile of the windrow can substantially improve microbial activity, and the mineralisation of nutrients. Quality controlled composts have the potential to improve the physical, chemical and biological properties of soil. To maximise the net present fertiliser value of the product, the end-point of the composting process must be dictated by the availability of nutrients and indicators of biological stability. Assigning carbon credits to the biologically active organic carbon fraction of a compost will enhance the adoption of composting for the sustainable recycling of organic waste.

Keywords: compost, organic waste, recycling, greenhouse gases, carbon credits

1. INTRODUCTION

Fifty years of agricultural specialisation and a reliance on inorganic fertilisers for crop production in Australia has resulted in a substantial decline in the organic carbon content of cropping soils (Dalal & Mayer, 1986). As a consequence the fertility and physical structure of the soil has declined (Golchin *et al.*, 1994), and the incidence of root diseases has increased (Pankhurst *et al.*, 1995).

Conversely, large stockpiles of organic waste have accumulated on-site at intensive rural industries such as piggeries, feedlots, cotton gins and sawmills. Loss of productive land used for stockpiling, the risks of spontaneous combustion and the eutrophication of local waterways has encouraged regulatory authorities to promote land application where possible, as a solution (Cameron *et al.*, 1997). However the application of sawdust or mulched yard trimmings to cropping lands is not appropriate due to the phytotoxicity of some organic constituents, and due to nutrient draw-down (Keeling *et al.*, 1994, Patriquin, 1986). Direct land application of grain-dust and cotton trash is equally undesirable, due to the potential spread of weed seeds, pesticide residues and plant pathogens.

Manure and other by-products of animal origin have attracted the most interest, because of their potential to replace inorganic fertilisers and to contribute to the organic carbon bank in soils. For example the nitrogen fertiliser equivalent in the excreta produced annually by

400,000 pigs in South Australia hypothetically could fertilise about 200,000 ha of cereal crops (Brechin & Mc Donald, 1994).

However the minerals contained in manure are predominantly bound in the organic fraction, unavailable for immediate plant use (potassium is an exception), and is highly variable in concentration. The feed formulation and the ability of animals to digest and absorb different chemical compounds determines the nutrient content of excreta (Tucker & Watts, 1993). For example pigs fed a high grain diet typically metabolise less than 40% of the phytate (form of organic phosphorus stored in seeds), with 60% excreted in the faeces.

Storage conditions and handling also substantially affect the nutrient composition of excreta (Sorensen, 1998). The nutrient content of slurry stored in anaerobic ponds is particularly variable, due to the diversity of aerobic and anaerobic microbial processes that may prevail. For example in a study of six anaerobic ponds receiving slurry from dairy sheds in New Zealand, no single management or environmental parameter was correlated with the nutrient content of the effluent (Hickey *et al.*, 1989).

Managing effluent irrigation to land in an environmentally sustainable manner is also difficult to achieve due to the inherent variability in the plant-available nutrient concentrations of such slurries. Stringent monitoring of the mineralisation rates of both nitrogen and phosphorus in the soil is required, to reduce applications of conventional inorganic fertilisers to ensure a balanced nutrient profile in the soil (Barkle *et al.*, 2000).

Hence most manures and slurries are used as ‘soil conditioners’, applied in addition to conventional inorganic fertilisers. Application rates are seldom based on crop or soil needs due to the extreme variability of nutrient concentrations, and a lack of information on rates of mineralisation (Cameron *et al.*, 1997). This increases the risk of uncontrolled growth flushes in subsequent crops, the risk of nutrient contamination of surface and/or underground water (Giusquiani *et al.*, 1998, Sharpley &Moyer, 2000), and the costs of crop production. Composting has the potential to improve the sustainability of reusing manures and other organic wastes, by stabilising the microbial biomass and accelerating the conversion of nutrients from the organic to the inorganic (plant-available) form.

2. COMPOSTING AND GREENHOUSE GAS EMISSIONS

Cellulose, the major constituent of plant cell walls, is the most abundant form of organic carbon on earth. Estimates of 7×10^{11} tonnes of cellulose exist at any one time in primary producers, with as much as 4×10^{10} tonnes replenished annually via photosynthesis (Coughlan &Mayer, 1991). Cellulose also accounts for 50% of the dry weight of the surpluses, residues and wastes generated by agriculture. Cellulolytic microorganisms are ubiquitous in the environment, eventually oxidising cellulose to carbon dioxide and water. However under anaerobic conditions, as much as 10% of the annual cellulose budget can be converted to methane and carbon dioxide.

The anaerobic conversion of organic carbon to methane and other gases is the principle behind the use of anaerobic ponds for waste management. However concern over greenhouse gas emissions and the unacceptability of odours that are also produced has encouraged the development of alternative waste management strategies. Aerobic composting is an increasingly popular alternative. Anaerobic microsites may still occur in a compost windrow, leading to the production of methane (Hellmann *et al.*, 1997, Hao *et al.*, 2001).

Managing composts to minimise methane is desirable for two reasons:

1) reducing greenhouse gas emissions and

2) reducing the likelihood of spontaneous combustion.

The total nutrient content of composts has the potential to replace conventional forms of inorganic fertilisers for crops, in addition to adding stable organic carbon to maintain soil structure and fertility. However the focus on the composting process has been on waste management, not on the production of a plant fertiliser. Too often this has lead to the production of immature composts, with little appreciation of how to maximise the biological value of the raw materials for agronomic use.

3. DEFINING THE COMPOSTING PROCESS

Composting is a bio-oxidative process with microbes converting biologically reactive organic matter into more stable humus compounds, carbon dioxide and water (Insam *et al.*, 1996). Ideally the water availability should be very high, with good oxygen diffusion favouring the activity of aerobic bacteria. However water availability and oxygen diffusion depend on the size of the voids between particles, and their water content. Feedstocks with many small pores will typically have a higher bulk density, with water-filled pores minimising oxygen diffusion.

Adding coarser bulking agents reduces the bulk density (Table 1), thereby increasing oxygen diffusion at higher water contents. The figure of 60% wet weight commonly recommended in composting publications, totally ignores the physical properties of the feedstock. In some composts 60% will be too dry for microbial growth, limiting the decomposition process (for example Smith, 1983). The more logical alternative is to improve passive aeration by reducing the bulk density of manures, using the concept of field capacity as the upper limit for water content.

Table 1: A comparison of the 60% wet weight and field capacity values for two windrow feedstocks differing in their bulk density (Pittaway unpublished)

Compost materials	Saturated water content (g/g)	Field capacity g/g)	60% wet weight (g/g)	Bulk density (gm/cm^3)
Feedlot manure only	0.75	0.68	0.45	0.68
Feedlot manure and sawdust	1.02	0.81	0.61	0.45

Actinomycete bacteria have a higher water requirement than fungi, and are extremely important in the degradation of phenolic compounds and in the control of plant and animal pathogens (Bollen &Volker, 1996). Temperatures maintained at between 50 and 65^0C selectively favour the activity of these preferred species. Above 65^0C many of these species are killed. Therefore managing the bulk density of a compost to achieve water contents close to field capacity and temperatures within the optimal range, maximises the biological value of the compost.

4. COMPOSTS, SOIL HEALTH AND SUSTAINABILITY

Too often the duration of composting is dictated by waste management objectives, not by plant growth requirements. In manures, most of the P and N is locked up in organic forms

(Van Kessel *et al.*, 2000). Plants cannot utilise these forms hence additional inorganic fertiliser must be applied to avoid nutrient draw-down. Over time the organic matter is mineralised by soil microbes, adding to pool of N and P available for crop use.

Excess nitrogen may pollute the environment via leaching to streams and nitrogen volatilisation (Barkle *et al.*, 2000). An imbalance of plant-available nutrients can also predispose plants to root disease (Pittaway *et al.*, 1999). Clearly, to be environmentally sustainable, the composting process must meet the requirements of plant growth and soil health: not just the economics of waste management.

The composting process consists of three stages:

1) Consolidation, during which the bulk density of a pile adjusts in accordance with the water content and height of the feedstock. Microbial activity may be limited by sub-optimal water availability, and by microbially toxic plant residues.

2) Active decomposition when water availability is optimal, and any toxic compounds have been degraded by microbial specialists, and

3) Curing phase by which time the readily degradable organic carbon has been used, with microbial activity reduced to a lower steady state.

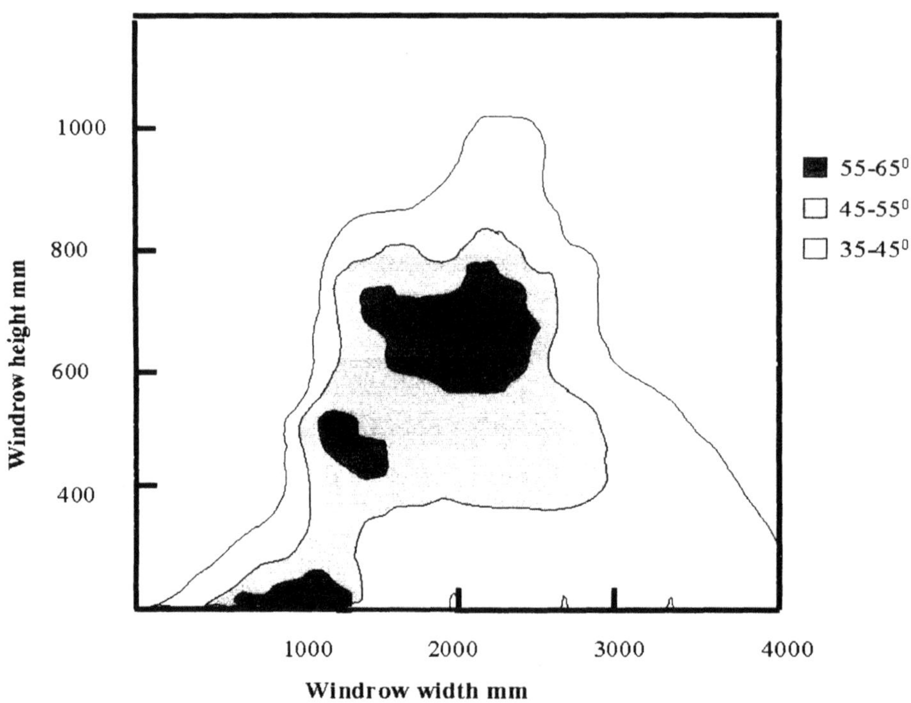

Figure 1. Temperature contours taken within a cross-section of a cotton trash and feedlot manure compost (2:1 v/v) windrow 4 weeks after consolidation. Optimal temperature range of 50-65^{0}C was maintained for more than 10 weeks during which time the windrow was turned only 3 times.

During the active phase heat evolution is greatest. By managing the bulk density of the feedstock and the height of the windrow, passive aeration and therefore temperature control can be achieved (refer figure 1). Accordingly the frequency of turning will be dictated by the need to pass all of the feedstock through the centrally heated core of the windrow. In our cotton trash composts and chicken manure composts passive temperature maintenance within 50-65^{0}C was achieved for over 10 weeks.

5. MANAGING COMPOSTS FOR AGRONOMIC USE

Steady-state microbial growth during the late active phase and curing phase results in a greater proportion of P in the available (mineralised) form (figure 2). In practice, the N and P mineralisation rates are faster in composts based on chicken manure, with feedlot manure requiring extended active and curing phases (Pittaway unpublished, Van Kessel *et al.*, 2000). Of the total K, most is in the available form except for that locked up in the microbial biomass (figure 3). In contrast most of the N is locked up in the organic matter. Animal requirements for P are two to three times greater than that of plants or microbes. Accordingly, a compost should be maintained well into the curing phase to ensure that at least 50% of the total P is available for immediate plant growth.

The 50% availability maximises the net present value of the compost as an inorganic fertiliser. At plant-available concentrations of less than 50% the net present fertiliser value is reduced (Table 2), reducing the marketing potential of the product.

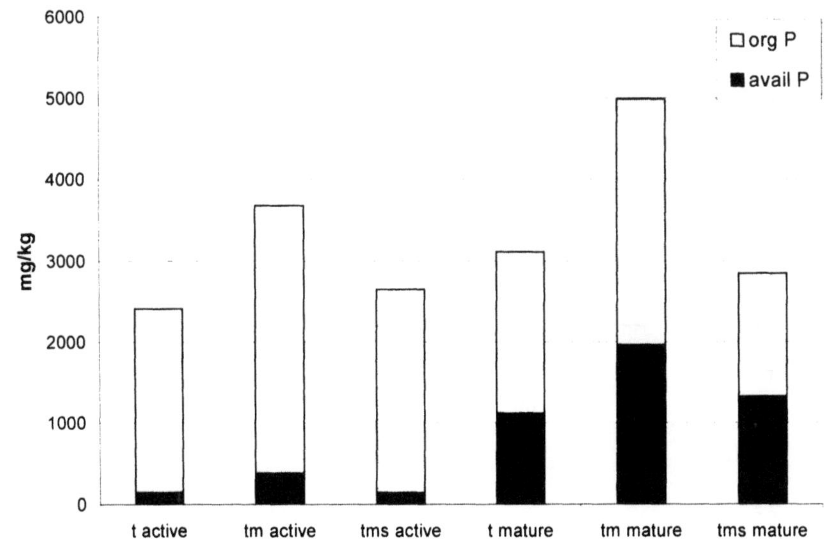

Figure 2 Concentration of phosphorus (P) in the organic and inorganic forms in cotton trash compost mixtures. Samples were from trash only (t), 2:1 v:v trash and feedlot manure ™, and 2:1:1 trash, feedlot manure and sawdust (tms)

Sustainable use depends on matching the available nutrient content of the compost with crop growth requirements (Table 2). Upper limits for application rates should be based on either the mineral P and/or K requirements for early crop growth. Application rates of conventional mineral P and K fertiliser should be reduced accordingly. Experience under southern Queensland environmental conditions with part composted feedlot manure indicates that N only becomes available 18 months after application. Conventional mineral N fertiliser will need to be applied to the first crop, with adjustments made on the basis of soil-available N for subsequent crops. To utilise the slow-release component over time sustainably, application of a compost as a fertiliser replacement should occur no more frequently than every second year.

Animal-based composts also contain high concentrations of calcium. In Australia gypsum is often applied to sodic soils, or to correct other forms of soil structural decline. However the addition of organic carbon and the formation of humic acids acts synergistically with gypsum to improve both the physical and chemical properties of soil (Chorom & Rengasamy, 1997). Estimates based on the calcium and organic carbon concentrations of a feedlot manure and grain offal compost suggest that compost applications of the order of 3.5tons per hectare (dry weight) may offset the rate of application of gypsum by 30 to 50% (Table 2).

Table 2: Estimation of the net present fertiliser value (NPV) of the feedlot manure compost applied at 5 tons per hectare (3.5 tons dry-weight basis) onto a Darling Downs soil for cotton production. QDPI Note on Cotton Growing on the Darling Downs was used to establish the conventional fertiliser application rates. The two asterisks indicate that calculations for gypsum equivalence assume that all the calcium in the compost is available (only total Ca data was available).

	Potassium	**Phosphorus**	**Nitrogen**	**Calcium**	**Total $**
Maximum fertiliser rates for Darling Downs cotton kg ha^{-1}	60	25	180		
Kg fertiliser in 3.5 (dry wt) tons ha^{-1} compost	62	7.7	0.1	49.2	
Aus$ NPV for inorganic fertiliser ex Toowoomba (ha^{-1})	(muriate potash) $59	(super phosphate) $27	(urea)	(gypsum)** $65	**$151**
Aus$ equivalence slow-release fertiliser investment ha^{-1}	$15	$162	$109		**$286**
Total $ fertiliser equivalent investment per hectare	**$74**	**$189**	**$109**	**$65**	**$472**

6. COMPOSTS AS CARBON CREDITS

Organic carbon is the corner-stone of primary production. The content and diversity of organic matter in the soil is also fundamental to the dynamics of nutrient recycling (Follett, 1987). However with the intensification of cropping systems and the dependence on inorganic fertilisers, the organic carbon content of both broadacre and intensive cropping lands has dropped substantially (Dalal &Mayer, 1986). The reduction and limited diversity of organic inputs has contributed to a decline in soil structure and an increase in soilborne diseases (Kennedy & Smith, 1995).

The most commonly recommended remedial strategy is replacing continuous cropping with one or more years of pasture (Dalal *et al.*, 1995). However the application of a mature compost at rates of 3 to 5 tons per hectare can replace the additional carbon input associated with a pasture phase. In the example provided in Table 2, at an application rate of 3.5 tons ha^{-1}(dry weight), the biologically active organic carbon contribution to the soil is 780 kg. In comparison the carbon contribution of one year of pasture above the contribution of continuous wheat is 650 kg ha^{-1}.

In the current climate of global carbon trading, the concept of gaining credits for the carbon content of soil is very attractive. However the turnover rate of carbon in soil is extremely variable, and soil composition is extremely heterogeneous. In contrast the carbon content of a defined compost mix is more homogeneous, and the microbial biomass, biologically available and more recalcitrant fractions of organic carbon can be reliably measured (Figure 3).

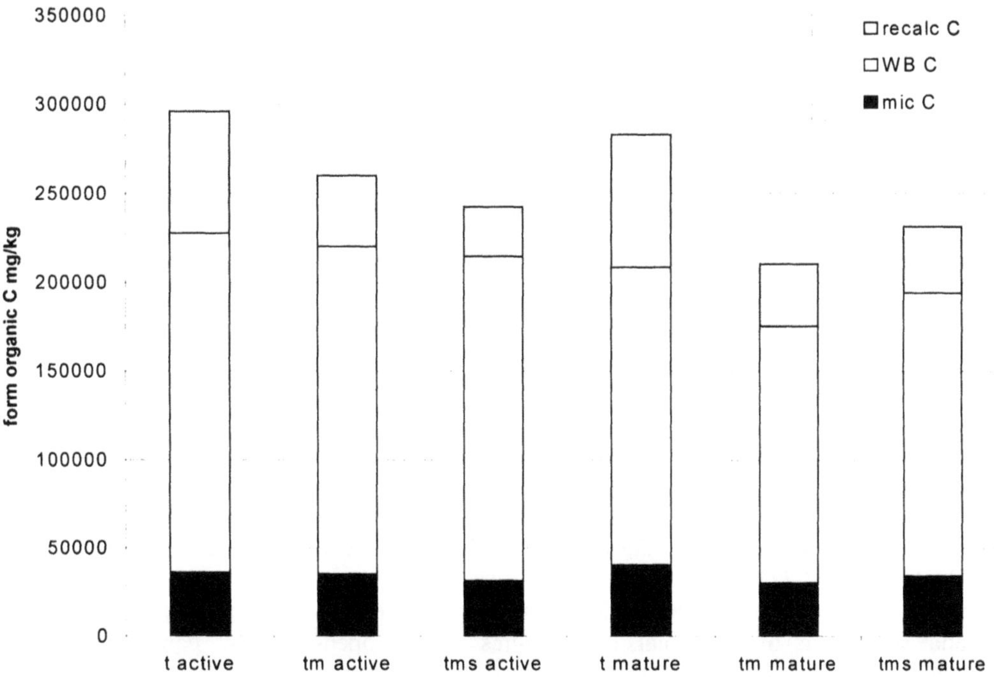

Figure 3. The microbial, Walkley-Black extractable and furnace organic carbon fractions in cotton trash compost mixtures. Samples were from trash only (t), 2:1 v:v trash and feedlot manure, and 2:1:1 trash, feedlot manure and sawdust

It is feasible to use the biologically active fraction of a compost (Walkley-Black method) as a carbon credit, thereby encouraging the application of composts to cropping soils. The costs of composting for plant growth requirements can therefore be offset by the P, K and Ca fertilser equivalence (Net Present Value: Table 2), and by the carbon credits assigned to the biologically available carbon fraction. On this basis, the recycling of intensive agricultural wastes will be both economically and environmentally sustainable.

REFERENCES

Barkle, G. F., R. Stenger, P. L. Singleton & D. J. Painter, 2000. Effect of regular irrigation with dairy farm effluent on soil organic matter and soil microbial biomass. Aust. J. Soil Res. 38: 1087-1097.

Bollen, G. J. & D. Volker 1996. Phytohygienic aspects of composting. In: The Science of Composting, de Bertoldi, M. Sequi, P., Lemmes, B. & Papi, T. (editors). Chapman and Hall, England. 233-246.

Brechin, J. & G. K. McDonald, 1994. Effect of form and rate of pig manure on the growth, nutrient uptake, and yield of barley (cv. Galleon). Aust. J. Exp. Agric. 34: 505-510.

Cameron, K. C., Di, H. J. & McLean, R. G., 1997. Is soil an appropriate dumping ground for our wastes? Aust. J. Soil Res. 35: 995-1035

Chorom, M. & P. Rengasamy, 1997. Carbonate chemistry, pH, and physical properties of an alkaline soil as affected by various amendments. Aust. J. Soil Res. 35: 149-161.

Coughlan, M .P. & F. Mayer, 1991. The cellulose-decomposing bacteria and their enzyme systems. In The Prokaryotes: a handbook on the biology of bacteria, Balows, A. (editor). Springer-Verlag, New York USA.

Dalal, R. C. & R. J. Mayer, 1986. Long-term trends in fertility of soils under continuous cultivation and cereal cropping in Southern Queensland. 111 Distribution and kinetics of soil organic carbon in particle-size fractions. Aust. J. Soil Res. 24: 293-300.

Dalal, R. C., W. M. Strong, E. J. Weston, J. E. Cooper, K. J. Lehane, A. J. King, & C. J. Chicken, 1995. Sustaining productivity of a Vertisol at Warra, Queensland, with fertilisers, no-tillage, or legumes. 1. Organic matter status. Aust. J. Exp. Agric. 35: 903-914.

Follett, R. F, 1987. Soil Fertility and Organic Matter as Critical Components of Production Systems. Soil Science Society America and American Society Agronomy, Madison, Wisconsin USA.

Golchin, A., J. M. Oades, J. O. Skjemstad & P. Clarke, 1994. Soil structure and carbon cycling. Aust. J. Soil Res. 32: 1043-1068.

Giusquaini, P. L., L. Concezzi, M. Businelli & A. Macchioni, 1998. Fate of pig sludge liquid fraction in calcareous soil: Agricultural and environmental implications. J. Environm. Quality 27: 364-371.

Hellmann B., L. Zelles , A. Palojarvl& Q. Bai, 1997. Emission of climate-relevant trace gases and succession of microbial communities during open-windrow composting. Appl. Environm. Microbiol. 63: 1011-1018.

Hickey, C. W., J. M. Quinn & R. J. Davies-Colley, 1989. Effluent characteristics of dairy shed oxidation ponds and their potential impact on rivers. N. Z. J. Marine Freshwater Res. 23: 569-584.

Hao, X., C. Chang, F. J. Larney & G. R. Travis, 2001. Greenhouse gas emissions during cattle feedlot manure composting. J. Environm. Quality 30: 376-386.

Insam H., K. Amor, M. Renner & Crepaz, 1996. Changes in functional abilities of the microbial community during composting of manure. Microb. Ecol. 31: 77-87.

Kennedy, A. C. & K. L. Smith, 1995. Soil microbial diversity and the sustainability of agricultural soils. In: The Significance and Regulation of Soil Biodiversity, Collins, H. P., Robertson, G. P. & King, M. J. (editors). Kluwer Academic Publishers. Netherlands. 75-86

Keeling, A. A., I. K.Paton & A. J. Mullett, 1994. Germination and growth of plants in media containing unstable refuse-derived compost. Soil Biol. Biochem. 26: 767-772.

Pankhurst, C. E., B. G. Hawke, H. J. McDonald, C. A. Kirkby, J. C. Buckerfield, P. Michelson, K. A. O'Brien, V. V. S. R. Gupta & Doube, B. M 1995. Evaluation of soil biological properties as potential biological indicators of soil health. Aust. J. Experim. Agric. 35: 1015-1028.

Patriquin, D. G., 1986. Biological husbandry and the 'nitrogen problem'. Biol. Agric. Hortic. 3: 167-189.

Pittaway P. A., N. Nasir & K. G. Pegg, 1999. Soil receptivity and host-pathogen dynamics in soils naturally infested with *Fusarium oxysporum* f. sp. *cubense*, the cause of Panama disease. Aust. J. Agric. Res. 50: 623-628.

Sharpley, A. & B. Moyer, 2000. Phosphorus forms in manure and compost and their release during simulated rainfall. J. Environm. Quality 29: 1462-1469.

Smith, J. F., 1983. The formulation of mixtures suitable for economic, short-duration mushroom composts. Sci. Hortic. 19: 65-78.

Sorensen, P., 1998. Effects of storage time and straw content of cattle slurry on the mineralization of nitrogen and carbon in soil. Biol. Fertil. Soils 27: 85-91.

Tucker, R. W. & P. J. Watts, 1993. Waste minimization in cattle feedlots by ration modification. Am. Soc. Agric. Engin. 36: 1461-1466.

Van Kessel, J. S., J. B. 111 Reeves & J. J. Meisinger, 2000. Nitrogen and carbon mineralization of potential manure components. J. Environm. Quality 29: 1669-1677.

Greenhouse Gases and Animal Agriculture
J. Takahashi & B. A. Young (Editors)

Status and prospects of manure management in Japan -Composting approaches

K. Nishizaki

Department of Agro-Environmental Science, Obihiro University of Agriculture and Veterinary Medicine, Inada-cho, Obihiro, Hokkaido, 080-8555, Japan

This paper reviews the progress and present status of manure management in Japan and introduces the newest development concerning composting machines. In addition, the prospective of manure management is briefly discussed.

Keywords: manure management, composting machines

1. INTRODUCTION

According to the latest statistics, the grassland area in Hokkaido is 540,000 ha. This value is equal to about half of the 1,200,000 ha of cultivated land in Hokkaido and covers 82% of grassland in Japan. The number of dairy cow farms is 10,000 in Hokkaido and corresponds to about 30% of all over the country. The number of dairy cows is 870,000, 49% of total dairy cows. The share of milk production occupies 42.5% of total milk produced and shows a tendency to increase. As mentioned above, dairy farming in Hokkaido performs an important part of milk production in Japan.

However, Hokkaido dairy farming also faces many problems demanding a rapid a rapid solution:

On one hand, a greater interest has to be put on the feed problem. The rate of self-supplying of feed in dairy farming is decreasing to about 38%. Forty percent of the total cost in milk production is occupied by the feed costs and the fluctuations of the feed cost directly affects farm management. To stabilize the dairy farm management the increase of self-sufficiency is an important subject.

On the other hand, manure management is also an important problem that needs a solution. The new agriculture fundamental law emphasized the need of a reasonable manure management and the importance of natural nutrient cycling.

The last problem is labor shortage. In a situation of decreasing numbers of farmers, dairy farmers also struggle with the enlargement of scale and the instability of milk price.

To clarify the background and the importance of manure management, in this paper the problems awaiting a solution of Hokkaido dairy farming are outlined.

2. THE REGULATIONS REGARDING MANURE DISPOSAL

The law for farm animal manure disposal was enforced on November 1, 1999. This law prohibits windrows/piles in the field and non-covered storage ponds for liquid manure, and after 5 years an offender is fined.

The Hokkaido government settled a promotion scheme utilizing manure disposal (Table 1).

Table 1. The promotion scheme utilizing manure disposal

Concentrated management system	400 farmers
Joint management system for several farmer	3,000 farmers
Individual management system	3,500 farmers
Simple management system using waterproof cover	3,500 farmers
Total	10,400 farmers

The Hokkaido government decided that composting facilities have to be introduced for all dairy farmers by 2004.

Moreover, a law for the recycling of food processing wastes was established on May 1, 2001. This law intends to utilize these wastes as feed and manure by recycling. A decrease in quantity is also considered as an appropriate way. But composting is regarded as a simple and cheap method.

Composting refers to a process of disposing biomass in the most efficient way to utilize it as a soil conditioner and an organic fertilizer.

However it is unclear, how manure management differs in the quality of finished composts and how manure should be utilized for crop productions. The difference in management affects the area required for the composting pad and it also influences the layout of the composting pad and the building expenses. Land area needs are based upon the composting system, the composting operation and the compost equipments selected. If the decomposition and composting period is rapid, less area is required.

Therefore, it relates to an effective utilization of the composting pad that the demand composting finishes rapidly by an appropriate management. The composting area required can be reduced to 1/2-1/3 depending on the quality of the finished compost and the management method. The height of the eaves can be the same, that is, the area and the height of eaves are related to the building expenses. The basic principle in designing a composting facility is to consider the composting operation and the final quality of manure desired. Dairy farmers also understand these points vaguely. But an appropriate manual in accordance with the purpose of composting is not yet made. This fact explains the farmer's hesitation and the delay in realization.

3. METHODS COMPOSTING AND COMPOST MACHINES

The purpose of composting is to change the properties of the biomass to alleviate the possibility of causing damage to the crops and to make it suitable for transport and storage. By composting, raw materials can be changed in a way that allows decomposition to proceed at a more favorable rate, making these materials easier to handle and more suitable as soil conditioners and organic fertilizers. The most important aspect of composting is to keep the conditions suitable for the activity of the microorganisms. It is necessary to maintain appropriate density, porosity, temperature and moisture content.

The most basic principle in composting is piling up raw materials. If the pilling is not sufficient, heat is lost from the surface of piles and sufficient decomposition cannot be expected. However, if the accumulation (windrow/pile height) becomes too high, the oxygen supply to the microorganisms becomes insufficient and there is a risk that the decomposition process stops. The appropriate moisture content of raw materials for sufficient decomposition is 60-65% (maximum permissible moisture content; 70-75%). As the materials for moisture content control is difficult to obtain, a level of 70-75% is used in the composting operation in general. By the arrangement of these conditions, a favorable composting process can take place. As a supplementary operating procedure, turning will increase the uniformity in the windrow/pile, and improve the porosity that often decreases due to swelling of the materials during composting.

To summarize, the main points in composting are; ① to keep appropriate porosity by turning, ② to avoid compression by excessive piling, ③ to maintain porosity by mixing and particle size reduction (well-mixed and porous), and ④ to keep sufficient surface area by low piling. There are at least five basic approaches to sustain the conditions that support the microorganisms doing the main work of composting:

1. The most popular method is a static piling approach, with no active management of the piles after initial mixing (Fig. 1). The windrow/pile is left alone for about three years. This method is now prohibited by the regulations.

Fig. 1. Static pile

2. The second is the loader-turned windrow approach. This approach minimizes new investments and is very similar to the static pile approach. The difference is that the piles are actively managed with the equipment. This equipment system will normally consist of a tractor with bucket (wheel) loader or a tractor, loader and manure spreader combination. But this system needs much space to maneuver the equipment. As mentioned earlier, the area and the height of eaves are related to the building expenses. Because most composting operations have to be done inside the buildings or under covered facilities, an effective utilization of the space is an important issue. Recently a simple compost turner was developed. This machine is a type of bucket loader, and a beater for mixing and particle size reduction is installed at the end of the bucket. Data from actual composting tests suggests that the bucket beater is effective (Fig. 2 and 3). This combined turner with a beater can process roughly 80-100 t of material per hour.

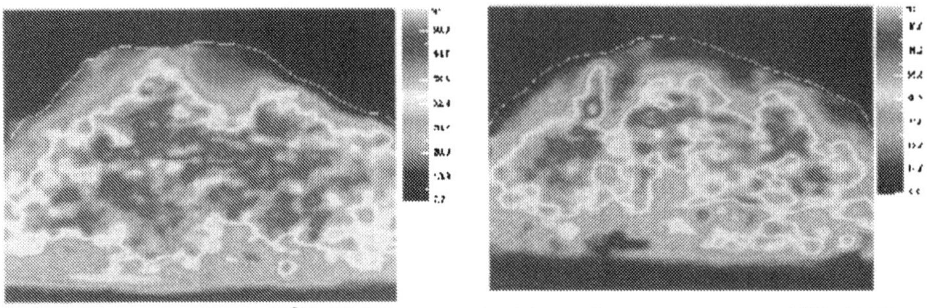

(with a beater, area above 40℃;75%) (without a beater, area above 40℃;40%)

Fig. 2. Temperature rise in the pile

Fig.3. Turning work by bucket loader with a beat

3. The third approach is a windrow system using specialized windrow turners. As the volume of the materials to be handled increases, the compost operations demand more land, more equipment, and more labor. Additional hired help may also be needed. Furthermore, farmers are facing to invest in specialized windrow turners (Fig.4. and 5). A newly developed compost turner is shown in Fig.4. This compost turner is attached to a tractor rear-side, and this turner has two beaters for pulverization and raking and a conveyer for discharge. Compost materials are raked in by 2 beaters and are conveyed rearward by the conveyer. These materials on the conveyer are discharged from the conveyer edge of height 2m. This turner can process about 600 tons per hour. Most windrow turners do a reasonably good job to reshape the windrow automatically. Most importantly, windrow turners can substantially reduce the amount of time that must be spent turning the pile.

Fig.4. Compost turner

Fig.5. Turning work (composting pad)

4. The fourth approach consists in aerated static pile systems. These depend on forced air blowers rather than mechanical windrow turners. A composting facility with aeration system is shown in Fig.6. Forced air blows through the piles from the pipe installed in the floor.
5. The fifth approach consists in in-vessel systems that use forced air and/or mechanical agitation (Fig.7). These systems use forced air and/or mechanical agitation to aerate waste in the confines of the vessel enclosed. The basic agitated bay technology is conceptually similar to a windrow technology. Enclosing the vessel into a structure also helps to further control or eliminate environmental impacts. In the case of introducing it in a cold area, the height of the piles (the depth of vessel) is an important factor for sufficient decomposition.

Fig.6. Composting facility with aeration system

Fig.7. In-vessel system

The duration of decomposition applying the second or third approach is roughly 1.5-2 months when used in buildings protecting against rain and with 2-3 times turning.

4. CONCLUSION

In this paper, the present status of manure management in Japan, and various methods of composting were briefly introduced, and the conditions to successfully use compost turners were described.

At present, the environmental problem has to be urgently solved and the application of solutions for dairy cow wastes is rapidly progressing. One solution for this problem is composting. As stated earlier, the composting activity differs in the quality of finished compost and the use of the finished compost. To initiate the composting process from the discharge of materials to the utilization of compost is the first step. An essential aspect of composting is to set up an appropriate design. If the arrangement of machinery and facilities is unfavorable, appropriate management is impossible.

REFERENCE

E.E.Fabian, T.L.Richard, D.Kay, D.Allee, & J.Regenstein, 1993., Agricultural Composting: A Feasibility Study for New York Farms. Cornell Composting homepage (http://www.cals.cornell.edu/dept/compost/feas.study.html

Greenhouse Gases and Animal Agriculture
J. Takahashi & B. A. Young (Editors)

Evaluation of the chamber system for gas measuring from the livestock waste compost

T. Osada & Y. Fukumoto

Lab. Animal Waste Management, National Institute of Animal Industry, Japan, Ikenodai 2, Kukizaki-machi, Inashiki-gun, Ibaraki-ken 305, Japan

A chamber system consisting of a chamber covering a composting mixture (3 m in diameter, 2.2 m in height, 13 m^3), a ventilator for suction of air into the chamber, and equipment to measure the gas composition and indicate temperature, was developed for evaluation of harmful gas emissions from the livestock waste composting. Fresh air was introduced through the space between the floor and the lower edge of the chamber, and exhaust gas was removed through an outlet placed on top of the chamber. NH_3, CH_4 and N_2O concentrations in exhaust air from the chamber were measured by Infrared Photoacoustic Detector (IPD, multi gas monitor type 1312, INNOVA, Copenhagen DK) at 5-min. intervals. The system was evaluated with standard gas of NH_3, CH_4 and N_2O. High recoveries of 98.5% (NH_3, SD 6.25), 96.6% (CH_4, SD 4.03), and 99.5% (N_2O, SD2.68) were obtained for each gas emission in the chamber over 17 min. The measured values of those gases obtained by the IPD method and conventional method at the time of a composting examination of swine waste were measured, and the differences were only a few percent of the total emissions.

Keywords: ammonia, nitrous oxide, methane, livestock waste, greenhouse gas

1. INTRODUCTION

In Japan, the composting process plays a central role in livestock waste treatment. Recycling of proper nutritive salts from livestock compost can't be completed only by circulation in an area where the livestock density per unit area is especially high. Thus, through the composting process, the livestock excrement can be changed into something easy to handle, and a product can be distributed over a wide area. Moreover, the "Law concerning the Appropriate Treatment and Promotion of Utilization of Livestock Manure" enacted in autumn of 1999, the composting process was recommended, and it is thought that those facilities should be increased.

We are already concerned about the environmental load gas generated by the composting process (Kuroda et al. 1996). This is because there is a risk of arising the serious pollution problem causing damage not only to neighbors environment but to wide range local environment (Hartung and Phillips, 1994). We used small composting examination equipment to examine the discharge gas which develops from the composting process (Osada et al. 2000). According to this result, NH_3 emission rates from composting one cubic meter of swine dungheap varied from 400 to 970 gNH_3-N m^{-3}. CH_4 and N_2O emission rates ranged from 0.6 to 385 g CH_4 m^{-3} and from 1.9 to 71.9 g N_2O-N m^{-3}, respectively. However, in actual cases in which the amounts generated were in a real processing facility, there was a great disparity in the values obtained (Husted, 1994, Czepiel et al. 1996). For the purpose of improving the calculation of emissions and developing more friendly composting technique, there is need for more extensive knowledge of the variable factors of greenhouse gases and NH_3 from the composting manure on the real scale.

For this reason, we developed a system for quantitative measurement of emission materials from composting using a large chamber in an experiment. The total amount measured as the recovery rate (i.e., the overall amount of materials measured in relation to the artificially generated (known) total amount) was considered, and the reliability of the data was confirmed. Moreover, the composting examination was conducted, and the measurement results with the conventional method and the IPD method were compared.

2. MATERIALS AND METHODS

2.1. Dynamic chamber system

The measurement system made consists of a dynamic chamber device to collect the exhaust gas from composting materials, and a measurement device (Fig. 1).

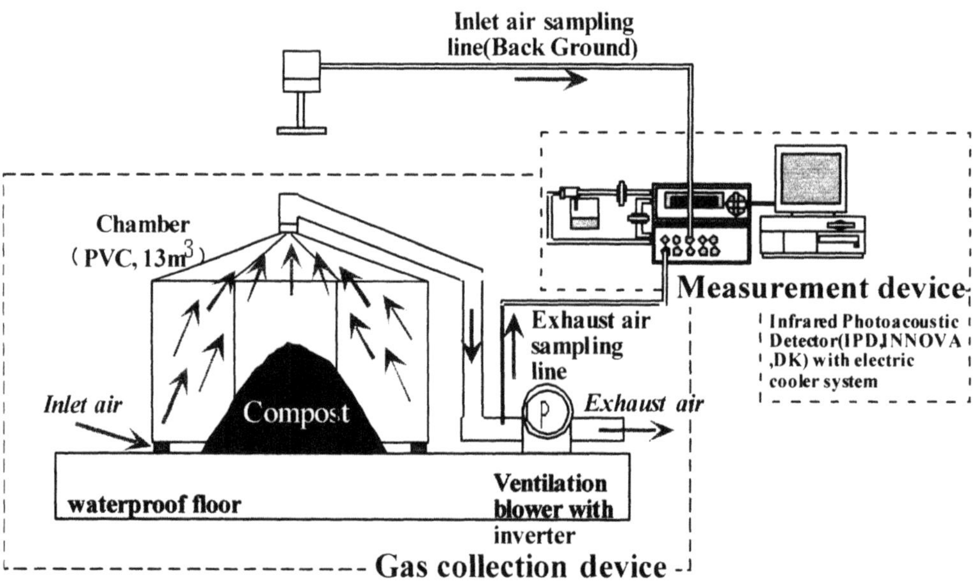

Figure 1 Measurement system for emissions from composting.

2.2. Calculation method for emission rate of each substances

The rate of emission (E) of each substance (NH_3, CH_4 and N_2O) was computed from the amount of ventilation and the concentration differences of each substance between the inlet and outlet air samples.

$$E\ (mg/5min.) = (Conc.\ of\ outlet\ air\ (mg/m^3) - Conc.\ of\ inlet\ air\ (mg/m^3)) \times 5/60 \times Ventilation\ rate\ (m^3/hour)$$

2.3. Recovery test

In order to evaluate recovery efficiencies, a fixed quantity of each substance (gas) was generated within the center of the chamber 15 cm high, and the total amount of emissions by this chamber system was calculated. The chemical cylinder filled with each pure substance (99.9% or more) was used for the emission source. Emission rate of NH_3, CH_4 and N_2O from the chemical cylinder was controlled 0.76 g/5min., 0.35 g/5min. and 0.51 g/5min, respectively, by a precision pressure regulator and the flow meter characteristics for each substance. These generation rates of three gases were determined from the most remarkable concentration detected in the composting facilities.

2.4. Comparison with values from conventional methods at composting examination

IPD is one of the Fourier Transform Infra Red (FTIR) spectroscopy detectors suitable for photoacoustic measurement of the absorption of infrared light from gas samples. The IPD data tended to overestimate both the CH_4 and the N_2O concentrations in the samples under high concentration of CO_2 or humidity condition just as composting (Osada et al 1997). For accurate evaluation of the gas emission, the IPD data were checked by means of conventional methods. A composting examination was performed, and the result of the IPD method used in the present study were compared with those by the sulfuric acid trap method (for NH_3), a flame ionized detector - Gas Chromatograph (FID-GC, for CH_4), and an electron capture detector - Gas Chromatograph (ECD-GC, for N_2O).

3. RESULTS AND DISCUSSION

3.1. Recovery test

Six-replication (N_2O) or seven-replication (NH_3 and CH_4) examinations were performed in this chamber system by each substance after around 17 min. generation. The average recovery of each substance was good based on the results of a field examination. NH_3, CH_4 and N_2O recoveries were 99.5% (S.D. 2.68), 98.5% (S.D. 6.25) and 96.6% (S.D. 4.03), respectively.

In the 130 m^3/h exchange volume condition for 17-minutes fixed amount of discharge of NH_3, the concentration of NH_3 within the chanber became elevated during 40 - 50 minutes. No great difference was found in movement of three sorts of substances. In a chamber experiment, recovery efficiency of substance is always a concern because the recovery percentage is directly related to the detection limit of that system. Especially with ammonia, the loss by adsorption on the chamber surface was suspected. According to this examination result, in this system, one can measure with sufficient accuracy not only CH_4 and N_2O, but also about NH_3.

3.2. Comparison with values from conventional methods at composting examination

A mixture of 660 kg of swine waste and sawdust was piled, the emission of the substances was measured by both the conventional methods and IPD method, and the results were the compared.

3.2.1. Ammonia.

Comparison of the values obtained by the sulfuric acid trap method with those by the IPD method when the composting examination was performed. By comparing the measured value in a usual sampling point (just before ventilation blower) with a sample in the chamber (exhaust port in the ceiling), the adsorption loss of ammonia at the time of suction was also examined.

In comparing these two methods the difference between them was 5% in each phase. Even during the 7 days after the composting was begun, these was only a 4% in the total amount. Moreover, the levels were compared between inside the chamber and the sampled portion (part removed from chamber), but the difference was around 3%. Since viscosity of ammonia is strong, it needs careful cautions for quantity measurement. This result thus means that with this measuring system the ammonia loss was sufficiently small.

3.2.2. Methane

The methane FID-GC measurement was conducted at 15-min. intervals using the automatic sampling system equipped on the GC. The results of both were compared in a time series measurement of the concentration, and the differences on the 6th day after composting was begun were considered to be very similar. When microbial activity is at its highest during composting, the CO_2 concentration reaches around 2000 mg m^{-3}. Such a high concentration may well have rather affected the methane measured values. The total amount of methane generated over 8 days following the start of the composting was 227g by the IPD method and 239g by the FID-GC method, and the difference was small at around 5%.

3.2.3. Nitrous oxide

In the measurement of nitrous oxide by GC, the analysis was performed manually and the two were compared. The nitrous oxide level in the composting examination is within this range of 0.3-1.4 ppm, and the atmospheric level of same is about 0.3 ppm. Great care is needed in the measurement since the absolute value of the nitrous oxide is so small. However, it was clear that the measurement of the nitrous oxide by the IPD method in the test yielded a value of sufficient reliability.

REFERENCES

Czepiel P., Douglas E., Harriss R., and Crill P. (1996) Measurement of N_2O from composted organic wastes. Environ. Sci. Technol. 30 (8) 2519-2525

Hartung J. and Phillips V.R. (1994) Control of gaseous emissions from livestock buildings and waste stores. J. Agric. Res. 57, 173-189

Husted S. (1994) Seasonal variation in methane emission from stored slurry and solid wastes. J. of Environ. Quality, 23, 585-592.

Kuroda, K., Osada T., Yonaga M., Kanematu A., Nitta T., Mouri S., and Kojima T.(1996) Emission of malodorous compounds and greenhouse gases from composting swine feces. Bioresouce Tech., 56,265-271

Law concerning the Appropriate Treatment and Promotion of Utilization of Livestock manure (LawNo. 112 of July28,1999) , Japan

Osada T., Kuroda K., and Yonaga M. (2000) Nitrous oxide, Methane and ammonia emissions from composting process of swine waste. The Japanese Society of Waste Management Experts 2 (1) 51-56

Osada T. & Fukumoto Y. (2001) Development of new dynamic chamber system for measuring harmful gas emission from composting livestock waste. Proceedings of 1st IWA International Conference on Odor and VOCs; Measurement, Regulation and Control techniques. Australia March 2001, p 163-170.

Greenhouse Gases and Animal Agriculture
J. Takahashi & B. A. Young (Editors)

Methane emission from dairy cow and swine manure slurries stored at 10°C and 15°C

D. I. Massé, F. Croteau, N.K. Patni & L. Massé

Agriculture and Agri Food Canada, Dairy and Swine Research and Development Centre. P.O. box 90, 2000 Road 108 east Lennoxville, Québec, Canada, J1M 1Z3.

The objective of this study was to determine the effect of slurry characteristics, temperature and storage duration on methane emissions from dairy cow and swine manure slurries. The manure was stored at 10°C and 15°C for a 272-day period in closed 250-litre plastic barrels simulating the anaerobic conditions of a full-scale storage tank. To reflect the range of manure solids contents found on Canadian livestock farms, tests included high and low total solids (TS) manure slurries.

As storage temperature was increased from 10°C to 15°C, total CH_4 emission increased for all types of manure slurries. During the 272-day storage period, total methane production from dairy cow manure ranged from 0.24 to 0.39 l CH_4 per l manure, except for barrels filled with low TS dairy cow manure (4.2% TS) and stored at 15°C. These barrels produced a total of 3.62 l CH_4 /l manure, which was mostly recorded between days 160 and 272 of storage, due to an increase in both biogas production and methane content of the biogas. Under similar storage conditions, swine manure produced more methane than dairy cow manure. Over the 272-day storage period, barrels filled with swine manure produced from 1.36 to 7.87 l CH_4 /l manure. Methane production rate from swine manure slurry was relatively constant over the whole storage period.

The results from this study showed that on typical Canadian farms, methane emissions from manure storage tanks over the late fall, winter and early spring period are negligible, because manure temperature remains substantially below 10°C. During the late spring, summer and early fall period, on the other hand, methane emissions could be substantially reduced by recommending storage periods shorter than 150 days and by using below ground storage tanks to maintain lower manure temperatures.

Keywords: methane emission, manure storage, swine, dairy cow, anaerobic conditions

1. INTRODUCTION

Methane is one of the well-recognised greenhouse gases that accumulate in the atmosphere due to human activity. Methane emissions from livestock manure are expected to increase in the coming decade, due to an increase in animal population and the use of liquid-based manure management systems. However, there are limited data on methane emissions from manure storage tanks. The aim of this paper was to measure CH_4 emissions from swine and dairy cow manure slurries at two total solids (TS) contents and stored at 10°C and 15°C.

2. MATERIALS AND METHOD

Manure type and storage conditions are presented in Table 1. All analyses were performed according to standard methods outlined in APHA (1992).

Table 1. Manure type and storage condition used in the experiment

Manure type	Total solids	Temperature	# of barrels per treatment
Dairy cow	9.2%	10°C and 15°C	4
	4.2%	10°C and 15°C	4
Swine	11.3%	10°C and 15°C	4
	4.9%	10°C and 15°C	4

3. RESULTS AND DISCUSSION

Manure characteristics are given in Table 2. Total chemical oxygen demand (TCOD), total nitrogen (TKN) and ammonia-N were approximately twice as high in swine than dairy cow manure. However the soluble COD (SCOD) and the volatile fatty acid (VFA) concentrations, which represent the readily available organic fraction for biological degradation, were similar in both manure types. TS contents are representative of the range of values found in manure from Canadian commercial farms.

Table 2. Average characteristics of the dairy cow and swine manure slurries.

Parameter	Pig manure				Dairy cow manure			
(g/l	High TS		Low TS		High TS		Low TS	
except pH)	Mean	s. e.†	mean	s. e.	mean	s. e.	mean	s. e.
TCOD	162.93	11.81	70.17	23.33	91.05	4.95	54.10	1.71
SCOD	34.19	7.25	15.44	8.02	29.92	1.76	14.97	1.49
TS	112.94‡	NA††	48.64	13.76	92.23	1.68	41.96	2.50
VS	83.07‡	NA††	35.78	11.10	77.48	1.72	33.44	1.41
pH	7.18	0.09	7.06	0.03	6.78	0.06	6.78	0.01
TKN	8.84	0.60	3.50	0.63	3.90	0.03	2.06	0.06
NH_4-N	5.74	0.35	2.48	0.44	2.12	0.01	1.12	0.04
Acetic	5.81	0.21	2.91	0.60	5.45	0.24	3.38	0.14
Propionic	2.73	0.23	1.25	0.21	1.71	0.03	1.04	0.06
Butyric	1.22	0.54	0.56	0.14	1.22	0.04	0.66	0.02

† Standard error of the mean
‡ Determined using average TCOD:TS and TCOD:VS ratios measured in the low TS swine manure
†† Not applicable (see note above)

Table 3 gives total methane production over two storage period lenghts (180 and 272 days) that are representative of manure management practices on Canadian livestock farms. For both manure types and TS contents, more methane was produced at 15°C than 10°C. However, the increase in methane production with temperature tended to be more pronounced for swine than dairy cow manure. During the 272-day storage period, total methane production from dairy cow manure ranged from 0.24 to 0.39 l CH_4 per l manure, except for barrels filled with low TS dairy cow manure (4.2% TS) and stored at 15°C. These barrels produced a total of 3.62 l CH_4 /l manure, which was mostly recorded between days 160 and 272 of storage, due to an increase in both biogas production and methane content of the biogas (Figs. 1 and 2). Under all storage conditions and duration, swine manure produced more methane than dairy cow manure (Table 3). Methane production from swine manure ranged from 1.36 to 7.87 l/l. Methane emissions from barrels filled with swine manure remained fairly constant over the whole storage period (Fig. 3). Methane content of the biogas was generally higher with swine than dairy cow manure (Figs. 2 and 4).

Table 3. Methane production from dairy cow and swine manure slurries for storage periods of 180 and 272 days.

Manure type	Solids content (%)	Storage time			
		180 days		272 days	
		10°C	15°C	10°C	15°C
		Litre of CH_4 per litre of manure slurry			
Dairy cow	9.2	0.28 ± 0.02	0.33 ± 0.03	0.31 ± 0.02	0.39 ± 0.01
	4.2	0.23 ± 0.02	0.42 ± 0.26	0.24 ± 0.02	3.62 ± 0.81
Swine	11.3	2.96 ± 0.69	5.50 ± 1.43	3.96 ± 0.85	7.87 ± 2.31
	4.9	0.95 ± 0.04	4.92 ± 1.48	1.36 ± 0.09	7.78 ± 1.34

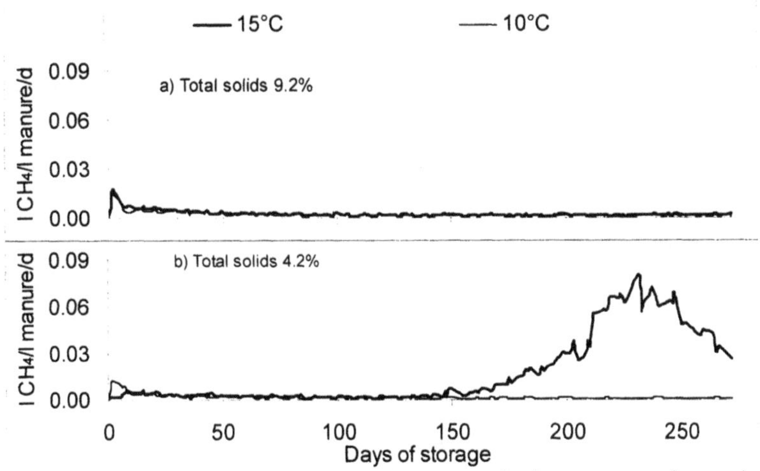

Figure 1. Methane production from dairy cow manure stored at 10°C and 15°C.

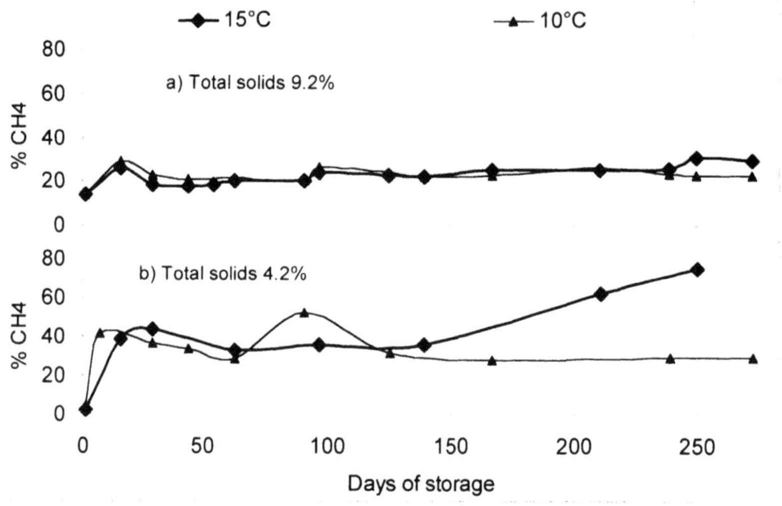

Figure 2. Methane content in biogas from dairy cow manure stored at 10°C and 15°C.

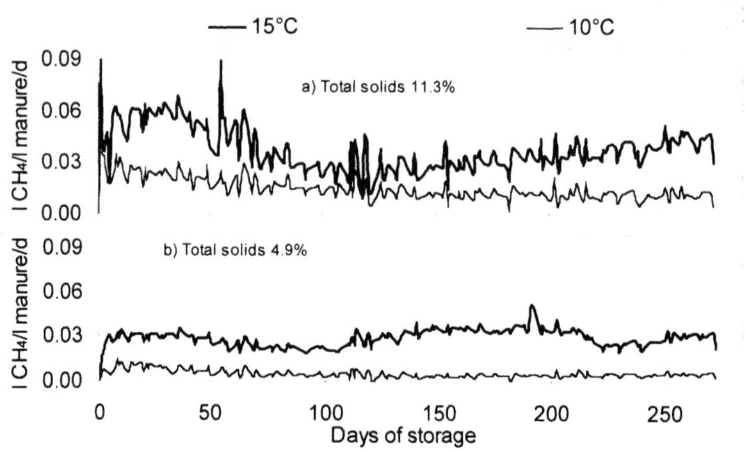

Figure 3. Methane production from swine manure stored at 10°C and 15°C.

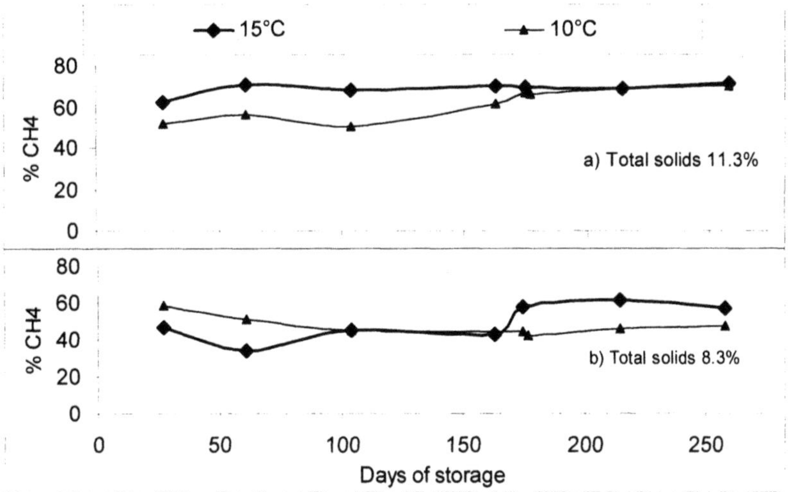

Figure 4. Methane content in biogas from swine manure stored at 10°C and 15°C.

4. CONCLUSION

The results of this study indicate that manure slurry characteristics, storage conditions and storage duration have an impact on methane emissions. It can be concluded that on typical Canadian farms, methane emissions from manure storage tanks over the late fall, winter and early spring period are negligible, because of manure temperatures substantially below 10°C. During the late spring, summer and early fall period, on the other hand, methane emissions can be substantially reduced by recommending storage periods shorter than 150 days and by using below ground storage to maintain lower manure temperatures.

REFERENCE

APHA, (1992), *Standard Methods for the Examination of Water and Wastewater*, 18th. ed. American Public Health Association, Washington, D.C.

Greenhouse Gases and Animal Agriculture
J. Takahashi & B. A. Young (Editors)

Ammonia assimilating microorganisms in animal manure treatment processes

Y. Nakai, H. Sasaki, J. Nonaka, T. Sasaki & M. Ohta

Laboratory of Animal Health and Management, Graduate School of Agricultural Science, Tohoku University, Aoba-ku, Sendai-shi, Miyagi, 981-8555 Japan

Ammonia is one of the most abundant malodorous substances in the animal manure and its treatment processes. We investigated microorganisms which assimilated ammonia in composting processes and lagoon treatment processes. Ammonia assimilating microorganisms were detected by nitrogen-limited medium which contained ammonia as a sole nitrogen source. Numbers of ammonia assimilating aerobes (Log CFU/ g) were 4.7 in the pre-fermentation stage of composting, 4.6 in the end of the primary fermentation and 5.1 in the end of the secondary fermentation on the culture plate incubated at 37°C. These numbers were 3.7, 3.2 and 3.2 in the respective stages on the plate at 55°C. From the lagoon, 3.4, 4.8, 5.0, 4.8 and 5.0 (Log CFU/ ml) of ammonia assimilating aerobes were detected on the culture plate incubated at 4, 10, 15, 20 and 25°C, respectively. Many isolates utilized ammonia at high rates when they were grown as pure cultures in the nitrogen-limited medium added with sterilized compost extract or lagoon extract. Many of them utilized ammonia even when they were cultivated in the media containing viable microbial flora of the compost or the lagoon. Analysis of DNA sequencing of 16S ribosomal DNA suggested that dominant species of microorganisms having high ammonia assimilating ability varied in the process of composting and lagooning.

Keywords: ammonia, animal waste, compost, lagoon, microorganism

1. INTRODUCTION

Ammonia is one of the most offensive odorous materials which are generated from the animal manure and its treatment processes. Control of ammonia volatilization is needed for preservation of the livestock environment and retention of nitrogen compounds in the compost or lagoon.

The major biological process of ammonia in the manure treatment is generally considered to be nitrification. However, nitrogen assimilation by microorganisms is one of important steps of ammonia removal from the wastewater (Nakai et al., 1999). Many species of microorganisms assimilate ammonia into glutamate in a stage of glutamine synthesis (Hu et al., 1999). Although ammonia assimilating microorganisms may exist in treatment systems of manure, details of their distribution and ability have been known little.

Then, we isolated the ammonia-assimilating bacteria from the composting processes and lagoons, and evaluated their ammonia-assimilating ability.

2. MATERIALS AND METHODS

Compost samples were obtained from the stage of the pre-fermentation, the end of the primary fermentation and the end of the secondary fermentation of composting of beef cattle waste. Lagoon samples were taken from the 30cm depth under the surface of basins of a lagoon for the wastewater from a paddock of dairy cattle.

Ammonia-assimilating microorganisms were counted by 10-fold dilution methods using the sterilized nitrogen-limited medium which contained ammonia as a sole nitrogen source. Inoculated media were incubated aerobically at various temperatures for 72h. After counted colonies, each single colony was isolated and pure culture was obtained by streaked twice.

Each isolate was inoculated to sterilized nitrogen-limited medium with 1% proportion added with compost extract or lagoon extract, and was aerobically incubated for 72h at the same temperature as it had been isolated. The ability of ammonia assimilation was determined by disappearance of ammonia in the culture medium. By measuring the nitrite concentration, the existence of the ammonia-oxidizing bacteria was confirmed. Isolates which showed especially high ammonia assimilation in the pure culture were cultivated in non-sterilized medium added with compost extract or lagoon extract by the same methods as described above.

Isolates which showed high ammonia assimilation in non-sterilized medium were identified by the analysis of 16S ribosomal DNA.

3. RESULTS AND DISCUSSION

The numbers of aerobes were decreased from 7.6 to 6.6 Log CFU/ g as the composting proceeded. Ammonia-assimilating microorganisms were at the range of 4.6 to 5.0 Log CFU/ g at 37°C. While at 55°C incubation the numbers were low, and they were at the range of 3.2 to 3.7 Log CFU/ g.

At 37°C incubation, in the pre-fermentation stage, we isolated 15 strains. 5 of them showed the consumption of more than 90% of ammonia in the medium. Especially, both CM-12 and CM14 showed about 99% of ammonia consumption. In the primary fermentation, we isolated 17 strains. Almost of them showed more than 50% ammonia consumption. Especially, IM-15 and IM-16 showed 96.4 and 95.9%, respectively. In the secondary fermentation, 13 strains were isolated. IIM-3 and IIM-5 showed 98.8 and 99.3% ammonia consumption, respectively.

At 55°C incubation, CT-1 showed high ammonia consumption (87.5%); however, other isolates showed relatively low ammonia consumption in the pre-fermentation stage. Among isolates from the primary fermentation, only IT-6 (99.3%) and IT-4 (71.5%) showed relatively high ammonia consumption. Isolates from the secondary fermentation showed high ammonia consumption. 80% of the isolates consumed more than 70% of ammonia in the media. Especially, IIT-2, IIT-5 and IIT-8 showed about 99%.

Abilities of ammonia assimilation of the isolates were observed in the mixed culture with microbial flora of each stage of the compost process. The isolates which showed high rates of ammonia consumption in the pure culture were inoculated into non-sterilized compost-extracted medium. CM-15 showed the highest ammonia consumption (46.0%) among the 37°C isolates. CT-1, IT-6, IIT-2, IIT-5 and IIT-8 were inoculated into non-sterilized compost-extracted medium and cultivated at 55°C for 72h. CT-1 isolated from the pre-fermentation stage showed 59.0% of ammonia consumption, and IT-6 from the end of primary fermentation showed 52.8%. However, three isolates from the end of secondary fermentation were lower ammonia consumption than the isolates from other stages.

Isolates which showed high ammonia consumption in non-sterilized compost-extracted medium were identified. CM-15 was *Zoogloea ramigera* (similarity: 100%), CM-17 showed 99% similarity to *Mycobacterium marinum*, *M. neoaurum*, *M. peregrinum* and *M. smegmatis*. Although a little arrangement of sequenced genes of IM-16 and IM-17 differed, both showed 97% similarity to *Pseudomonas* sp.. Moreover, arrangement of sequenced genes of IIM-4 and IIM-6 was completely the same, and both of them were *Streptomyces scabies* (100%). Isolates from 55°C were identified as genera of thermophile or endosporogenous. CT-1 showed 90% similarity to *Roseiflexus castenholzii*, both IT-6 and IIT-6 showed 99% similarity to *Bacillus smithii*.

From the lagoon, 3.4, 4.8, 5.0, 4.8 and 5.0 (Log CFU/ ml) of ammonia-assimilating microorganisms were detected on the culture plate incubated at 4, 10, 15, 20 and 25°C, respectively.

We isolated totally 34 strains from these culture plates. 15 of them showed the consumption of more than 95% of ammonia in sterilized medium added with lagoon extract.

Isolates which showed high rates of ammonia consumption in the pure culture were inoculated into non-sterilized medium including lagoon extract. L1002 isolated at 10°C and L2510 isolated at 25°C showed relatively high ammonia consumption, 24.4% and 25.8%, respectively.

Isolates which showed high ammonia consumption in non-sterilized medium including lagoon extract were identified. L2510 was identified as enterobacteriaceae, which showed 99% similarity to *Buttiauxella agrestis*, *B. brennerae*, *Enterobacter nimipressuralis* and *Obesumbacterium proteus*. L1002 showed 100% similarity to *Pseudomonas cedrella*, *P. fluorescens*, *P. gessardii*, *P. libaniensis* and *P. putida*.

Changes of substances and concentrations of nutrients and shifts of the temperature in the process of composting and lagooning may affect the ability of ammonia assimilation; consequently, dominant species of ammonia-assimilating microorganisms may be selected under the influence of these complex environmental conditions.

ACKNOWLEDGMENT

This study was supported in part by a grant from the Livestock Technology Association.

REFERENCES

Abeliovich, A. 1985. Nitrification of ammonia in wastewater. Field observations and laboratory studies. Water Research. 19: 1097-1099.

Hu, P., T. Leighton, G. Ishkhanova & S. Kustu. 1999. Sensing of nitrogen limitation by *Bacillus subtilis*: Comparison to enteric bacteria. Journal of Bacteriology. 181: 5042-5050.

Nakai, Y., T. Abe, C. Kohda & T. Ando. 1999. Water characteristics and microbial flora in a lagoon system for wastewater from a paddock of dairy cattle. Animal Science Journal. 70: 38-42.

Greenhouse Gases and Animal Agriculture
J. Takahashi & B. A. Young (Editors)

Traffic effects on greenhouse gas impact of organic waste injection

J. N. Tullberg

University of Queensland, Gatton, Qld4343, Australia

Soil injection of animal and municipal wastes is becoming increasingly popular in Europe, despite the high level of cost, energy input and degradation of soil structure. This degradation is a consequence of of tillage under high soil moisture conditions and field traffic by heavy slurry tankers. Godwin (1990) has demonstrated improvements to the system when field tankers are replaced by "umbilical hose" systems to supply waste to the injector.

Soil injection has major advantages in terms of nutrient recycling, odour and pathogen control, but it has important negatives in terms of greenhouse gas emissions. The high fossil fuel energy requirement itself contributes to GHG emissions, but the subsequent soil structural degradation reduces soil aeration and contributes to waterlogging, which in turn increase methane and nitrous oxide emissions. Just as importantly, this can be corrected in the short-term only by tillage under better soil moisture conditions, involving more fossil fuel, and further emissions due to soil disturbance.

In controlled traffic cropping systems soil structure is enhanced, and the energy requirements of tillage and traffic dramatically reduced by restricting the wheels of all field equipment to permanent, compacted traffic lanes. This paper provides a summary of the impact of tractor wheels on soil, and the effects of controlling field traffic. The potential advantages of precision interrow slurry injection within controlled field traffic crop production system are discussed, together with the results of a preliminary experiment.

Keywords: greenhouse gases, traffic control, organic waste

1. INTRODUCTION

Wheel traffic affects soil by applying load to the surface. When this load exceeds soil strength, soil deforms until the increase in soil strength or area is sufficient to resist the load. Load and deformation are transmitted from surface to subsurface layers. The surface loading applied by transport wheels is primarily vertical and compressive, but traction wheels also apply horizontal shear loads of up to 40% of vertical load. These loads produce vertical and horizontal deformation, examples of which have been illustrated by Horn (2000). Soil is not a simple solid easily defined in terms of elasticity etc, and soil response to load is difficult to predict, even in idealised conditions.

Engineers and agronomists (e.g. 6, 10) take a distinctly different approach to the study of soil response to management inputs. Soil response to traffic is reviewed from a number of perspectives in Soane and Van Ouwekerk (1994). Despite the complexity of the situation some general principles can be distilled from the literature. While there might be important exceptions, the author has found the ideas listed below useful in understanding the impact of traffic on soil.

- Most of the damage caused by any given traffic event occurs on the first pass.
- Tyre pressure determines mean contact stress and the potential for surface damage.
- Axle load largely determines the potential for subsoil degradation.
- Soil damage occurs largely due to plastic deformation, in both compression and shear.
- Reduced porosity and pore connectivity are the major symptoms of damage.
- Greater load and deformation result in damage to progressively smaller pores.
- Soil strength generally decreases with moisture content, and increases with depth.
- Field operations often occur when surface soil is trafficable, but subsoil is still wetter.

Energy dissipation in the soil by a tractor/implement system is simple to estimate, because power is energy per unit time. Tillage implement energy input to the soil can be calculated from power input to the implement, and the area over which it is dissipated. Tractor traffic energy input to the soil can be found given estimates of power output, tractive efficiency, and area wheeled/unit time.
With modern farm equipment, traffic energy input values are typically around 10kJ/m^2. Tillage energy inputs to soil disturbance, on the other hand, are typically in the range 3 -- 6kJ/m^2 for broadacre planting and tillage. Only in deep (subsoiling) operations does tillage energy input to the soil approach traffic energy input, and Tullberg (2000) has shown that the draft of tillage tines following tractor and implement wheels can be approximately twice that of other tines.

People are often surprised to find that tractor and implement wheels normally dissipate more energy per unit area of soil, over a greater depth of the soil profile, than most tillage or planting implements. This energy input occurs regardless of whether or not its damaging effects are totally repaired, partially rectified, or simply disguised by subsequent operations. Beneath the disturbed layer soil structural damage is often visible, and only in this sub-tillage layer is "soil compaction" commonly recognised. As depth of damage increases, mechanical rectification becomes more expensive, and natural amelioration processes occur more slowly.

Approximately half the total power output of a tractor can be dissipated in the process of creating and undoing the traffic effects of its own wheels. When traffic is controlled, tractor size and fuel requirements are typically reduced by 50%, regardless of other changes. A long-term investigation of traffic and tillage effects (Tullberg 2001) also demonstrated major effects on soil permeability, plant available water, soil organic matter and soil health. Infiltration and soil health effects are illustrated in Figure 1 and 2.

Controlled traffic reduces or eliminates the formation of large aggregates or clods, and hence reduces the timeliness constraints and the surface disturbance that occurs during some soil-engaging operations. The impact on planting, for instance, is illustrated in Figure 3 in terms of the proportion of aggregates passing through a 12 mm sieve. Controlled traffic also requires greater precision in field operations, to ensure that all heavy wheels remain on the permanent traffic lanes and the large-scale adoption of controlled traffic systems in Australia

(Yule, 1998) is largely responsible for the rapid development of high-precision DGPS tractor guidance systems.

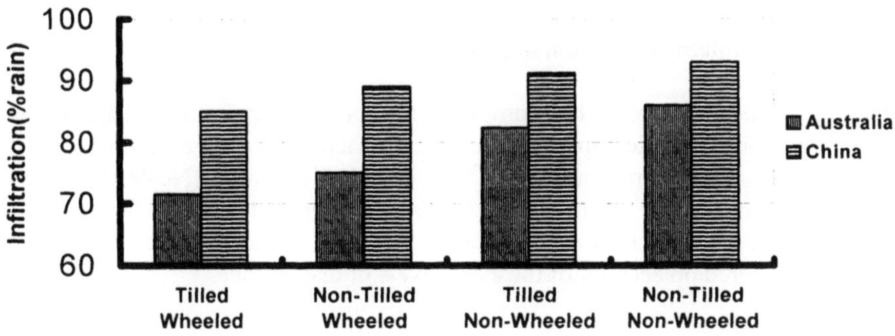

Fig 1. Cumulative effect of wheels and tillage in Australia and China (from Li et al. 2001.)

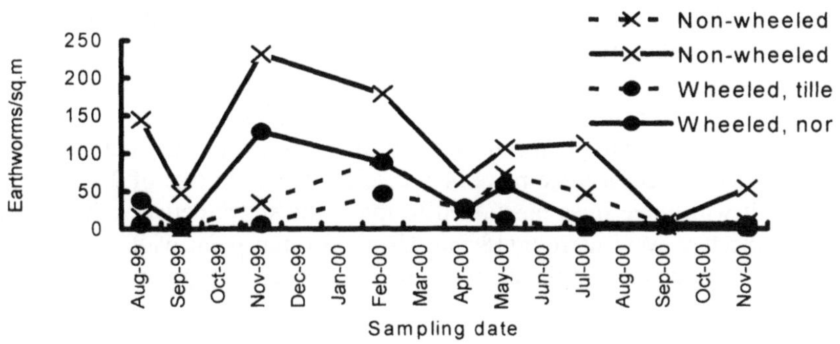

Fig 2. Traffic and tillage effects on earthworm population (U. Pangnakorn pers. com.)

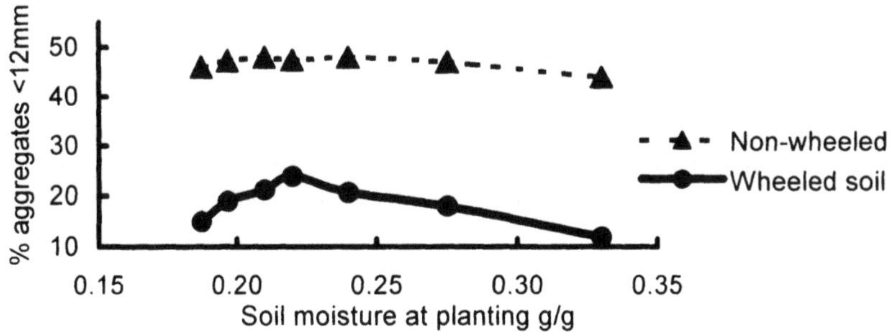

Fig 3. Wheeling effects on aggregate size in planting zone (T. Tapaevalu pers. com.).

In terms of potential for soil injection of nutrient-rich waste, precision controlled traffic provides:

- Improved wet-weather traffic access on permanent traffic lanes.
- Better soil aeration, to reduce methane and nitrous oxide emissions.
- Reduced soil disturbance and energy requirements of injection.
- The opportunity for interrow injection in perennial row crops.

This technology appears particularly appropriate for forage crops such as lucerne, where soil compaction by random traffic appears to be a major factor reducing crop productive life, and earlier work demonstrated improved yield under controlled traffic (Neale and Tullberg, 1998). The experimental work summarised here was designed to assess the use of soil injection in precision controlled traffic production of lucerne, a situation for which it would not normally be considered on account of the excessive soil disturbance.

2. METHODOLOGY

Lucerne was planted in August 1999 on an area of self-mulching heavy clay soil in the UQG Field Crop Teaching Research Unit. An older, conventionally trafficed lucerne plot was available nearby as a control treatment.

All injection work was carried out with a 70 kw two wheel drive tractor modified to work on a 3 m controlled traffic system (ie 2.5 m of lucerne crop plus 0.5 m compacted wheel track). The same tractor was used to plant controlled traffic plots and to carry out the injection operation. Lucerne was planted in three rows (4 x 0.75 m) at a seed rate of 2.5 kg/ha. An injector unit was built with three tines positioned to inject the slurry between these lucerne rows. The tines themselves were narrow (18 mm shank, 35 mm point width) designed for zero tillage planting. They were equipped with 20 mm injection pipes delivering slurry immediately behind the point.

For the simple injector, the slurry was delivered by gravity from a tank mounted above the injector unit. Rather than dilute animal manure, potassium rich fermenter waste was used as the injection material, and injection rate varied by changing forward speed. Four comparisons were included in this experiment :

Soil condition –	controlled traffic and conventional random traffic.
Injection depth –	100 mm and 150 mm.
Injection rate –	1.5 l/m^2 and $3l/m^2$.
Row spacing –	250 mm and 350 mm.

Three replicates of each were set out in a completely randomised block design.

The injecting tractor was towed by another tractor to evaluate the energy requirement of injection. Soil disturbance was assessed using a field profile meter, and cone index used to assess soil compaction differences between 100 mm and 150 mm injection. Gravimetric soil moisture was taken at five 10 cm depth increments down the profile.

3. RESULTS AND DISCUSSION

3.1. General

The injection unit appeared to work satisfactorily, providing relatively easy adjustment of tine spacing and depth, using conventional depth wheels. In all cases, the injected material was not visible from the surface after injection, however this might be a factor of the low soil moisture content (Fig. 2) at the time of these measurements (the maximum value, 18% gravimetric is close to wilting point for this soil).

3.2. Soil disturbance

The soil disturbance was recorded photographically, and assessed with the profile meter. While there was a clear impression that soil profile was smoother when injecting within the controlled traffic system, it has not been possible to back this up with statistical evidence. Cone penetrometer measurements also demonstrated that deeper injection substantially loosened soil only in the immediate injection area. These results are not very meaningful considered on their own, but looked at together with the surface profile measurements and visual evidence, they confirm the practicability of subsurface injection of wastes with minimal soil disturbance.

3.3. Energy

Energy results demonstrated apparent energy reductions between 17 and 37% in the controlled traffic situation. This energy effect was much smaller than expected, probably because of the draft measurement methodology. Draft measurements in controlled traffic plots can often be increased when implement depth control wheels run in the permanent tractor wheel tracks so operating depth is increased compared with other treatments.

To a

Other comparisons (see Table 1, Tullberg 2000) indicate that the energy requirement of injecting

Table 1. Mean traffic penalty with varying tillage units, locations and wheel treatments

Tillage Unit and Country	Tillage Depth mm	Wheel Weight Range 5-10kN	20-25kN
		Traffic Penalty	
Australia –sweep tillage	75	1.69	1.92
Australia –sweep tillage	100	1.82	1.96
Australia –chisel tillage	155	2.54	2.54
China - planter points	75	1.85	1.85
China - planter points	100	2.20	2.80

*traffic penalty is the ratio of tillage draft in wheeled soil to tillage draft in non-wheeled soil.

into controlled traffic lucerne would around 50% of that in conventional practice. Other factors being equal, this would translate directly into reduction in fuel use and greenhouse gas emission resulting from the energy input into waste injection.

3.4. Discussion

In the Australian context, National Greenhouse Gas Inventory(1) data can be used to estimate that mean annual tillage energy use contributes approximately 1200Gg CO_2 equivalent per annum of greenhouse gas (GHG) emissions. This is only approximately 20% of the GHG emissions resulting from soil disturbance, waterlogging and nitrogenous fertilizer use. An unknown, but significant proportion of these should however be attributed to the soil disturbance and structural degradation caused by tillage and traffic.

It is probably reasonable to assume that emissions from soil disturbance are directly related to energy input to that process. In this case the major reduction in energy input achieved by precision interrow controlled traffic injection of organic waste will provide a substantially greater reduction in the overall greenhouse gas impact of animal manure recycling. Further investigation of the process is currently being undertaken.

4. CONCLUSION

This brief experimental program confirmed the feasibility of interrow injection of organic wastes and the advantages of precision controlled traffic systems for such operations. Systems of this type are clearly capable of accepting organic wastes even under moderately wet weather conditions, without major negative effects on crop productivity, without the creation of odour and pathogen problems, and at small direct energy cost.

ACKNOWLEDGEMENTS

Much of the work reported here was supported by the Australian Centre for International Agricultural Research under projects 9209 and 96143 -- "Sustainable Mechanised Dryland Grain Production" -- with China Agricultural University.

REFERENCES

Australian Greenhouse Office. 2001. National greenhouse gas inventory, 1999.

Godwin, R.J.,Warner, N.L. and Hann, M.J. 1990. A Comparison between Tanker-Mounted and Umbilical Systems for the Injection of Animal Slurry. Agric.Eng, 45: 2, 45-50

Horn, R. 2000. Proceedings, Fourth International Conference on Soil Dynamics, Adelaide.

Li, Y., Tullberg, J.N. and Freebairn, D.M. 2000. Traffic and Residue Cover Effects on Infiltration. Australian Journal of Soil Research 39. 2392-47.

Neale,T.J. and Tullberg, J.N. 1988. Controlled Traffic in Lucerne Hay Production. Proc. Second Controlled Traffic Conference, University of Queensland Gatton, p 169.

Passioura, J.B. 2000. Proceedings, Fourth International Conference on Soil Dynamics, Adelaide.

Soane, B.D. and van Ouwerkerk, C. (Eds), 1994. Soil Compaction in Crop Production. Elsevier Science B.V. Amsterdam.

Tullberg, J.N. 2000. J. Agricultural Engineering Research. 75(4) 375-382. Elsevier, Amsterdam.

Tullberg, J.N., Ziebarth, P.J. and Li Yuxia. 2001. Tillage and Traffic Effects on Runoff. Aust. J. of Soil Research. 39:249-257

Way, T.R., Erbach, D.C., Bailey, A.C., Burt, E.C. and Johnson, C.E. 2000. Proceedings, Fourth International Conference on Soil Dynamics, Adelaide, S.A.

Yule, D.F., 1998. Proceedings Second National Controlled Traffic Conference, University of Queensland Gatton, Qld. pp 6-12.

Greenhouse Gases and Animal Agriculture
J. Takahashi & B. A. Young (Editors)

Factors of soil-machine resistance and soil compaction at the operation of animal effluent application

T. Kishimoto[1], T. R. Way [2], H. Jun[3] & K. Nishizaki[1]

[1] Dept. of Agro-Environmental Science, Obihiro University of Agriculture and Veterinary Medicine, Inada-cho, Hokkaido, 080-8555, Japan, [2] National Soil Dynamics Laboratory, USDA, ARS, Auburn, Alabama 36832, USA, [3] Department of Bio-Mechatronic Engineering, Sung-Kyun-Kwan University, Suwon, 440-746 Korea

Some slurry handling and spreading systems have been introduced to ensure that the intended amount of nutrients in slurry is applied to the field. Factors affecting soil-machine resistance and soil compaction and their impacts on reducing energy requirements of slurry spreading and injecting implements are discussed. Soil stresses were determined at a 250 mm depth beneath the centerline of tires for estimating soil compaction. Soil-machine resistance and soil compaction are thought to be relatively great for deep injectors on upland fields. A shallow injector or a trailing shoe spreader produces little resistance to forward motion and generates little soil compaction on grassland. The band spreader generally causes soil compaction but has no soil-machine resistance other than the motion resistance of the tanker.

Keywords: spreader, injector, motion resistance, tillage resistance, soil compaction

1. INTRODUCTION

Several systems for animal effluent application have been introduced in European countries because there are some regulations to protect the environment from contamination. Some of these systems, the use of a slurry tankers and spreaders or injectors, have recently been introduced into Hokkaido, Japan. However, they are still in the trial use stage so suitable slurry handling and spreading systems for local use and their performance have not been obtained.

In order to design or select a system suitable for Hokkaido, the factors of soil-machine resistance by the slurry tanker or the spreader/injector must be quantified to reduce the energy requirement. Implement energy requirement is also very important and related to the choice of an adequate size of a tractor for a selected system. Animal effluent application systems are operated both on upland fields and grassland. The discussion that which parts of the machine are mainly producing soil-machine resistance is conducted under these field conditions.

Soil compaction affects the soil physical environment for crop production in agriculture (Gill and Vanden Berg, 1967). Excessive soil compaction reduces infiltration of water into soil thereby increasing the potential for runoff and erosion. Excessive soil compaction also prevents crop roots from extending into soil for water and nutrients. Soil compaction research has been conducted mainly for agricultural tractor drive tires. Dynamic load and inflation pressure have been shown to affect soil stresses and rut depths (Bailey et al., 1996). Way et al. (1995) measured soil stresses beneath a tire in soft soil (sandy loam and clay loam) after rotary tillage. In the section on the soil compaction in this study, the soil stresses in structured soil are measured in conditions similar to grassland, and are compared with soil stresses in soft soil similar to an upland field.

2. SLURRY HANDLING AND SPREADING SYSTEMS

It is important that adequate machinery is selected to give accurate slurry application. Chambers et al. (2001) recommended selecting the right handling and spreading system to ensure that the intended amount of nutrients from slurry have been applied to the crop. The following are the four main types of slurry transport and four main types of slurry distribution systems (Chambers et al. 2001).

2.1. Slurry transport systems

2.1.1. Vacuum tanker

Slurry is sucked into the tanker by using an air pump to evacuate the air from the tank to create a vacuum. The tanker is emptied using the air pump to pressurize the tanker. The slurry is discharged by these operations. This system can be used for most slurry transporting.

2.1.2 Pumped tanker

Slurry is pumped into and from the tanker using a slurry pump, either a centrifugal or positive displacement pump. This generally provides better spreading precision than a vacuum tanker. More maintenance is required for a positive displacement pump.

2.1.3. Umbilical hose

Slurry is fed by a drag hose to a tractor equipped with a distribution system. The hose is supplied with slurry, usually directly from the slurry store, by a centrifugal or positive displacement pump. This method has the potential for crop damage due to the hose dragging across the ground. Hose damage and wear can be a problem on abrasive or flinty ground.

2.1.4. Irrigator

This system has a self-traveling machine with flexible or reeled-in hoses usually fed from a network of underground pipes. A centrifugal or positive displacement pump which is situated near the slurry store is used for supplying the slurry. This system is suitable for semiautomatic operation, but anti-pollution safeguards are required.

2.2. Slurry distribution systems

2.2.1. Broadcast spreader (splash plate or nozzles)

Figure 1 shows a typical broadcast spreader. The slurry is forced under pressure through a nozzle, often onto an inclined plate to increase the sideways spread.

2.2.2. Band spreader

Figure 2 shows a typical band spreader. The spreader boom has a number of hoses connected to it, distributing the slurry close to ground in strips or bands. It is fed with slurry from single pipe, thus relying on the pressure at each of the hose outlets to provide even distribution. Advanced systems use rotary distributors to proportion the slurry evenly to each outlet.

2.2.3. Trailing shoe spreader

Figure 3 shows a typical trailing shoe spreader. This has a configuration similar to the band spreader with a shoe added to each hose allowing the slurry to be deposited under the crop canopy onto the soil. It is also possible to spread in growing crops without contaminating the crops.

2.2.4. Injector

Slurry is injected under the soil surface. There are various types of injectors but each fits into one of two categories. One of them is open slot shallow injection for up to 50 mm depth. The other is deep injection for depths greater than 150 mm. Figures 4 and 5 show the shallow and deep injectors respectively.

Figure 1. Typical broadcast spreader (splash plate).

Figure 2. Typical band spreader.

Figure 3. Typical trailing shoe spreader.

Figures 4. Typical shallow injector.

Figures 5 Typical deep injector

3. SOIL-MACHINE RESISTANCE AND SOIL COMPACTION BY SLURRY DISTRIBUTION SYSTEMS

3.1. Factors of soil-machine resistance

In this section, factors of producing resistance between soil and machinery for slurry distribution systems are discussed.

3.1.1. Broadcast and band spreaders

Figure 6 shows a schematic diagram of external forces acting on the slurry distribution system with a band spreader and a tanker. As there is no soil-machine contact at a spreader, the system with a broadcast spreader and a tanker is a similar condition. Soil-machine resistance of these systems is mainly the motion resistance of the tanker tires as the spreader components do not engage the soil. Broadcast spreaders are usually operated on grassland. The motion resistance ratio on the grassland is shown as 0.10 (JSAM, 1996) so the operation of broadcast spreaders on grassland is not thought to be difficult. When the dynamic load of the 12000 L tanker filled with liquid is 150 kN, the motion resistance is estimated at 15 kN. Band spreaders are usually used on upland fields after crop harvesting. In this case, the motion resistance of the tanker increases to more than twice the resistance on grassland as the motion resistance ratio on upland fields is 0.20 to 0.30 (JSAM, 1996).

The motion resistance of the tanker tires depends on the dynamic load of the tires and tire size. A suitable choice of the tanker tire could reduce the motion resistance.

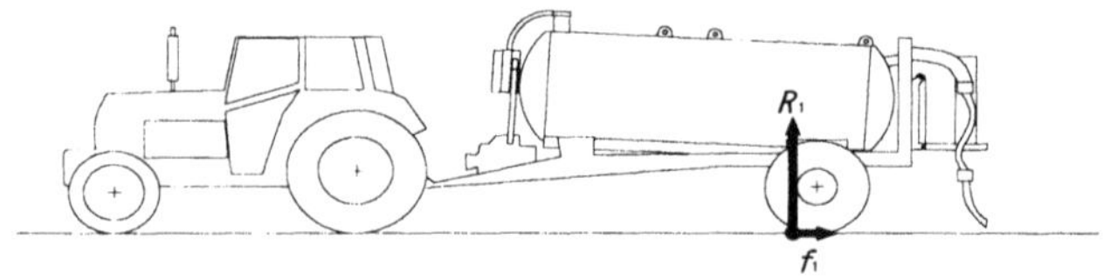

Figure 6. Schematic diagram of external forces acting

3.1.2. Trailing shoe spreader

Figure 7 shows the schematic diagram of external forces acting on the system with a trailing shoe spreader and a tanker. The shoes of the trailing spreader come in contact with the soil surface. The resistance produced by the friction between the shoes and soil or grass is greater, compared to the band spreader system.

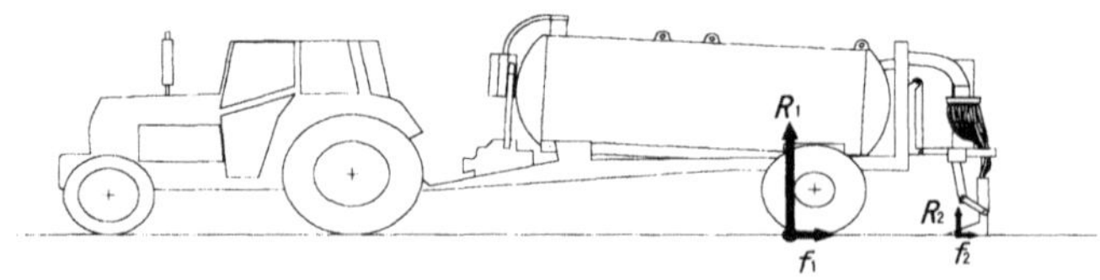

Figure 7. The schematic diagram of external forces acting on a trailing shoe spreader and a tanker.

The coefficient of the friction between the steel and the soil is in the range of 0.3 to 0.6 depending on the soil type. As the trailing shoe spreaders are usually operated on grassland, the coefficient between the steel and the grass is thought to be smaller

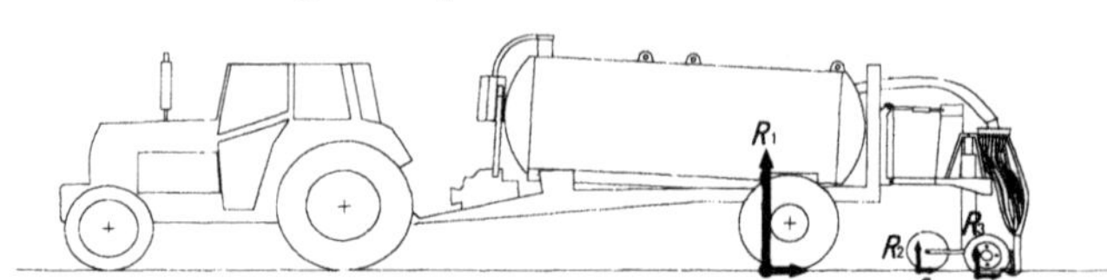

Figures 8-a. The schematic diagram of external forces acting on a shallow injector and a tanker.

than that of steel on soil. The resistance due to the friction has not been measured in Japan. It is important to obtain the coefficient of friction to achieve a suitable shoe design.

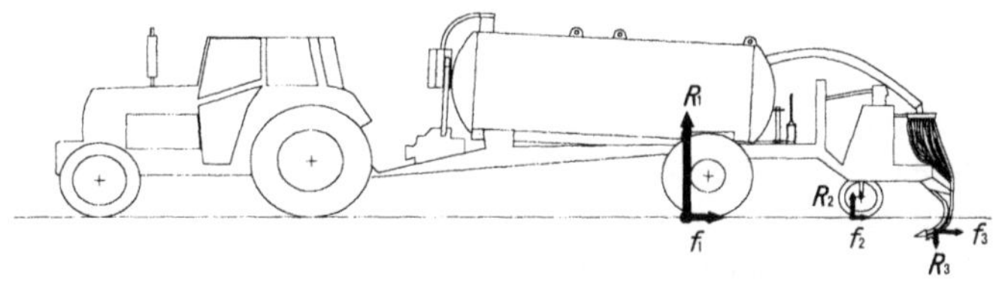

Figure 8-b The schematic diagram of external forces acting on a deep injector and a tanker.

3.1.3. Injector

Figures 8-a and 8-b show the schematic diagrams of external forces acting on a shallow injector and tanker, and a deep injector and tanker, respectively. The trailing guides and slot cut disks of a shallow injector come in contact with grass as the injector is usually operated on the grassland. The resistance produced by the friction between these parts and grass is greater, compared to the band spreader system. The injector is sometimes equipped with gauge wheels instead of trailing guides which can reduce the resistance of trailing guides.

A deep injector has several chisels to inject slurry under the soil for upland fields. This type of injector is not usually used for grassland because it disturbs the field surface. As the chisel point goes beneath the soil surface, the resistance becomes greater than the resistance produced by the trailing shoe or the shallow injectors. Therefore, the width of the injector is usually narrower than the trailing shoe or the shallow injectors. Gauge wheels also produce motion resistance, but the magnitude varies depending on the tire and the soil conditions.

3.2. Soil compaction

The tanker of a slurry injector or a spreader is quite heavy when the tank is filled-up before field operation. A trailer type slurry tanker of 12000 L capacity has been introduced on the market, even in Japan. One of these types of tankers is trailed at the drawbar of the tractor and supported by tandem axles. In this case, the static loads of the front and rear axles of the tanker and on the drawbar are about 65, 72 and 11 kN (6.7, 7.4 and 1.1 t) respectively (Nishizaki, 2000). If the tanker has one axle, the static load on the axle becomes 137 kN. Soil compaction by slurry tanker tires may be significant when heavy slurry tankers are used in fields.

In this section, the difference in soil stresses beneath a tire on an upland field and on grassland is discussed. Way et al. (1995) measured soil stresses beneath a tire in soft soil. As the tire size used by Way et al. (1995) is the same as in the experiments in this study, the authors conducted experiments on structured soil and compared the stresses between soft and structured soils.

3.2.1. Procedures

The experiment was conducted at National Soil Dynamics Laboratory (NSDL), a facility of the USDA Agricultural Research Service in Auburn, Alabama, using the NSDL single wheel Traction Research Vehicle (TRV) operating in a soil bin (Burt et al., 1980). An Armstrong 18.4R38 radial-ply tractor drive tire was used in structured soil. The experiment was designed with three combinations of dynamic load and inflation pressure (13.2 kN-41 kPa, 19.8 kN- 83 kPa and 25.3 kN-124 kPa). Each combination uses the correct inflation pressure for the corresponding dynamic load on the 18.4R38 tire. Five replications were used in the experiment.

The soil used in this study was the Hiwassee Clay soil in an outdoor soil bin. The composition of the soil was 59.6 % clay, 17.2 % silt, and 23.2 % sand. The soil has not been tilled for several years so the soil had developed structure. The mean cone index for a 0-150 mm depth range was 850 kPa. This condition is thought to be similar to grassland.

Six directional soil pressures were measured by a stress state transducer (SST) as shown in Fig. 9. Nichols et al. (1987) described the details of the SST. Soil stresses were determined at a 25 cm depth in the soil beneath a lug at the centerline of the tire. Soil was augured at 45 degree from the opposite side of the approaching wheel. The bottom part of the augured hole was cut using a specially shaped knife to attain good contact of the pressure sensors with the soil. Figure 10 shows the location of tire tread relative to the SST.

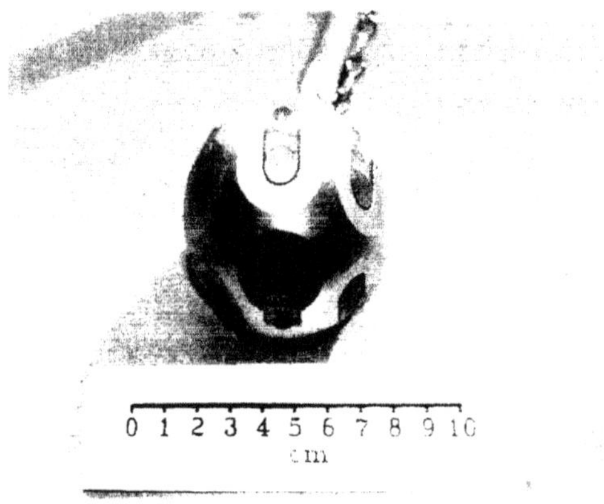

Figure 9. Stress state transducer (SST)

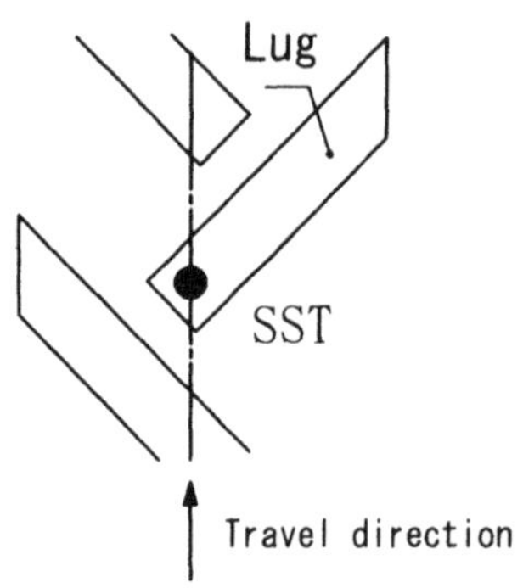

Figure 10. Location of SST.

4. RESULTS

Table 1 shows the means of peak maximum values of σ_{oct} and mean corresponding τ_{oct} at-the hardpan (252 mm sandy loam, 288 mm clay loam) by Way et al. Table 2 shows the means of peak maximum values of σ_{oct} and mean corresponding τ_{oct} at the hardpan in the structured soil.

The mean of the peak maximum values of σ_{oct} for the 25.3 kN-124 kPa treatment in structured soil is similar to the stresses in both soft soils. These results indicate that the soil stresses increase as the dynamic load increases when the inflation pressure is set at the correct inflation pressure corresponding to the dynamic load. When the dynamic load decreased, the stresses in structured soil decreased significantly. These results indicate the possibility that significant soil compaction does not occur in grassland if the dynamic load applied to the tanker axle does not exceed particular values.

Table 1. Means of maximum σ_{oct} and mean corresponding τ_{oct} at hardpan (Way et al.)

Norfork sandy loam		Decatur clay loam	
σ_{oct} (kPa)	τ_{oct} (kPa)	σ_{oct} (kPa)	τ_{oct} (kPa)
113	157	93	164

110 kPa inflation pressure, 25.0 kN dynamic load

Table 2. Means of maximum σ_{oct} and mean corresponding τ_{oct} at hardpan (Hiwassee clay)

Combination of dynamic load and inflation pressure	σ_{oct} (kPa)	τ_{oct} (kPa)
13.2 kN-41 kPa	14.1	48.3
19.8 kN-83 kPa	38.4	77.2
25.3 kN-124 kPa	93.4	186.5

The soil stresses under a tractor tire increase as the dynamic load increases at constant inflation pressure, and also increase as the inflation pressure increases at constant dynamic load [Bailey et al. (1996) and Way et al. (1995)]. Soil stresses caused by slurry tanker tires are estimated to be twice or more than those of tractor drive tires. This is because the static load of the tanker tire is more than twice that of the tractor tire and the tanker tires, for example 12000 L capacity tanker, are sometimes overinflated for easy operation on the road. Appropriate selection of tire size and adjustment of inflation pressure are important factors in reducing soil compaction and motion resistance of tires.

5. CONCLUSION

Factors affecting soil-machine resistance and their impacts on reducing energy requirements of slurry spreading and injecting implements were discussed. Soil stresses in the structured soil were determined in clay and compared with stresses in soft soil. The following conclusions were drawn from the experiments

1. The band spreader generally causes soil compaction but has no soil-machine resistance other than the motion resistance of the tanker.
2. A shallow injector or a trailing shoe spreader produces little resistance to forward motion and generates little soil compaction on grassland.
3. Soil-machine resistance and soil compaction are thought to be relatively great for deep injectors on upland fields.
4. The mean of peak maximum values of σ_{oct} in structured soil is similar to stresses in a soft sandy loam and a soft clay loam soil, but the soil stresses decreased significantly as dynamic load decreased.
5. Motion resistance and soil compaction of tanker tires are thought to be large. Proper selection of tire size and adjustment of tire inflation pressure may reduce these factors.

REFERENCES

Bailey, A. C., R. L. Raper, T. R. Way, E. C. Burt & C. E. Johnson, 1996. Soil stresses under a tractor tire at various loads and inflation pressures. J. Terramech. 33(1): 1-11

Burt, E.C., C.A. Reaves, A.C. Bailey & W. D. Pickering. 1980. A machine for testing tractor tires in soil bins. Transactions of the ASAE 23(3): 546-547,552.

Chambers, B., N. Nicholson, K. Smith, B. Pain, T. Cumby & I. Scotford, 2001. Managing Livestock Manures Booklet 3 Spreading systems for slurries and solid manures. Ministry of Agriculture, Fisheries and Food. London, UK

Gill, W. R. & G. E. Vanden Berg, 1967. Soil dynamics in tillage and traction. Agriculture Handbook No. 316. USDA. Washington, DC., USA.

Japanese Society of Agricultural Machinery, 1996. Bio-production Machinery Handbook, Omiya, Japan. 332, 408

Nichols, T. A., A. C. Bailey, C. E. Johnson & R. D. Grisso, 1987. A stress state transducer for soil. Trans. ASAE 30(5), 1237-1241

Nishizaki, K., 2000. Safety for trailed vehicles and the necessity of trailer brake. J. JSAM. 62(1): 13-17

Way, T. R., A. C. Bailey, R. L. Raper & E. C. Burt, 1995. Tire lug height effect on soil stresses and bulk density. Transactions of the ASAE 38(3): 669-674

Greenhouse Gases and Animal Agriculture
J. Takahashi & B. A. Young (Editors)

Fertilizer value of anaerobically co-digested dairy manure and food processing wastes

K. Umetsu, R. Kondo, M. Tani & T. Hayashi

Dep. of Agro-environmental science, Obihiro University of Agriculture and Veterinary Medicine, Inada-cho, Obihiro, Hokkaido, 080-8555 Japan

Anaerobic co-digestion of dairy manure and food processing wastes under thermophilic(55℃) and mesophilic(35℃) temperatures were investigated using laboratory scale digesters. Two types of influent feedstocks were utilized: 100% dairy manure (A) and a 7:3 mixture (wet weight basis) of dairy manure and food processing wastes (B). The content of the food processing wastes, as the feedstock was a 3:3:3:1 mixture of cheese whey, animal blood, used deep-fat frying oil and fried potato residue. The four continuous digestions are abbreviated as follows: 35A, 55A, 35B and 55B. Co-digestion of dairy manure and food processing wastes under thermophilic (55℃) in the 55B digester increased the methane production per unit digester volume. However, co-digestion in the 35B digester was inhibited. While total Kjeldahl N of these wastes was 73.1 to 91.9% recovered after digestion, NH_4^+-N was recovered in excess of 100% (169.6 to 171.0% for A and 275.4 to 278.4% for B digesters) due to the mineralization of influent organic N. The food processing wastes mixture had greater recoveries of NH_4^+-N during digestion compared to the dairy manure only, reflecting its greater organic N degradability. The extractable/total ratios of influent K, P, and Mg were slightly reduced after digestion, possibly due to the incorporation of these components into the solid fraction. Centrifugation distributed NH_4^+-N and K between supernatant and sludge fractions in proportions that paralleled the effluent total mass distribution between those fractions and distributed P, Ca and Mg in proportions that paralleled effluent TS distribution. Since supernatants accounted for most of the mass, the majority of the effluent NH_4^+-N (53.4 to 58.7%) and total K (66.0 to 73.1%) was present in the supernatants. Since the sludges accounted for most of the TS, the majority of effluent total P (72.4 to 87.1%), total Ca (63.8 to 75.2%) and total Mg (78.7 to 79.6%) was present in the sludges. Thus, the co-digested effluent could be a readily available fertilizer for N and K, while be a slow release fertilizer for P, Ca, and Mg.

Keywords: organic fertilizer, dairy manure, co-digestion, biogas, methane, anaerobic fermentation

1. INTRODUCTION

Many modern large-scale manure based biogas plants have been constructed recently in Japan (Kimura, Y. and Umetsu, K., 2000). The benefits include renewable energy production, reduced greenhouse gas emissions, efficient organic waste recycling, and improved utilization of the manure as fertilizer. In Denmark, all biogas plants take in different types of organic waste, primarily from the food processing industry in order to increase the biogas production (Danish Institute of Agricultural and Fisheries Economics, 1999). Co-digestion offers the opportunity to deal with such types of waste, and ensures the recycling of nutrients back to the agricultural sector. Some types of waste containing high concentrations of oil/grease or proteins have a high potential to produce methane, but must be treated with care and fed to the process at a controlled rate. Moreover, the addition of such types of waste to dairy manure increases the fertilizer value of the effluent. A unique attribute of the production of methane from agricultural wastes by biogas plants is the high recovery of nutrients important to plant growth. Total Kjeldahl N (TKN) is usually partially lost by conventional lagoon treatment systems. However, manure N is conserved and NH_4^+-N composition of manure is increased during the anaerobic digestion process (Field *et al.*, 1984 and 1985; Converse *et al.*, 1977; Hashimoto, 1983; Umetsu *et al.*, 2000).

The objectives of this study were to determine recovery, changes in extractability, and distribution between effluent supernatant and sludge fraction of N, P, K, Ca, and Mg following thermophilic and mesophilic digestion of dairy manure and a mixture of dairy manure and food processing wastes.

2. MATERIALS AND METHODS

2.1. Materials

Four 10 L continuous digesters were constructed from stainless steel vessels. The digesters were placed in temperature-controlled water baths. A schematic of the experimental digester was shown in a previous paper (Umetsu *et al.*, 1998). Two types of influent feedstocks were utilized: 100% dairy manure (A) and a 7:3 mixture (wet weight basis) of dairy manure and food processing wastes (B). The manure (feces only) from ration-fed dairy cows was collected from concrete floors at the Obihiro University Farm. The dairy cows were fed a diet of hay, plus supplement feeds including dent corn silage and beet pulp. No antibiotics were incorporated into the feed. The feedstock of the food processing wastes was a 3:3:3:1 mixture of cheese whey, animal blood, used deep-fat frying oil and fried potato waste. Characteristics of experimental materials as feedstocks are summarized in Table 1.

Table 1. Characteristics of experimental materials as feedstocks

		Dairy manure	Dairy manure + Food processing wastes
TS	(%)	8.1	17.9
VS	(%)	6.1	16.5
TS /VS	(%)	75.3	92.2
pH		6.8	6.4
TVA	(mg/L)	1189.08	1600.69
FA	(mg/L)	18.18	0
AA	(mg/L)	885.85	1247.46
PA	(mg/L)	185.06	235.33
BA	(mg/L)	99.99	8.58

2.2. Operation

Each feedstock was operated with two temperature ranges (35℃ and 55℃). The four continuous digestions are abbreviated as follows: 35A, 55A, 35B and 55B. The operations of the methane digesters are summarized in Table 2. Influent samples were collected from the onset of the experiment. Effluent collection began 20 days following the onset of the experiment after achievement of steady gas yield.

Table 2. Operation of methane digester

Digester	Temp (℃)	Material type	Loading (L)	Loading rate (g·VS/L/day)	HRT (day)
35A	35	DM	0.5	3.05	20
55A	55	DM	0.5	3.05	20
35B	35	DM + FPW	0.5	8.25	20
55B	55	DM + FPW	0.5	8.25	20

HRT: Hydraulic retention time

2.3. Analytical methods

The daily volume of gas produced was measured by wet gas meters. All gas measurements were expressed at 0°C and a pressure of one atmosphere. The composition of the produced gas was analyzed every day by a Shimadzu gas chromatograph (GC-4C) with a thermal conductivity detector. A stainless steel column (3m × 2m) packed with 80/100 mesh silica gel was used with Helium as the carrier gas at a flow rate of 28ml/min. The injector and detector temperatures were 80°C and 120°C, respectively. The column temperature was 80°C. Total volatile acids (TVA) were determined by a Shimadzu HPLC (LC-10A), using a Shim-pack SGR-102H. The method of analyses for TVA was described in detail in a previous paper (Kimura *et al.*, 1994).

The sample suspensions were centrifuged at 3000 g for 15 min, and the supernatant fraction and the sludge fraction were separated. The original suspension and the supernatant fraction were analyzed as below:

2.3.1. Weight measurement of each fraction

The weights of the original sample suspensions and supernatant fractions obtained by centrifugation were measured. The weight of the supernatant fraction and the sludge fraction was calculated as below:

$$\text{Msl (g)} = \text{Mor (g)} - \text{Msu (g)}$$
$$\text{Fsu (\%)} = \text{Msu} / \text{Mor} \times 100$$
$$\text{Fsl (\%)} = \text{Msl} / \text{Mor} \times 100$$

Msu: weight of supernatant fraction
Msl: weight of sludge fraction
Mor: weight of original suspension
Fsu: fraction percent of supernatant
Fsl: fraction percent of sludge

2.3.2. Solid and organic matter contents

The suspension was dried at 105℃ for 24 h, and solids content was calculated from the difference of weights before and after drying. The dried matter was heated at 550℃ for 4 h, and organic matter content was calculated from the loss on ignition.

2.3.3. Total nitrogen

A 5 g sample was weighed into a Kjeldahl digestion tube, and 10 ml of conc. H_2SO_4 and accelerator (Kjeldahl catalyst tablets including K_2SO_4 and $CuSO_4$) were added. The tube was placed in a Kjeldahl digestion apparatus (Sibata, B-412) and allowed to be heated for 2 h. The digestion was then carried out for another 30 min to ensure conversion of all nitrogen to ammonium sulphate. After cooling, the digested sample was transferred into 100-ml volumetric flasks and diluted with distilled water. A 20 ml volume of the digested solution was analyzed for ammonium by using Kjel-Auto (Mitamura, MRK automatic nitrogen / protein measurement apparatus 037800 DTP-3).

2.3.4. Inorganic nitrogen (ammonium, nitrate, and nitrite)

A 5 g sample and 50 ml of 1 mol l^{-1} KCl solution were placed into a centrifugation tube. After shaking for 1 h with a reciprocating shaker, the suspension was centrifuged at 3000 g for 15 min and the supernatant was passed through a filter paper. A 5 ml volume of the filtrate was analyzed for ammonium-N (NH_4^+-N) by using the Bremner steam-distillation apparatus. The distillate was titrated against 0.01mol l^{-1} H_2SO_4. After distillation of ammonium, powdered Devarda's alloy was added, and the filtrate was re-distillated and then the distillate was titrated again to measure the sum of nitrate-N and nitrite-N.

2.3.5. Total Ca, K, Mg, and P

A 5 g sample was weighed into a tall-beaker (300 ml), and 20 ml of conc. HNO_3 and 10 ml of $HClO_4$ were added. The beaker was covered with a watch-glass and digested at 170℃ on a hot-plate. The digestion was carried out until the fume changed to completely white. After cooling, the digested sample was filtered and diluted in a volumetric flask to 100 ml with distilled water. The solution was analyzed for Ca, K, and Mg by using an atomic absorption spectrometer. Before P measurement, the solution was agitated with charcoal activated for 1 h and filtered to remove the colored constituents, which could interfere the absorbance measurement. P in the solution was analyzed by a molybdate-blue colorimetric method.

2.3.6. Extractable Ca, K, Mg, and P

A 5 g sample was weighed into a polypropylene vessel and 20 ml of 0.05mol l^{-1} HCl + 0.0075 mol l^{-1} H_2SO_4 mixture was added. After shaking for 5 min, the suspension was passed through a filter paper and diluted in a volumetric flask to 100 ml with distilled water. Ca, K, Mg, and P in the solution were analyzed as mentioned above. The analyses were carried out on the original suspension and the supernatant fraction. The concentration of the sludge fraction was calculated as below:

$$Csl = (Cor - (Csu \times Fsu)) / Fsl$$

Csu: concentration in supernatant fraction
Csl: concentration in sludge fraction
Cor: concentration in original suspension

$$Dsu = Csu \times Fsu / Cor \times 100$$

$$Dsl = 100 - Dsu$$

Dsu: distribution percent of supernatant
Dsl: distribution percent of sludge

3. RESULTS AND DISCUSSION

3.1. Methane gas yields

Methane gas yields are presented in Table 3. The average yields were greatest from the 55B digestion. The 35B digester was inhibited as evidenced by the lack of gas production and low effluent pH as shown in Table 4. The inhibition suggests that the rate of FPW additions combined with the rate of cell wash out at 20 days HRT may have been too rapid for maintenance of a stable methanogenic population. FPW addition to dairy manure in the 55B digester increased the methane production per unit digester volume. The pH increased in all digesters except the inhibited 35B digester as shown in Table 4. The pH of the 35A and 55A digesters increased by 0.9 and 1.3 pH units, to 7.7 to 8.1, respectively, and in the 55B digester by 0.9 pH units to 7.3.

Table 3. Methane gas yields

	CH_4 L/digester L/day	CH_4 L/g VS/day
35A	0.53±0.080	0.15±0.025
35B	Ceased	Ceased
55A	0.53±0.079	0.14±0.024
55B	1.48±0.022	0.17±0.027

3.2. TS and VS

The concentrations of influent TS and VS of feedstocks are presented in Table 1. The concentrations of influent TS and VS in material B are higher than those of material A. Addition of food processing wastes to dairy manure increased the TS and VS contents of material B influent. The distribution of mass and solids between supernatant and sludge fractions of the influents and effluents are presented in Table 5. The proportion of solids present in the material B sludge (28.8%) was less than the proportion in the material A sludge (54.6%). The percentages of solids distribution into the supernatant in material B were higher than those in material A. TS and VS reduction rates are presented in Table 4. For a given influent type, TS and VS reduction rates of 55℃ digesters were high compared to rates observed from 35℃ digesters.

3.3. Inorganic and total nitrogen

After the digestion, the concentration of nitrate and/or nitrite was considerably lower than that of NH_4^+-N in the effluent and almost ignorable as constituents of inorganic nitrogen. Kirchmann and Lundvall (1993) also found that nitrate and nitrite were extremely rare and ammonium was abundant in the anaerobically fermented or digested pig and cattle slurries. These should be reasonable because the digestion was carried out under anaerobic condition in which the nitrification process could not be occurred. The NH_4^+-N/TKN ratio of the B influent (Table 6) was low compared to ratios observed for dairy manure(A). The low ratio can be partially explained by the low NH_4^+-N/TKN ratio of the food processing wastes. The recoveries of NH_4^+-N (Table 7) from the B digesters were greater than those from A digesters. This suggests that material B had a more readily degradable organic N like as amino acids and proteins. The mineralization of influent organic N from the 55℃ digesters were slightly higher than those of the 35℃ digesters. The mineralization of organic N from the 35B digester was slightly low than that of the 55B digester. Thus, the inhibition of the 35B digestion affected organic N mineralization to a much lesser extent than it affected CH_4 yield. The measured recoveries of TKN (Table 7) in A ranged from 91.9 to 85.7%, which were close to 100%. The measured recoveries of TKN (Table 7) in B ranged from 78.1 to 73.1%.

Table 4. Characteristics of digested dairy manure effluent

		35A	55A	35B	55B
TS	(%)	7.0	6.1	14.3	10.4
VS	(%)	5.2	4.5	12.8	8.9
pH		7.7	8.1	5.7	7.3
TVA	(mg/L)	998.4	323.6	2822.2	805.1
FA	(mg/L)	0	0	0.3	0.4
AA	(mg/L)	387.3	260.1	1273.9	490.0
PA	(mg/L)	595.9	61.6	1000.4	294.5
BA	(mg/L)	15.2	1.9	547.6	20.2
NH_4^+-N	(mg/L)	1653.6	1639.5	2344.9	2369.9
TKN	(mg/L)	3906.4	3644.3	4700.0	4396.6
NH_4^+-N/TKN	(%)	42.3	45.0	49.9	53.9
TS reduction	(%)	13.9	24.3	20.3	42.2
VS reduction	(%)	15.6	26.6	23.2	45.8

TS: Total solid VS: Volatile solid TVA: Total volatile acid
FA: Formic acid AA: Acetic acid PA: Propionic acid BA: Butyric acid
NH_4^+-N: Ammonium nitrogen TKN: Total Kjeldahl nitrogen

The distribution of NH_4^+-N and TKN between supernatant and sludge fractions of the influent and effluent are presented in Table 9. The proportions of effluent TKN present in the sludges were less than the proportions of solids (Table 9) and greater than the proportions of mass present in the sludges. This is because some of the N was in the soluble NH_4^+-N form and not associated with solids.

Table 5. Distribution of mass and total solids

		Mass		Total solids	
		Supernatant (%)	Sludge (%)	Supernatant (%)	Sludge (%)
35A	Influent	71.5	28.5	45.4	54.6
	Effluent	68.9	31.1	33.5	66.5
55A	Influent	71.5	28.5	45.4	54.6
	Effluent	67.6	32.4	27.5	72.5
35B	Influent	74.9	25.1	71.0	29
	Effluent	43.7	56.3	24.5	75.5
55B	Influent	74.9	25.1	71.2	28.8
	Effluent	61.2	38.8	37.2	62.8

Table 6. Ammonium N, TKN and extractable and total K, P, Ca, and Mg of the influent and effluent

	NH_4^+-N mg/L	TKN mg/L	Extr K mg/L	Total K mg/L	Extr P mg/L	Total P mg/L	Extr Ca mg/L	Total Ca mg/L	Extra Mg mg/L	Total Mg mg/L
Influent										
A	996.9	4250.5	3728.1	3982.1	601.9	847.4	1105.1	1356.9	991.6	1030.2
B	851.3	6015.1	3021.1	3158.2	472.1	624.5	831.1	984.9	764.7	785.7
Effluent										
35A	1653.6	3906.4	2926.8	3328.1	522.1	723.9	1170.7	1324.5	860.6	910.5
55A	1639.5	3644.3	2781.5	3242.2	489.8	696.9	1196.1	1297.9	818.5	918.1
35B	2344.9	4700.1	2306.6	2794.1	498.9	669.1	921.1	1104.8	824.8	860.5
55B	2369.9	4396.6	2250.1	2689.2	429.2	606.8	792.7	986.7	713.5	725.1

The 55B effluents which had the greatest NH_4^+-N/TKN ratio, therefore, had the greatest portion of their effluent TKN in the supernatants. The proportions of effluent NH_4^+-N present in the supernatants approximately paralleled the proportions of mass present in the supernatants. This is because the mass was composed mostly of liquid and the NH_4^+-N was highly associated with the liquid component of the effluents. Consequently, most of the NH_4^+-N of the effluents was distributed in the supernatant fractions. It implies that the co-digested effluents should play a critical role as ready available nitrogen fertilizes when they would be applied to the arable land.

Table 7. Recovery of nutritional elements (%)

	TKN	NH_4^+-N	Total Ca	Extra-Ca	Total K	Extr-K	Total Mg	Extr-Mg	Total P	Extr-
35A (%)	91.9	171.0	97.6	105.9	83.6	78.5	88.4	86.8	85.4	86.7
55A (%)	85.7	169.6	95.6	108.2	81.4	74.6	89.1	82.5	82.2	81.4
35B (%)	78.1	275.4	112.2	110.8	88.5	76.4	109.5	107.9	107.1	105.7
55B (%)	73.1	278.4	100.2	95.4	85.2	74.5	92.3	93.3	97.2	90.9

3.4. Potassium

The concentrations of extractable and total K are presented in Table 6. The extractable/total K ratios of the influent (Table 8) indicate that most of the K of dairy manure and food processing wastes is extractable. On the contrary, the extractable/total K ratios of the effluent decreased slightly. The distributions of extractable and total K between fractions of the influent and effluent are presented in Table 9. The proportions of both extractable and total effluent K present in the sludge and supernatant fractions paralleled the proportions of mass distributed in these two fractions. This was due to the same reasons given for NH_4^+-N. Consequently, large proportions of the effluent K were present in the supernatant fractions and could be readily available as a fertilizer. Recoveries of total K were approximately over 80%.

3.5. Phosphorus

The concentrations of extractable and total P are presented in Table 6. The extractable/total P ratios of the influents indicate that a large component of manure P is unextractable comparing to the other nutritional elements. The effluent extractable P concentrations of all digesters were reduced compared to the influent concentration except 35B digester. This indicates that a component of extractable P in the influent became unextractable during digestion. This might be due to the incorporation of dissolved P into the solid fraction and converted to the organically-bound phosphate fraction. Consequently, extractable/total P ratios of the influents decreased slightly following digestion (Table 8). The concentrations of both extractable and total P in the sludge fractions of the effluents were greater than the concentrations in the supernatants as shown in Table 9. The distributions of extractable and total P between fractions of the effluent are presented in Table 9. The proportions of P distributed between fractions of the effluent paralleled trends observed for the TS distributions as shown in Table 5. Differing from the case of N and K, the effectiveness of the co-digested effluent as a ready available phosphate fertilizer could not be counted. However, the combination of the digested slurry with phosphate fertilizers could lead to the integrated P availability in soils (Dravid and Biswas, 1996), since it affected as organic manure to release phosphorus from the soils and as slow release phosphate fertilizer.

Table 8. NH_4-N/TKN, Extractable/Total Ca, K, Mg and P of the influent, effluent, supernatant and sludge

	NH_4^+-N/TKN	Extr-/Total Ca	Extr/Total K	Extr/Total Mg	Extr/Total P
	%	%	%	%	%
Influent					
35A	22.7	81.4	93.6	96.3	71.0
55A	22.7	81.4	93.6	96.3	71.0
35B	14.2	84.4	95.7	97.3	75.6
55B	14.2	84.4	95.7	97.3	75.6
Effluent					
35A	42.3	88.4	87.9	94.5	72.1
55A	45.0	92.2	85.8	89.2	70.3
35B	49.9	83.4	82.6	95.9	74.6
55B	53.9	80.3	83.7	98.4	70.7
Supernatant					
35A	49.0	99.4	101.2	98.6	55.3
55A	53.8	97.6	94.0	97.7	57.1
35B	59.0	97.8	101.4	108.5	90.3
55B	55.0	101.0	93.9	83.2	76.3
Sludge					
35A	35.5	84.6	51.8	93.4	75.4
55A	36.7	103.6	63.1	87.0	72.2
35B	44.9	78.1	67.8	87.5	66.4
55B	52.7	68.6	63.7	102.5	68.6

3.6. Calcium

The concentrations of Ca in the influents and effluents are presented in Table 6. The extractable/total Ca ratios of the influents indicate that a large component of manure Ca is extractable (greater than 80%). Recoveries of extractable and total Ca from the digesters were close to 100%. The concentrations of both extractable and total Ca in the sludge fractions of the effluents were greater than the concentrations in the supernatants as shown in Table 9. Therefore, paralleling the trends observed for P, Ca was also highly associated with the solids of the effluents. The proportions of Ca distributed between fractions of the effluent paralleled trends observed for the TS distributions as shown in Table 5.

Table 9. Distribution of nutritional elements between supernatant and sludge fractions of the influent and effluent

	Influent		Effluent	
	Supernatant (%)	Sludge (%)	Supernatant (%)	Sludge (%)
TKN				
35A	69.3	30.7	50.7	49.3
55A	69.3	30.7	48.5	51.5
35B	68.1	31.9	35.5	64.5
55B	68.1	31.9	52.3	47.7
NH_4^+-N				
35A	53.2	46.8	58.7	41.3
55A	53.2	46.8	58.0	42.0
35B	71.2	28.8	42.0	58.0
55B	71.2	28.8	53.4	46.6
Total Ca				
35A	56.8	43.2	25.8	74.2
55A	56.8	43.2	24.8	75.2
35B	59.5	40.5	26.7	73.3
55B	59.5	40.5	26.7	63.8
Extr-Ca				
35A	52.2	47.8	29.0	71.0
55A	52.2	47.8	15.5	84.5
35B	54.4	45.6	31.4	68.6
55B	54.4	45.6	45.5	54.5
Total K				
35A	69.3	30.7	73.1	26.9
55A	69.3	30.7	73.4	26.6
35B	74.7	25.3	43.8	56.2
55B	74.7	25.3	66.0	34.0

Table 9. Distribution of nutritional elements between supernatant and sludge fractions of the influent and effluent

	Influent		Effluent	
	Supernatant (%)	Sludge (%)	Supernatant (%)	Sludge (%)
Extr-K				
35B	73.7	26.3	53.9	46.1
55B	73.7	26.3	74.1	25.9
Total Mg				
35A	39.1	60.9	21.2	78.8
55A	39.1	60.9	20.4	79.6
35B	74.9	25.1	39.8	60.2
55B	74.9	25.1	21.3	78.7
Extr-Mg				
35A	31.5	68.5	22.1	77.9
55A	31.5	68.5	22.4	77.6
35B	71.7	28.3	45.1	54.9
55B	71.7	28.3	18.0	82.0
Total P				
35A	38.8	61.2	16.5	83.5
55A	38.8	61.2	12.9	87.1
35B	71.1	28.9	34.3	65.7
55B	71.1	28.9	27.6	72.4
Extr-P				
35A	27.2	72.8	12.7	87.3
55A	27.2	72.8	10.5	89.5
35B	72.9	27.1	41.5	58.5
55B	72.9	27.1	29.8	70.2

3.7. Magnesium

The concentrations of Mg in the influents and effluents are presented in Table 6. The extractable/total Mg ratios of the influents indicate the high extractability of manure Mg as was also observed for Ca and K. The concentrations of both extractable and total Mg in the sludge fractions of the effluents were greater than the concentrations in the supernatants as shown in Table 9. Therefore, as was observed for P and Ca, Mg was also highly associated with the solids of the effluents. The distributions of extractable and total Mg between fractions of the effluent are presented in Table 9. Consequently, 78.7 to 79.6% of total Mg and 77.6 to 82% of the extractable Mg in the effluents were present in the sludge fractions (ignoring 35B). Thus, the co-digested effluent would be expected to act as a slow release fertilizer for Ca and Mg.

3.8. Coliform bacteria

The changes in colony forming units of Coliform bacteria are given in Table 10. Coliform bacteria in the digested slurries declined to low or undetectable numbers after the digestion process (ignoring 35B).

Table 10. The numbers of coliform bacteria

		Influent	Effluent
35A	log CFU/mL	7	1
55A	log CFU/mL	7	0
35B	log CFU/mL	7	3
55B	log CFU/mL	7	0

CFU : Colony forming units

4. CONCLUSION

Total N added to the digester was recovered to a large extent in the effluent. Large increases in the concentrations of NH_4^+-N during digestion were attributed to the mineralization of organic N. The FPW mixture had greater recoveries of NH_4^+-N during digestion compared to the DM, reflecting its greater organic N degradability. The extractable/total ratios of influent K, P, and Mg were slightly reduced after digestion.

Fractionation of the effluent indicated that distributions of effluent NH_4^+-N and K were related to the distribution of mass between supernatants and sludges. Since effluent mass was composed mostly of water, the majority of these nutrients were present in the liquid associated supernatants. For these constituents, the digested effluents should be counted as ready available fertilizer.

Distributions of total P, Ca, and Mg were related to the distribution of TS between the supernatant and the sludge fractions of the effluents. Since the majority of the TS from the effluents was present in the sludges, a majority of the effluent total P, Ca, and Mg was also present in the sludges. Liquids and solids fractions of anaerobically digested manures are often handled separately. Therefore, some indication of the proportions of nutrients present in the separate fractions can be estimated by measuring the mass and TS distributions.

ACKNOWLEDGEMENT

We thank Dr. Thomas R. Way, USDA, ARS, National Soil Dynamics Laboratory, for reviewing and comments of the paper.

REFERENCES

Convers, J.C., R.E. Graves and G.W. Evans. 1977. Anaerobic degradation of dairy manure under mesophilic and thermophilic temperatures. Transactions of ASAE 20(2):336-340.

Danish Institute of Agricultural and Fisheries Economics, 1999. Centralized Biogas Plants-Integrated Energy Production, Waste Treatment and Nutrient Redistribution Facilites-

Dravid, M.S. and Biswas, D.R. 1996. Effect of phosphorus, poultry manure, biogas slurry and farmyard manure on dry matter yield and utilization of applied P by wheat. Journal of Nuclear Agriculture and Biology. Vol. 25, No.2: 89-93.

Field, J.A., J.S. Caldwell, S. Jeyanayagam, R.B. Reneau, W. Kroontje and E.R. Collins, Jr. 1984. Fertilizer recovery from anaerobic digesters. Transactions of ASAE

Hashimoto, A.G. 1983. Conversion of straw-manure mixtures to methane at mesophilic and thermophilic temperatures. Biotechnol and Bioeng. 25:185-200.

Kirchmann, H. and Lundvall, A. 1993. Relationship between N immobilization and volatile fatty acids in soil after application of pig and cattle slurry. Biology and Fertility of Soils. Vol. 15: 161-164.

Kimura, Y., Umetsu, K., Takahata, H. 1994. Effects of methane fermentation on seed survival of Broadleaf dock. Journal of Japanese Society of Grassland Science. Vol. 40, N0.2: 165-170.

Umetsu, K. Neu, Y. Hoshiba, H. Takahata, H. 1998. Effect of plastic media on mesophilic methane fermentation of separated dairy manure slurry at high solids concentration. Journal of the Society of Agricultural Structures, Japan. Vol. 29, No.2: 57-67.

Umetsu, K. and Y., Kimura. 1999. Effluent handling and management in Japan. Geenhouse gases in animal husbandry: Monbusho international scientific research program. Obihiro University and The University of Queensland. The University of Queensland ISBN 1864993626: 43-46.

Umetsu, K., Takahata H. and Takeuchi, Y. 2000. Farm scale anaerobic digester of dairy manure slurry in a cold region. Journal of the Society of Agricultural Structures, Japan. Vol.31, No.3: 179-185.

Greenhouse Gases and Animal Agriculture
J. Takahashi & B. A. Young (Editors)

Current situation on cattle wastes and their efficient use for reducing the risk of environmental pollution from dairy farming in Hokkaido, Japan

T. Matsunaka

Dept. of Dairy Science, Faculty of Dairy Science, Rakuno Gakuen University, Ebetsu, Hokkaido, 069-8501 Japan

Dairy farming in Hokkaido has been developing, with a large increase in the number of dairy cattle since 1960. The number of dairy farms decreased constantly during this period and both the number of the cattle per farm and per hectare of grassland has increased dramatically. At present in Hokkaido, around 900 thousand dairy cattle are raised and they produce 16 million metric tones of waste, such as feces and urine, per year. It is well recognized that the wastes are a good source of plant nutrients for crop growth but can also be a potential cause of environmental pollution, if misused. Complete management of the wastes is a first step to avoid direct leakage of the nutrients from the wastes to environment such as atmosphere, soil and ground water. The capacity for storing the wastes, however, is often insufficient compared with the volume produced by the cattle and is one of the causes of the nutrient leakage form the wastes to wider environment. The most efficient application method of the wastes such as cattle slurry to grassland in Hokkaido was as follows; application rates from 20 to 30 t/ha should be applied both in autumn (from early September to late October) and in the following Spring (until middle May). In order to use the wastes with low environmental impact, establishment of the guideline for the number of cattle raised per unit land is an urgent priority, because if there is no sufficient land area to be available for the application at correct rates of the wastes, they cannot help being applied at excessive high rates that cause environmental pollution. Within the guideline it will also be important to promote the efficient use of the wastes.

Keywords: animal wastes, efficient application, environmental pollution, nitrogen, raising density.

1. BRIEF HISTORY ON DEVELOPMENT OF DAIRY FARMING IN HOKKAIDO

Different types of agriculture have developed as a result of different climatic conditions in Hokkaido, and dairy farming is concentrated mainly in northern and eastern regions because of their unfavorable climatic conditions for general arable crops. In the eastern region called Konsen, a national project was initiated, The Konsen Pilot Farm Project, with financial support from the World Bank in 1956. The project, which was completed in 1967, promoted large-scale dairy farming where grass silage and hay were the main feedstuffs for dairy cattle. This type of farming was the first experience as grassland farming in Japan.

The number of dairy cattle increased with time from 182,810 in 1960 to 878,200 heads in 1999, accounting for 48% of the total in Japan, while the number of dairy farms in Hokkaido decreased continuously from 63,690 in 1960 to 10,300 in 1999 (Fig. 1 & 2). Since the decrease in the number of dairy farms in Japan was greater than that in Hokkaido, the dairy farms in Hokkaido accounted for 29% of the total in Japan in 1999. Hence the number of dairy cattle per farm in Hokkaido increased surprisingly from 2.9 head in 1960 to 85.3 head in 1999 (Fig. 2). The milk production per year was also steadily increasing from 397,150 ton in 1960 to 3,633,723 ton in 1999, accounting for 43% of the national share. The annual milk yield per milking cow was 7,370 kg in 1999 and it was slightly greater than 7,220 kg of the Japanese average.

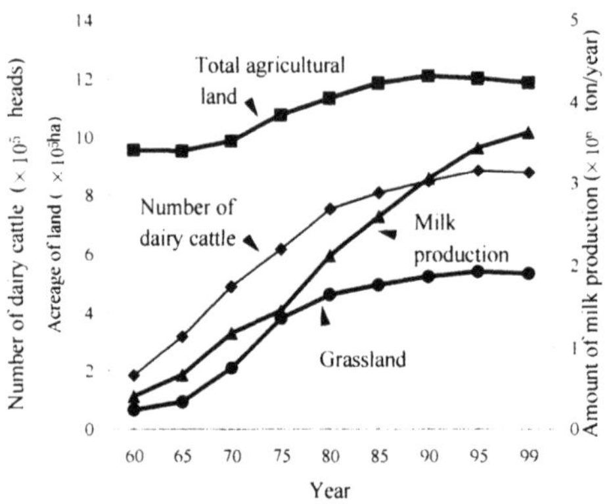

Figure 1. Change in acreage of agricultural land and grassland, milk production and number of dairy cattle in Hokkaido with years.

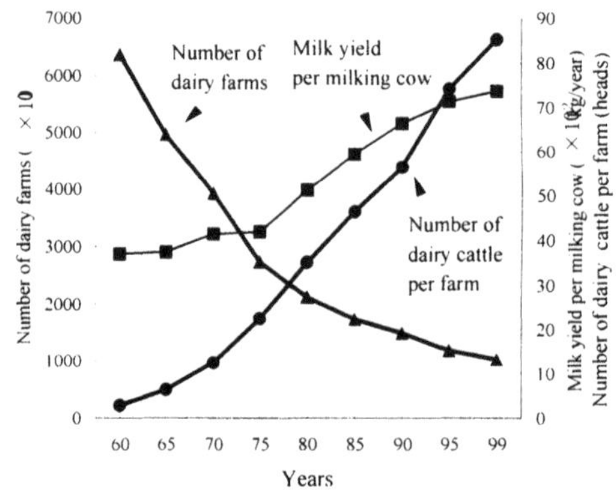

Figure 2. Change in number of dairy farms and of dairy cattle per farm, and milk yield pre milking cow in Hokkaido with years.

As shown in Fig. 1, total area of grassland in Hokkaido also increased rapidly with the number of dairy cattle from the 1960s to the 1970s and then the increase slowed down from the 1980s. The area in 1999 was about 533,500 ha, accounting for 82% of the total grassland in Japan. These facts show that Hokkaido is now the main location for dairy farming in Japan. It is likely, however, that this rapid growth of the dairy farming in Hokkaido results inevitably in environmental pollution problems, which are not serious yet.

2. MANAGEMENT OF THE WASTES TO CONTROL NITROGEN LEAKAGE FROM THE DAIRY FARM TO THE WIDER ENVIRONMENT

2.1. Quantity of the wastes from dairy cattle and nitrogen load

Naturally enough, the quantity of wastes from dairy cattle, i.e. feces and urine, has increased in line with the numbers raised in Hokkaido. This increase in the quantity can be a potential risk of environmental pollution, although the wastes are also a good source of plant nutrients for crops. In particular, nitrogen (N) and phosphorus (P) in the wastes are the primary causes for the pollution derived from the wastes.

The 878,200 dairy cattle raised in Hokkaido in 1999 produced approximately 12 million tons of feces and 4 million tons of urine. The total N content of these wastes is estimated to be about 84,000 tons per year. In addition, the 413,800 beef cattle and 549,200 swine in Hokkaido excreted 20,000 and 7,000 tons of N, respectively. The total N load from main livestock groups in Hokkaido is therefore approximately 111,000 tons per year. The N load in Hokkaido is greater than that applied to all of the agricultural land as inorganic fertilizer, which is estimated to be approximately 100,000 tons.

Assuming that the N derived from these livestock could be evenly applied to all the agricultural land in Hokkaido, the application rate of N is estimated to be 94 kg/ha. In the case of an application of N derived from dairy cattle wastes only to the grassland, the annual application rate is about 158 kg/ha. The N load in both cases is therefore less than 200 kg/ha that is recognized as the recycling capacity of N on grassland to minimize environmental pollution from animal wastes (Research Group on New Agricultural Policy, 1992). These scenarios suggest that on average in Hokkaido, the dairy cattle wastes are not a potential cause of environmental pollution at the moment. This situation is in contrast to that in the other regions of Japan. For example, many prefectures in Honshu, Kyushu and Shikoku islands have already shown that annual N load derived from livestock wastes substantially exceeds the recycling capacity for N (Tsuiki and Harada, 1997) and causes increase in nitrate nitrogen (NO_3-N) concentration in ground water in the regions (Kumazawa, 1998).

2.2. N budget for dairy farms

In the past, nutrients were efficiently recycled through the soil-plant-cattle system in low input grassland farming. The intensification in dairy farming in recent years, however, has resulted in huge quantities of wastes being produced on the farm. In addition, nutrients such as purchased concentrates and chemical fertilizers are imported into each dairy farm. These input flows of the nutrients have resulted in a nutrient accumulation on dairy farms. N accumulation and its leakage from farms to the wider environment are of particular concern.

As a case study to determine the N budget for a dairy farm, the author had ever measured the annual N flow through a typical dairy farm in northern Hokkaido (Matsunaka, 1999). The farm has 45 dairy cows and 60 heifers. The total area of the dairy farm is 73 ha, including 70 ha of grassland and 3 ha of building area (cattle barn and its surrounding area, manure facilities etc.). Annual milk production of the farm was 350 tons per year. This production is slightly more than the average production per farm in the region.

From the results of the study, total annual input N derived from the purchased concentrates, fertilizer and N fixation by Rhizobia was estimated at 8,590 kg, while total annual output N was apparently only 1,740 kg in the milk. Since accumulated N in the body of heifers was estimated as 460 kg and unknown N was 260 kg, the N balance between input and output was 6,130 kg. On average, therefore, N accumulation on the farm was 84 kg/ha/year. This N accumulation becomes a potential worry for environmental pollution derived from the dairy farming.

2.3. Water pollution by N accumulated in a dairy farm

N accumulation in a dairy farm is the balance between input and output N. The N accumulation, however, is not evenly distributed throughout the farm. For example, the average N accumulation per year for the case study farm described above was 84 kg/ha, while the rate of accumulation for grassland for silage and/or hay was 61 kg/ha, and for the grazing pasture was 93 kg/ha (Fig. 3). We should also pay special attention to the 2,490 kg-N derived from the wastes distributed on the 3 ha immediately around the cattle barn. In this area of the farm the N accumulation rate was, surprisingly, 830 kg/ha/year.

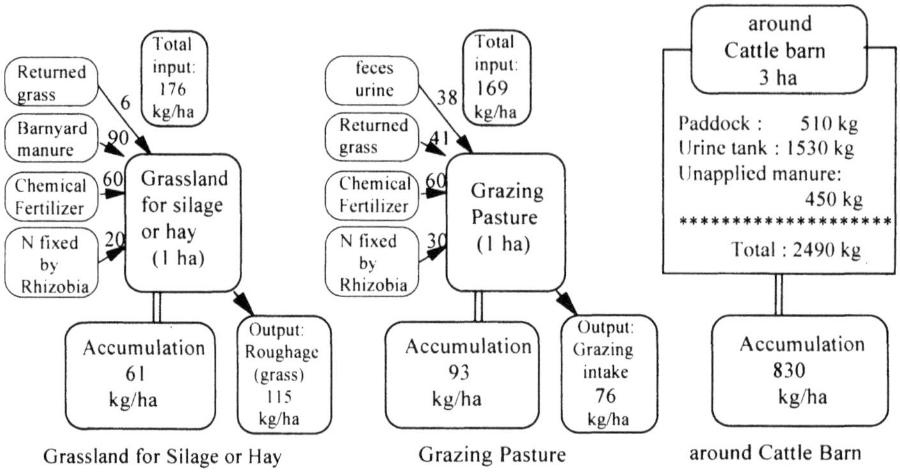

Figure 3. Distribution of N accumulation in the dairy farm surveyed as case study.

The extra surplus N around the barn is potentially the primary cause of environmental pollution from the dairy farm. In fact, Matsumoto *et al.* (1993) and Ishiwata *et al.* (1998) reported an extremely high soil mineral N content in the vicinity of a barnyard manure store. Clearly, these results indicated that seepage and run-off effluent from the unroofed storage facility of barnyard manure had polluted the soil and probably the underground water. The possibility of water pollution by dairy farms was supported by the results of Ohmura (1994) and Hojito (1998) who demonstrated that the N concentration in rivers increased after the rivers passed through dairy farming areas. In addition, Hojito (1998) showed that nitrate nitrogen (NO_3-N) concentration in well water in dairy farming areas in Hokkaido were generally less than 5 mg-N/L, but in some cases exceeded 10 mg-N/L (the maximum permitted concentration for drinking water). Direct leakage of urine from its storage tanks to all watercourses is the most significant risk of causing the pollution. Consequently it is very important to manage the nutrients, particularly N, in the wastes at all stages, viz. housing, storage and landspreading, in order to reduce the N leakage to the environment.

According to a report by the Hokkaido government (1995), however, 10% of the dairy farms surveyed had no storage facilities for barnyard manure, and even if farms had the manure storage facilities, 90% of the facilities had no roof (Table 1). Also the capacity of the facilities did not meet the amount of the manure produced by the dairy cows raised on the farm (Hara, 1995). The situation for urine was worse than that for barnyard manure (Hokkaido government, 1995). Only 73 % of the farms surveyed had any storage facilities for urine and 17 % of the farms discharged the urine with no treatment (Table 1). In addition, Hara (1995) pointed out that none of the storage tanks for urine had sufficient capacity for the number of the raised cattle and that the storage capacity of urine was not related to the number of the cattle raised on a farm.

These results indicate that the storage capacity for cattle wastes has not kept pace with the increasing numbers of cattle raised per dairy farm in Hokkaido. It is therefore feared that environmental pollution caused by dairy farming will become a serious problem in the near future, although on average the pollution in Hokkaido is not yet serious at present.

Complete management of the wastes in a dairy farm is a first step to prevent direct N leakage from the farm. This problem should be solved before discussing the efficient use of the wastes to grassland. When the farms intend to introduce the storage facilities of the wastes, however, financial costs should be also considered. The current cost for the facilities was more than 2 million Japanese yen per year at present (Research Project Team, 1999). The expense is considerably heavier to the farms. This is one of the causes for restricting the introduction of proper storage facilities of the cattle wastes into the farm.

Table 1. Current situation of equipping the storage facilities for the cattle wastes on dairy farms in Hokkaido. *(Hokkaido Government, 1995)*

Regions in Hokkaido	The number of respondents	The percentage of the farms equipping the storage facilities							
		Barnyard manure		The roof covering the storage facility for the barnyard manure			Urine tank		
		Yes	No	Yes	partial yes	No	Yes	Tentative tank	No
Southern part	328	93	5	2	2	91	74	6	15
Central part	300	89	9	2	1	91	80	4	12
Arable farming	512	89	9	1	1	92	77	7	12
Dairy farming	698	82	17	0	1	88	60	7	28
All of Hokkaido	1,838	88	10	1	1	90	73	6	17

3. EFFICIENT USE OF THE WASTES FROM DAIRY FARM

3.1. Time and rate of application of dairy wastes to grassland

Efficient use of the wastes ensures that a high proportion of plant nutrients are recycled through the soil-plant-cattle system and only removed from the dairy farm in the grass or milk rather than lost to the watercourse and atmosphere. This requires control over the rate and time of the application to land.

The application time for the wastes, however, is usually restricted in Hokkaido due to long winter and heavy snow cover. Consequently the main periods of the wastes application to grassland are October to November before wintering and May after wintering. The author *et al.* (1988a) showed that the application time of cattle slurry (mixture of feces, urine of dairy cattle and water) had a significant effect on dry matter yield. Since the N content of the slurry applied each time in the experiment was different, it is possible that the results were attributed to the different rate of N supplied from the slurry. We therefore calculated the efficiency of dry matter production per unit N (*Ne*) applied in the slurry at each time (Fig. 4).

There were no great differences in *Ne* in the slurry applied in early September, in late October and in the following middle of May. However the *Ne* in the middle of December, in the following late May and in the middle of July were clearly lower than those in the application times described before. Low *Ne* in the slurry applied in late May and the middle of July was due to the low efficiency of dry matter production per unit N absorbed by the grass (*Ndm*), while that in the middle of December was caused by a low efficiency of N absorption (*Nab*). The low *Nab* of the application in December suggests N loss from the system. In December the soil was completely frozen and the grass could not efficiently absorb N in the applied slurry. The slurry-N remained on the surface of the grassland as a risk of loss to the aqueous environment via runoff. From these results, the recommended application time of the slurry to grassland in Hokkaido was the period from early September to late October and the following middle of May.

Even if slurry is applied to grassland at the recommended time, excessively high rates of application of the wastes can lead to accumulations in the soil and subsequent loss or leaching of surplus N as nitrates. For example, as shown in Fig. 4, *Nab* of the application of cattle slurry in autumn (September to October) was relatively lower than that in spring (Middle of May). It is therefore likely that the application of the wastes in autumn poses a slightly greater risk to the environment than in spring. At present, both application at the rate of 20 to 30 t/ha of the slurry in autumn (September to late October) and in the following middle May is recommended as a safe application rate to grassland (Matsunaka *et al.*, 1988b), while the rate is dependent on the botanical composition such as the ratio of legume to grass on the grassland (Research Project Team, 1999).

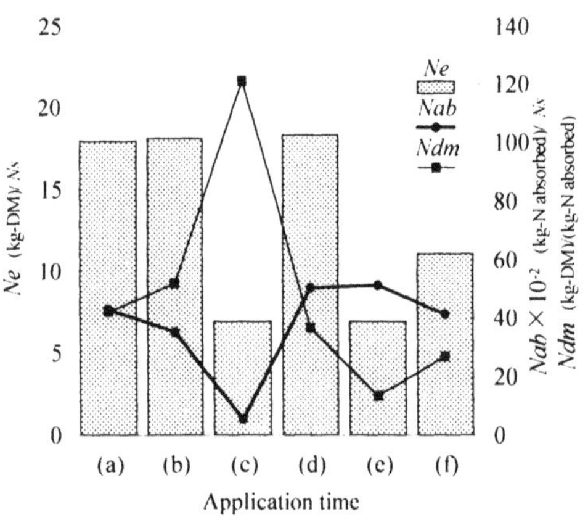

Figure 4. Nitrogen efficiency of the cattle slurry applied in different times.

Ne (Nitrogen efficiency): (Yt-Yn)/Ns, Nab (Efficiency of nitrogen absorption by grass): (Nt-Nn)/Ns, Ndm (Efficiency of dry matter production by the absorbed N): (Yt-Yn)/(Nt-Nn), where Ns is the amount of applied N, Yt and Yn are the dry matter yield of the treatment of the cattle slurry application and of no slurry application, respectively. Nt and Nn are N absorbed by the grass in the treatment of the cattle slurry application and of no slurry application, respectively. Application times are as follows; (a): early September, (b): late October, (c): middle of December, (d): middle of May, (e) late May, (f) middle of July.

3.2. Gaseous losses of N from the wastes

The wastes in dairy farming, such as barnyard manure, cattle slurry and urine, are generally spread onto grassland in Hokkaido. Surface spreading of the wastes, however, is a major source of complaints from the public due to the offensive odor and emission of Ammonia (NH_3) to atmosphere. NH_3 emission not only results in significant reductions in the value of the wastes as fertilizer but following transport and deposition it is associated with the eutrophication of natural and semi-natural ecosystems and also with the acidification of weakly buffered soils. In fact, NH_3 emission following the surface spreading accounted for 20 to 35 % of NH_4-N in the applied slurry (Fig. 5). The concentration of NH_4-N in the snow and rain near the storage facility of barnyard manure was higher than that on an arable field to which the distance from the facility is about 600m (Matsunaka, 1999). This fact suggests that the emitted NH_3 is re-deposited to the soil surface in wet deposition. Since the wet deposition contributes to environmental damage such as eutrophication, soil acidification and the disturbance of nutrient balance in the tree and in the soil, we should control the NH_3 emission during and following spreading the slurry and establish methods to minimize the emission.

Nitrous oxide (N_2O) emission following land spreading of the wastes poses another risk of polluting the atmosphere, because the gas is of key importance in relation to the greenhouse effect and global warming. Compared with many studies in European countries (Fowler *et al.*, 1997), research on N_2O emission following the application of the wastes to grassland in Hokkaido only began recently. Kouda and Hojito (1999) confirmed N_2O emission was increased following the application of barnyard manure and inorganic fertilizer to grassland. Takakura and the author are now developing a new system to measure automatically N_2O emission following the application of the digestive cattle slurry that is produced after methane fermentation of the wastes under anaerobic condition (Takakura and Matsunaka, 2001). We also recognized clearly N_2O emission from the grassland. Further research in Hokkaido should therefore focus on practical recommendations to reduce N_2O emission following the application of the wastes to grassland.

The N_2O emission has a great influence in the greenhouse effect. The quantity of N loss as N_2O from the grassland following the application of the wastes, however, is very less than the loss by NH_3 emission. The case study of the N flow through a dairy farm showed that the N loss as a form of NH_3 was estimated at 14% of total input N, although it as a form of N_2O was only 1% (Konsen Agric. Expt. Stn., 1999). In addition, the study found out that the loss by NH_3 emission was the greatest of the N loss from the dairy farm to environment and it accounted for 46% of the total N load to the environment (Konsen Agric. Expt. Stn., 1999). It is therefore very important for reducing N loss from dairy farm to minimize NH_3 emission.

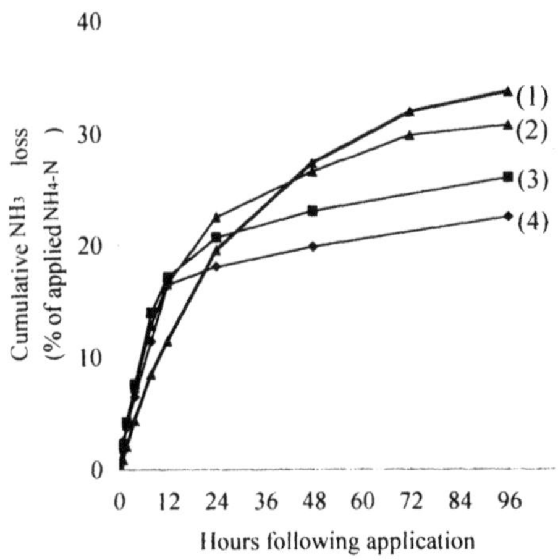

Figure 5. Cumulative ammonia emission following application of several kinds of cattle slurries in some trials.

The application rates and pH of the slurries were as follows; Trial A; (1) High rate (120t/ha), no aeration and pH7.1, (2) Standard rate (60t/ha), no aeration and pH 7.1, Trial B; (3) Standard rate (60t/ha), no aeration and pH7.6, (4) Standard rate (60t/ha), aeration for 4 weeks before application (pH8.9)

3.3. Application techniques for minimizing NH_3 emission

When the animal wastes come into contact with soil, e.g. through injection of the slurry into soil or incorporation immediately after spreading, NH_3 emission is reduced considerably. This indicates that the technique for slurry application can be used to control NH_3 emission. New slurry application techniques for grassland have been developed in recent years (Huijsmans *et al.*, 1997; Morken and Sakshaug, 1998). These techniques include direct ground injection (injection by using a pump to pressurize the slurry), shallow injection (cutting a shallow slit in the sward into which slurry is applied) and band spreading (application of the slurry in bands on the soil surface using a trailing foot implement). This development of the techniques is progressing slowly in Hokkaido compared with in European countries. The development, however, is urgently required to ensure the environmentally efficient use of the animal wastes in Hokkaido.

4. THE NEED FOR AN INTEGRATED APPROACH TO N LEAKAGE

There is a need for an integrated approach to the environmentally friendly use of the animal wastes. For example, preventing the NH_3 emission from the wastes means that the NH_3 is conserved in the soil where it may increase the risks of NO_3-N leaching to the ground water and N_2O emission from the soil. Without the integrated approach, there is a risk that reducing one form of the N leakage from the wastes management may simply cause an increase in another form, and/or an increase in the same form of the leakage at another stage of nutrients cycle.

5. GUIDELINE FOR THE NUMBER OF CATTLE RAISED PER UNIT LAND

As described above, since Hokkaido has a comparatively larger land area per dairy cattle than the other regions in Japan, the environmental problems associated with the wastes of dairy cattle have not yet been serious. In recent decades, however, the raising density of the dairy cattle per unit land area in Hokkaido has been constantly increasing and this trend results in an increase in N load from the wastes per unit land area. Even in Hokkaido, therefore, the environmental problems will be realized in near future, if we leave the trend as it is and take no remedial action.

We can effectively use the wastes as a valuable source of nutrients for plant growth without environmental risks, when the wastes can be exploited by careful control of time, rate and method of application to land. This requires sufficient storage capacity for the wastes and sufficient land area to be available for application at correct rates and with an appropriate machine. It is likely that the most difficult of these requirements to meet in intensive dairy farming is to ensure the sufficient land area. There is no official guideline for the raising density of the dairy cattle per unit land area in Hokkaido and Japan, but according to the author's trial calculation based only on prevention of water pollution by N derived from the wastes, the suitable raising density was about 2 heads of milking cows pre hectare (Matsunaka, 2001).

When the guideline for the raising density is clearly established through the research findings, the land area of the farm restricts the number of cattle raised on the farm, and the number of cattle will dictate the capacity of the storage facilities. Within the condition under the guideline for raising density it will be necessary to promote the environmentally friendly use of the wastes based on discussion in this paper. Under the guideline, the dairy farms that have sufficient land area have an advantage over those that have less land, because the larger land area can support a greater number of cattle. Administrative support will be required for the disadvantaged farms, when the guideline becomes effective as a official regulation.

6. CONCLUSION

In light of the above discussion, the following points should be considered in order to reduce the risk of environmental pollution from dairy farming in Hokkaido; a) Elucidate how many cattle per unit land area can be raised with least environmental pollution, b) Establish the guidelines for the number of cattle raised per unit land, c) Adopt the integrated approach to the environmentally friendly use of the wastes and d) Promote efficient use of the wastes within the guidelines under the integrated approach.

ACKNOWLEDGMENTS

The author thanks Dr D. Chadwick of Institute of Grassland and Environmental Research, North Wyke, UK and Prof. J. W. Snape of John Innes Centre, UK for their critical reading and correction of this manuscript.

REFERENCES

Fowler, D., U. Skiba & K. J. Hargreaves, 1997. Emissions of Nitrous Oxide from Grasslands. In: Gaseous Nitrogen Emissions from Grasslands, S. C. Jarvis & B. F. Pain, CAB International, Wallingford, UK. 147-164.

Hara, Y., 1995. Current situation and problems on management of animal wastes. Journal of livestock management, Hokkaido, 31: 83-95. (in Japanese)

Hojito, M., 1998. Actual circumstances regarding water pollution in dairy farming areas in Hokkaido. In: Proceedings of International Workshop on Environmentally Friendly Management of Farm Animal Waste, T. Matsunaka, Civil Engineering Research Institute, Sapporo, 73-77.

Hokkaido Government, 1995. Survey Report of current situation of animal waste management and intention of dairy farmers, Sapporo, 1-181. (in Japanese)

Huijsmans, J. F. M., J. M. G. Hol & D. W. Bussink, 1997. Reduction of Ammonia Emission by New Slurry Application Techniques on Grassland. In: Gaseous Nitrogen Emissions from Grasslands, S. C. Jarvis & B. F. Pain, CAB International, Wallingford, UK. 281-285.

Ishiwata, T., N. Takamiya & T. Ishida, 1998. Development of dairy farming in Hokkaido and issues regarding dairy farm waste. In: Proceedings of International Workshop on Environmentally Friendly Management of Farm Animal Waste, T. Matsunaka, Civil Engineering Research Institute, Sapporo, 1-9.

Konsen Agricultural Experiment Station, 1999. A case study for nitrogen flow through a dairy farm. Hokkaido Konsen Agricultural Experiment Station, Nakashibetsu, 1-41.

Kouda, H. & M. Hojito, 1999. Change in nitrous oxide flux on timothy swards in Konsen region of Hokkaido. In: Abstract of the Annual Meeting, Japanese Society of Soil Science and Plant Nutrition, 45: 338. (in Japanese)

Kumazawa, K., 1998. Present state of nitrate pollution in groundwater, Jpn. J. of Soil Sci. Plant Nutri., 70: 207-213. (in Japanese)

Matsumoto, T., Y. Koide, T. Yoshikawa & M. Noshiro, 1993. Current situation and problems of cattle waste management in Nemuro region, Hokkaido. Hokuno, 60: 380-384. (in Japanese)

Matsunaka, T., J. Koseki & H. Kondo, 1988a. Efficient application time of the cow slurry to the timothy-legume mixture sward in Konsen district, Hokkaido. Hokuno, 55, (8): 17-30. (in Japanese)

Matsunaka, T., J. Koseki & H. Kondo, 1988b. Efficient application rate of the cow slurry to the timothy-legume mixture sward in Konsen district, Hokkaido. Hokuno, 55, (9): 30-44. (in Japanese)

Matsunaka, T., 1999. Nutrients flow in dairy farming and environmental pollution. In: Soil Science and Plant Nutrition in Hokkaido Agriculture 1999, Hokunokai, Sapporo, 248-251. (in Japanese)

Matsunaka, T., 2001. How many cattle per unit land area can the environmentally friendly dairy farming raise? In: Discussions on the dairy farming in 21st century, Extension Center of Rakuno Gakuen University, Ebetsu, 106-111. (in Japanese)

Morken, J. & S. Sakshaug, 1998. Direct Ground Injection – a new system for slurry injection. In: Proceedings of International Workshop on Environmentally Friendly Management of Farm Animal Waste, T. Matsunaka, Civil Engineering Research Institute, Sapporo, 219-224.

Ohmura, K., 1994. Circulation of N, P in the dairying area at the Hokkaido and water conservation. Jpn. J. of Soil Sci. Plant Nutri., 65: 573-577. (in Japanese)

Research Project Team, 1999. Handbook of Animal Waste Management and Utilization in Hokkaido 1999, Hokkaido Government and Hokkaido Prefectural Agricultural and Animal Husbandry Experiment Stations, 8-12 & 53-58. (in Japanese)

Research Group on New Agricultural Policy, 1992. New policy of food, agriculture and rural community, Chikyusya, Tokyo, 386-429. (in Japanese)

Takakura, K. & T. Matsunaka, 2001. An examination of continuously automatic measurement for nitrous oxide flux from grassland by using infrared absorption method. In: Abstract of the Annual Meeting, Japanese Society of Soil Science and Plant Nutrition, 47: 230. (in Japanese)

Tsuiki, M. & Y. Harada, 1997. Current circumstance in farm animal wastes in Japan ant its problems. In: Environmental conservation and New Animal Husbandry, M. Nishio, Norinsuisan Gijutsu Kyokai, Tokyo, 15-29. (in Japanese)

Greenhouse Gases and Animal Agriculture
J. Takahashi & B. A. Young (Editors)

Soil rehabilitation using livestock and organic wastes

G. Xiao, L. T. Yu & L. Ming

ZhengZhou College of Animal Husbandry Engineering, 450008, P R China

This paper put forward effective methods for soil rehabilitation using livestock and organic waste .by the establishment of integration of cereal crop—economic plant—forage plant farming system and Conversion of stalk (straw) into organic fertilizers by feeding animals based on save grain typical husbandry. The results showed these comprehensive measures can not only improve degraded land and realize soil rehabilitation, but also enhance ecological benefit and economic benefit , and realize good recycling of agriculture and husbandry production with sustainable development.

Keywords: rehabilitation, waste, livestock, recycling

1. INTRODUCTION

In recent years agriculture and animal husbandry in China are developing rapidly and China has become one of the most important countries in animal production in the world. But because of over-grazing, over-cultivation, deforestation, mismanagement of irrigation systems, water resources irrational utilization and rapid population increase has resulted an abrupt decrease in land productivity as well as the destruction of fertility, especially in rural area of China, in soil erosion, salination and desertification. By 1990, the area of degraded land amounted to over 40% of the whole territory, Rehabilitation of soil has been limited sustainable development of the whole nation. But in recent years, many measures have been taken to solve soil rehabilitation, among them the establishment of integration of cereal crop—economic plant—forage plant farming system and development of save-grain husbandry in rural area. These measures proved to be feasible and have bright promise.

2. MATERIALS AND METHODS

The studies was mainly conducted in Henan Province in central China and began in 1991 and lasted 10 years. The concrete methods employed were the following :

2.1. The establishment of Integration of cereal crop—economic plant—forage plant farming system

Farmers were accustomed to develop agriculture production only by way of integration of cereal crop—economic plant farming system, the results caused soil deterioration because of over-cultivation, rational application of chemicals and lower multiple crop index. In fact, many grasses and legumes have greater primary productivity and greater potential to sustain and improve degraded soil. It is necessary to combine forage plants (grasses and legumes)with integration of cereal crop—economic plant farming system and form integration of cereal crop—economic plant----forage plant farming system (three agricultural component structure),which increase multiple crop index and the proportion of granular structure in soil through rational rotation in order to produce more and forage for animal production and promote soil rehabilitation. The forage plants chosen for three agricultural component structure were comprised mainly of alfafa ,white clover, red clover arabian cron,. sweet clover, rumex ,prairie milkvetch, forage maize, the main rotation planting pattern used in the central China was showed as following:

(1) wheat + alfalfa→wheat—alfalfa→alfalfa→wheat + cotton→wheat = cotton→prairie milkvetch→forage maize+ soybean→wheat + alfalfa.(2) wheat + sweet clover→wheat= cotton—sweet clover—wheat + cotton→wheat—cotton→red clover→wheat + sweet clover.(3)wheat→wheat—soybean→corn—sweet clover→rumex+ sweet clover—wheat → wheat—sweet potato→corn=soybean—wheat.(4)prairie milkvetch→prairie milkvetch→wheat —sweet potato→alfalfa→Arabian corn + cabbage→wheat= cotton→wheat—soybean→wheat —sweet potato→prairie milkvetch.(5) wheat + alfalfa→wheat→forage maize +soybean→ wheat—sweet potato→alfalfa→alfalfa + cabbage→wheat .(6)wheat+ sweet clover→wheat= cotton—sweet clover→corn= soybean →wheat.(7)wheat +pea→wheat—pea—cotton→corn + pepper→alfalfa→alfalfa + wheat= cotton→wheat. (→ **rotating planting between two years, — rotating planting in a year, = interplanting , + intercropping)**

2.2. Conversion of stalk (straw) into organic fertilizers by feeding herbivorous animals based on save grain system

China is the largest developing country with the highest population in the world, so it is of vital importance to develop save grain husbandry. Herbivorous animals not only consume less grain than swine and poultry , but also can make good use of agricultural byproducts such as stalk and straw. China is the largest agricultural country in the world with abundant stalk (straw) resources and has a greater potential to produce roughage. But because of lower nutritive value and utilization rate of raw materials, stalk (straw) was rarely used as roughage directly, and it is estimated that less than 5%of stalk and straw were used as roughage 15 years ago , most of which was regarded as waste to be burned out or rotted away slowly in soil that produced severe pollution and waste of forage resources. Since 1990s there has been a wide adoption of techniques of ammoniated straw and stalk silage in rural areas, and the nutritive value and edibility of stalk and straw was significantly improved, stimulating more agricultural stalk (straw) to be used as roughage to raise herbivorous animals. Many other organic wastes such as beet pulp (top), vine, peanut shell, potato pulp and bran can also be utilized as concentrate feeding stuff.

2.3.Manure disposal for organic fertilizer

Animal waste can usually be utilized as organic fertilizers in natural conditions because recycling of the elements is frequently adequate in a vast agricultural land space, although the volume of feces produced was twice that of urine. In feedlot fecal excreta does not create a serious problem because it can be properly dealt with by composting. But urine has a pungent smell and often pollutes the environment with ammonia, methane gas etc, and the odors can have significant adverse health effects, and high sulfide concentrations can be deadly for animal and humans, so it is necessary to choose right materials to establish a manure disposal system in animal farms for absorbing urine and avoid forming effluent in order to be taken away for composting, such as straw, clay, sawdust and dried leaves of shrubs or trees. Animal farms usually adopt the gravity drain pit system to collect manure. The mixture of feces and urine in the pit is a slurry, and the organic loading rate is higher. It was necessary to maintain control of the water content of fecal excreta to promote organic matter decomposition and exterminate seed and pathogens. The optimum range for organic matter composting of fecal excreta water content is kept about 65%, and sawdust hay and dried straw were used to adjust the water content of fecal excreta because it has a higher water absorption capacity. Many alternatives such as dried straw, dried leaves of shrubs and trees and raw or processed rice hulls also can be used to enhance the water absorption potential. Aerobic composting accelerates organic matter decomposition and result in a higher temperature necessary for seed and pathogen destruction and moisture evaporation. Aerobic composting also minimizes the potential for nuisance odor. In the end, livestock manure became organic fertilizer by composting.

2.4. Application of organic fertilizer for land rehabilitation

Organic fertilizer can promote the formation of soil granular structure and enhance soil fertility. Because of the short of trace elements in soil, such as zinc, molybdenum, and boron. So it was necessary for application of $ZnSO_4$, $(NH_4)MoO_7$, H_3BO_3 with organic fertilizers together. Organic fertilizers are usually used as base fertilizer for the development of agricultural production.

3. RESULTS

3.1. The analysis of land productivity under the establishment of integration of cereal crop-economic plant-forage plant farming system (three agricultural component structure)

Based on analysing the three agricultural component structure in Henan Province, situated in central China as well as the whole nation, multiple crop index and green coverage in agriculture production were raised, agricultural grain output and byproduct yield increased by 12.5% and 46% respectively since 1990 because of rational rotation with highly productive forage plants, the level of protein of roughage from stalk (straw) and forage plants increased by 25%~30%. Because of fixation of nitrogen of Phizobium in perennial legumes, quick-acting nitrogenous fertilizer in soil raised by 17.8%.

3.2. The analysis of effects of save grain

Unlike swine and poultry, herbivorous animals don't create any grain competition with human beings. Ammoniated straw and stalk silage were substituted for roughage, and processed peanut or cottonseed cake and bran were used as concentrate instead of grain, so the production cost was much lower. Since 1992, over 1.2×10^7metric tons were saved for population consumption each year, which relieves pressure on arable land and avoids land degradation. .

3.3. Agriculture byproduct utilization and the development of animal production

Henan Province led in China in the development of ammoniated straw and stalk silage.

Compared with 1990, the utilization ratio of stalk and straw used for roughage through ammonia and silage reached 29.5% in all stalk and straw, and was 1.4 times that of China, an increase by 370% and 147%, respectively in Henan Province and China(Figure 1). In the meantime, the number of herbivorous animals increased by 58% and 29% for Henan Province and China, respectively. In addition, cattle, goats and sheep increased by 40.6%,14.8% and 68% respectively in Henan Province(Figure 2), and was significantly higher than mean increasing rate of the whole nation.

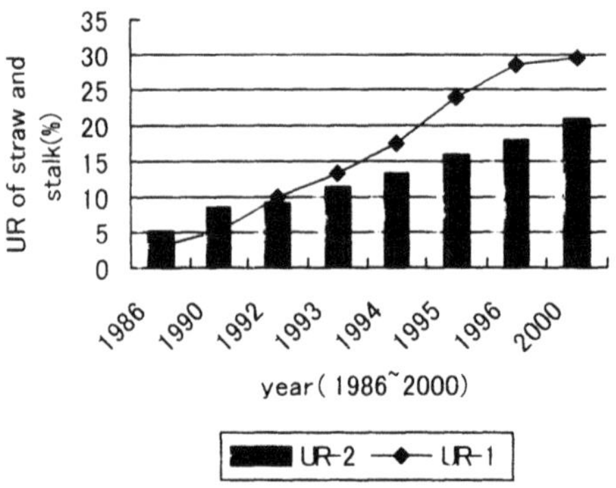

Figure 1 *Utilization rate of stalk and straw in Henan Province(1) and China(2)*

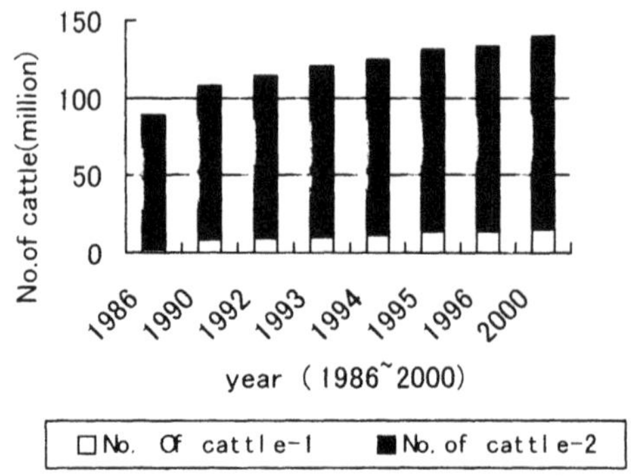

Figure 2 The comparison of No. of cattle Between Henan Province(1) and China(2)

3.4. The utilization of manure and soil rehabilitation

Since 1990, the manure increased by 50%~65% for Henan Province, and 25%~30% for the entire country with the development of herbivorous animal production. Because many feedlots usually adopted the gravity drain pit system to collect manure for it provided convenience of manure disposal and needed the least labor and expenses which favored the conversion of manure and other organic wastes into organic fertilizers, organic fertilizer applied for agriculture production increased 15% and degraded land area decreased 5.6%. Organic fertilizer from the manure was estimated to increase by over 3.5% annually all over the country, Composting liquid or solid materials with livestock manure is beneficial for the plants as nutrient sources and soil conditioner. In addition, because of fixation of nitrogen of Phizobium in perennial legumes, quick-acting nitrogenous fertilizer in soil increased 1.1×10^6 metric tons annually. Based on manure and waste, soil conditions were dramatically improved, and the degraded arable land declined significantly, in Henan Province since1990 (Figure 3). Organic matter and granular structure in soil rose by 7.8%, 13.4% respectively, land fertility was obviously enhanced (Figure 4). It is assured that soil rehabilitation will speed up by further using livestock and organic wastes as well as extensive development of the three agricultural component structure in future.

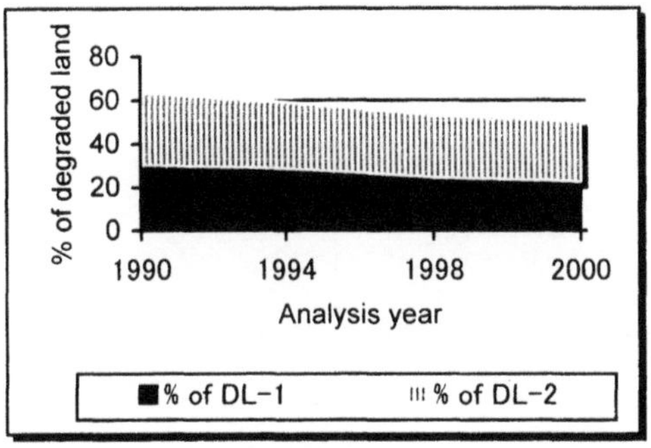

Figure 3 *The tendency of % of degraded arable Land(DL) from 1990~2000 in China (2) and Henan Province(1)*

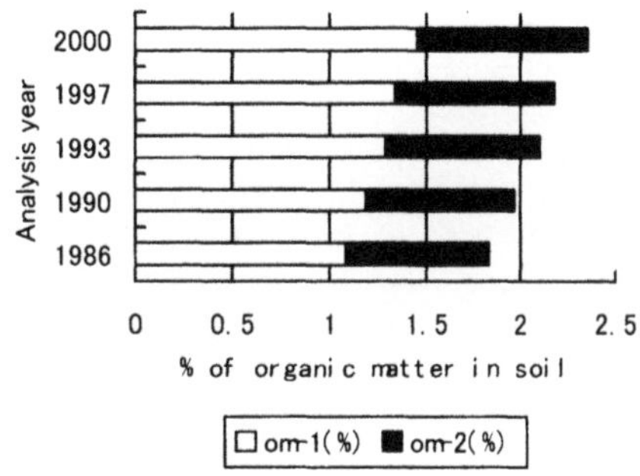

Figure 4 *the temporal variation of organic matter(OM) in Henan Province(1) and the whole nation(2)*

REFERENCES

Chen, Y.C. 1996. Integration of livestock-fish farming system in China. In Proc. FAO/AAAP Symposium on integrated Systems of Animal production in the Asian Region.p.77~82

Peart,R.M. and Puckett, H.B. 1982 Feed processing and Animal Feeding , Management by Exception. Agriculture England, 63:18~19

Levins,R.A.,M.A.Schmitt,and D.W.Richardson. 1996. Extension programming for teaching manure management to farmers.Review of Agricultural Economics 18(2):250~320

Sweeten,J.M. 1996. Cattle feedlot manure and wastewater management practices for water and air pollution control .p.60~80.

Tamminga,S. 1996. A review on environmental impacts of nutritional strategies in ruminants.J.Anim. Sci. 74:3120~3124

Bolsen,K.K.,J.E.Baylor and M.E.McCullough.1991.Field guide for hay and silage management. Natl. Feed Ingredients Association.pp.13~32

AUTHOR INDEX

SUBJECT INDEX

www.ingramcontent.com/pod-product-compliance
Ingram Content Group UK Ltd.
Pitfield, Milton Keynes, MK11 3LW, UK
UKHW050948290926
15470UKWH00005B/1702